Machine Learning and Data Science in the Oil and Gas Industry

*Dedicated to my parents Aline and Kurt Bangert
who supported me always and instilled a good work ethic, and
my teachers John Lammin, David Tovee,
and Mitchell Berger who taught me boundless
curiosity and precision.
Thank you*

Machine Learning and Data Science in the Oil and Gas Industry

Best Practices, Tools, and Case Studies

Edited by

Patrick Bangert

Artificial Intelligence Team, Samsung SDSA, San Jose, CA, United States;
algorithmica technologies GmbH, Küchlerstrasse 7, Bad Nauheim,
Germany

Gulf Professional Publishing
An imprint of Elsevier

Gulf Professional Publishing is an imprint of Elsevier
50 Hampshire Street, 5th Floor, Cambridge, MA 02139, United States
The Boulevard, Langford Lane, Kidlington, Oxford, OX5 1GB, United Kingdom

Notices
Knowledge and best practice in this field are constantly changing. As new research and experience broaden our understanding, changes in research methods, professional practices, or medical treatment may become necessary.

Practitioners and researchers must always rely on their own experience and knowledge in evaluating and using any information, methods, compounds, or experiments described herein. In using such information or methods they should be mindful of their own safety and the safety of others, including parties for whom they have a professional responsibility.

To the fullest extent of the law, neither the Publisher nor the authors, contributors, or editors, assume any liability for any injury and/or damage to persons or property as a matter of products liability, negligence or otherwise, or from any use or operation of any methods, products, instructions, or ideas contained in the material herein.

Library of Congress Cataloging-in-Publication Data
A catalog record for this book is available from the Library of Congress

British Library Cataloguing-in-Publication Data
A catalogue record for this book is available from the British Library

ISBN: 978-0-12-820714-7

For information on all Gulf Professional Publishing publications
visit our website at https://www.elsevier.com/books-and-journals

Publisher: Joe Hayton
Acquisitions Editor: Katie Hammon
Editorial Project Manager: Timothy Bennett
Production Project Manager: Prasanna Kalyanaraman
Designer: Matthew Limbert

Typeset by Thomson Digital

Working together
to grow libraries in
developing countr

www.elsevier.com • www.bookaid.org

Contents

6. Getting the Most Across the Value Chain
Robert Maglalang

7. Project Management for a Machine Learning Project
Peter Dabrowski

8.   The Business of AI Adoption
Geoffrey Cann

Contributors

Patrick Bangert, Artificial Intelligence Team, Samsung SDSA, San Jose, CA, United States; algorithmica technologies GmbH, Küchlerstrasse 7, Bad Nauheim, Germany

Geoffrey Cann, MadCann Alberta Inc., British Columbia, Canada

Jim Crompton, Reflections Data Consulting, Colorado Springs, CO, United States

Peter Dabrowski, Wintershall Dea, Hamburg, Germany

Guoqing Hana, China University of Petroleum, Beijing, China

Peter Kronberger, Wintershall Dea Norge AS, Stavanger, Norway

Robert Maglalang, Value Chain Optimization at Phillips 66, Houston, TX, United States

Arnold Landjobo Pagou, China University of Petroleum, Beijing, China

Long Peng, China University of Petroleum, Beijing, China

Wu Qing, Department of Science & Technology Development, China National Offshore Oil Corporation, Beijing, China

Jin Shua, China University of Petroleum, Beijing, China

Foreword

We live in a changing world. Not only this, the pace of change is itself changing. We have entered an era of incredible universal transition that will influence various aspects of our lives. Today, the energy and technological worlds are witnessing drastic changes that are influenced by intertwined causes of growth: energy demand, rapid transition of energy systems, political factors, technological evolution, and societal and cultural implications.

These changes simultaneously depend on and influence the way we live, the way we work, and the way we move. There is nothing more crucial than bringing together leadership power, domain expertise, knowledge, and the many lessons of past civilizations and industrial revolutions.

In the energy sector, the petroleum industry is still the most important source of energy and the provider of primary material to many industries. This industry is also influenced by the transformative technologies that offer tremendous potential to redraw not only the physical infrastructure but also the intellectual climate, business models, and asset management systems. This petroleum industry is very different from other industries. It is among the few industries that are perceived as generators of big data through the various phases from exploration to abandonment. It is an industry that has been treasuring all sorts of data for decades as it relies heavily on these data in managing its assets and running its operations. Companies in the petroleum industry are risk averse and have strict safety standards due to the nature of their business. None of these companies can afford to operate without appropriate decision-making processes that rely greatly on adequate technologies, experience, and data analytics. The recent advancements in the digital world including the increase in computer power and information processing, the speed of communication, and the reduction in cost of data storage; most of the companies in the petroleum industry expedited and extended their digital business with great ambitions to extract or generate new value and identify potential opportunities from both their old and new data augmented with either existing or new business models.

Many of these opportunities can be identified if the underlying mechanism is exposed in numerical format and encapsulated in the language of mathematical formulas, the language of data science. It is the purpose of data science to obtain, clean, and curate the data necessary and the purpose of machine learning to produce the formulas (i.e., new knowledge). With this new knowledge, managers of assets are expected to do better.

The authors of this book have stepped back from debating what might be achievable and instead they started with the more fundamental exploration of the potential of machine learning and demonstrated how it can help the petroleum industry. They have clearly indicated that machine learning in itself is not an objective but a means. No one is expected to implement a technology for the sake of it or because of hype around it, but for the benefits it may bring.

Today, machine learning is among the most widely used data analytics techniques in various industries including health care, education, manufacturing, automotive, retail, finance, logistics, transportation, and energy including oil and gas. No one denies the value that this technology brings if applied appropriately and at large scale. I can see that there will eventually be no industry that will be left untouched by machine learning, as it possesses the potential to empower humans to do what may be perceived today as impossible. In the oil and gas industry, machine learning has already been implemented or tested in a wide range of business sectors and the industry is very ambitious about extending its applications and very optimistic about its positive impact.

Although many of the challenges in the petroleum industry are global, many solutions, assessment studies, and remedial actions are almost local. Only few companies are effective at transferring and exchanging experience and learning from their successes and failures. This book is a clear testimony of the desire of its authors to demonstrate the meaning and importance of sharing local knowledge globally.

This book presents an overview of the field of machine learning and points toward the many used cases in the oil and gas industry where these have already solved problems. The hype surrounding machine learning will hopefully be cleared up as this book focuses on what can realistically be done with existing tools. It aims to present much of the fundamental understanding of the petroleum industry and machine learning in a way that is comfortable for all to follow. It summarizes the fundamental concepts of various applications of machine learning in the oil and gas industry and does not pretend to present a comprehensive review of the details though references to books providing such reviews are included where appropriate. While this book will not make you into a machine learner, it will provide enough knowledge to conduct yourself effectively with data scientists and also to be able to reflect on the overall quality and results of a project.

The book addresses four main groups of readers: Oil and gas professionals, machine learners and data scientists, students and researchers, and the general public.

The primary audience is any person working in the oil and gas industry who wants to understand what machine learning is and how it applies to the industry.

Machine learners and data scientists will learn about the oil and gas industry and its complexities as well as the use cases that their methods can be put to in this industry. They will learn what an oil and gas professional expects to see from the technology and the final outcome. The book puts into perspective some of the issues that take center stage for data scientists, such as training time and model accuracy, and relativizes these to the needs of the end user.

This book also targets graduate students and researchers at the cutting edge of investigations into the fundamentals and applications of machine learning in the oil and gas industry.

For the general public, this book presents an overview of the state-of-the-art in applying a hyped field like machine learning to an old-world industry.

The book can be divided into two parts. The chapters of the first part discuss the petroleum industry and machine learning with associated management challenges. The second half focuses on practical case studies that have been carried out by oil companies and report on what has been done already as well as what the field is capable of. In this context, the reader will be able to judge how much of the marketing surrounding machine learning is hype and how much is reality.

This book will have served its purpose if an oil and gas company uses the advice given here in facilitating a machine learning project or toolset to drive value. It is meant partially as an instruction manual and partially as an inspiration for oil company managers who want to use machine learning and artificial intelligence to improve the industry and its efficiency.

As such, this book is deliberately designed to be a passionate discussion starter for what I consider to be one of the digital world's most important debates. I believe the authors were successful in their mission to open up and stimulate further discussions. In a rapid transition of energy systems, such debates are essential and can only be achieved by fundamental research, free from economic and political constraints, and by people free to rethink our current challenging business environment.

I truly believe that a main reasons for societies and industries to continue, grow, and flourish is the existence of distinguished members who generously share their valuable experience, knowledge, and inspiration.

I am deeply grateful to the leading author, Patrick Bangert, and the other authors for writing this book. Moreover, I hope that this book inspires you to think deeply about the opportunities machine learning can create, and I do invite you to contribute to this journey by sharing your story.

Saeed M. Al Mubarak
Chairman, SPE Digital Energy Technical Section (2017–23)
Chairman of the Standards Committee and Chairman of
the P&FDD Academy Technical Program, Saudi Aramco

Chapter 1

Introduction

Patrick Bangert

Artificial Intelligence Team, Samsung SDSA, San Jose, CA, United States; algorithmica technologies GmbH, Küchlerstrasse 7, Bad Nauheim, Germany

1.1 Who this book is for

This book will provide an overview of the field of machine learning (ML) as applied to industrial datasets in the oil and gas industry. It will provide enough scientific knowledge for a manager of a related project to understand what to look for and how to interpret the results. While this book will not make you into a machine learner, it will provide everything needed to talk successfully with machine learners. It will also provide many useful lessons learned in the management of such projects. As we will learn, over 90% of the total effort put into these projects is not mathematical in nature and all these aspects will be covered.

An ML project consists of four major elements:

1. *Management*: Defining the task, gathering the team, obtaining the budget, assessing the business value, and coordinating the other steps in the procedure.
2. *Modeling*: Collecting data, describing the problem, doing the scientific training of a model, and assessing that the model is accurate and precise.
3. *Deployment*: Integrating the model with the other infrastructure so that it can be run continuously in real-time.
4. *Change management*: Persuading the end-users to take heed of the new system and change their behavior accordingly.

Most books on industrial data science discuss mostly the first item. Many books on ML deal only with the second item. It is however the whole process that is required to create a success story. Indeed, the fourth step of change management is frequently the critical element. This book aims to discuss all four parts.

The book addresses three main groups of readers: Oil and gas professionals, machine learners and data scientists, and the general public.

Machine Learning and Data Science in the Oil and Gas Industry.
http://dx.doi.org/10.1016/B978-0-12-820714-7.00001-7

Oil and gas professionals such as C-level directors, plant managers, and process engineers will learn what ML is capable of and what benefits may be expected. You will learn what is needed to reap the rewards. This book will prepare you for a discussion with data scientists so that you know what to look for and how to judge the results.

Machine learners and *data scientists* will learn about the oil and gas industry and its complexities as well as the use cases that their methods can be put to in this industry. You will learn what an oil and gas professional expects to see from the technology and the final outcome. The book will put into perspective some of the issues that take center stage for data scientists, such as training time and model accuracy, and relativize these to the needs of the end user.

For the *general public*, this book presents an overview of the state-of-the-art in applying a hyped field like ML to an old-world industry. You will learn how both fields work and how they can work together while the industry is transitioning to a new way of supplying energy to the world.

One of the most fundamental points, to which we shall return often, is that a practical ML project requires far more than just ML. It starts with a good quality data set and some domain knowledge, and proceeds to sufficient funding, support and most critically change management. All these aspects will be treated so that you obtain a holistic 360-degree view of what a real industrial ML project looks like.

The book can be divided into two parts. The first 8 chapters discuss general issues of ML and relevant management challenges. The second half focuses on practical case studies that have been carried out in real industrial plants and report on what has been done already as well as what the field is capable of. In this context, the reader will be able to judge how much of the marketing surrounding ML is hype and how much is reality.

1.2 Preview of the content

The book begins in *Chapter 2* with a presentation of data science that focuses on analyzing, cleaning, and preparing a dataset for ML. Practically speaking, this represents about 80% of the effort in any ML project if we do not count the change management in deploying a finished model.

We then proceed to an overview of the field of ML in *Chapter 3*. The focus will be on the central ideas of what a model is, how to make one, and how to judge if it is any good. Several types of model will be presented briefly so that one may understand some of the options and the potential uses of these models.

A review of the status of ML in oil and gas follows in *Chapter 4*. While we make no attempt at being complete, the chapter will cover a large array of use cases that have been investigated and provides some references for further reading. The reader will get a good idea of what is possible and what is hype.

In *Chapter 5*, Jim Crompton addresses how the data is obtained, transmitted, stored, and made available for analysis. These systems are complex and diverse and form the backbone of any analysis. Without proper data collection, ML is impossible, and this chapter discusses the status in the industry of how data is obtained and what data may be expected.

Management is concerned with the business case that Robert Maglalang analyzes in *Chapter 6*. Before doing a project, it is necessary to defend its cost and expected benefit. After a project, its benefit must be measured and monitored. ML can deliver significant benefits if done correctly and this chapter analyzes how one might do that.

ML projects must be managed by considering various factors such as domain expertise and user expectations. In a new field like ML, this often leads to shifting expectations during the project. In *Chapter 7*, Peter Dabrowski introduces the agile way of managing such projects that has had tremendous successes in delivering projects on time, in budget and to specifications.

Many projects get stuck in a proof-of-concept phase and do not get rolled out. This state of purgatory is presented by Geoffrey Cann in *Chapter 8*. It can be resolved by clearly communicating needs and expectations on both sides. If the expectation on the operator side is "to learn something," then this is also legitimate and can be factored into the project so that a roll-out is not expected on the software side.

The next several chapters discuss concrete use cases where ML has made an impact in oil and gas.

In *Chapter 9*, Wu Qing presents many applications that ML has been put to in China National Offshore Oil Corporation (CNOOC) and focuses primarily on exploration and refining.

Environmental pollution such as the release of NOx or SOx gasses into the atmosphere while operating a gas turbine is harmful. With ML, physical pollution sensors can be substituted by models. These are not only more reliable, but they allow model predictive control and thus are able to lower pollution. Shahid Hafeez presents this in *Chapter 10*.

The simple algorithm of principal component analysis finds a great use in *Chapter 11* where Long Peng predicts the failure of electric submersible pumps. This addresses the most famous use case in industrial ML: predictive maintenance.

A forecasting and classification methodology for rod pumps is presented in *Chapter 12* where problems can be diagnosed in advance of a failure and a maintenance measure planned and scheduled in a timely manner.

The forecasting of slugging events in gas-lift wells is presented by Peter Kronberger in *Chapter 13*. These events are upsetting and can be mitigated if choke valves are closed at the right time. This intricate advanced process control application is powered by a time-series forecasting model.

1.3 Oil and gas industry overview

Millions of years ago, when sea creatures and plants died and collected at the bottom of the ocean or swamps, they sometimes found themselves in oxygen starved environments and covered by sediment. Over time, the deposits above grew, the pressure rose and with it the temperature. Under these conditions, the organic material was slowly converted into a collection of hydrocarbon molecules. This is, of course, a very simple presentation of a complex process and there are other mechanisms for the creation of oil.

As the continents moved and the Earth's surface changed, the structure of these sedimentary layers was changed. This forced some of these deposits to flow elsewhere. Some locations were special in that they attracted these deposits from far and wide and in that they were covered from above by an impenetrable layer of rock. These places are the storehouses of hydrocarbons and they are called *reservoirs*. Finding reservoirs is a difficult task as they are often far underground beneath terrain that is hard to get to. This task is called *exploration*.

A reservoir is not (usually) an underground lake of oil but rather it is a porous rock the pores of which contain the oil. Depending on the structure and its history, the internal pressure of the reservoir may be high enough, that if it is punctured by *drilling* a well, the oil will flow out of it naturally. In other cases, the internal pressure is not enough to transport the oil to the Earth's surface and the process must be assisted by a pump. This is a process known as *artificial lift*.

A well may be hundreds of meters deep and only half a meter in diameter. Such a fragile structure would collapse if it were left to its own devices and so it must be lined, often while being drilled. When the well has been drilled, and oil has been found at its bottom, it must be closed off at the top and attached to a system of pipelines that will transport the oil from the well to locations of further processing. This process is known as *completions*.

Managing the well from the start of operations, through its lifetime that may extend over several decades, is called *production*. During production, the equipment at the well must be monitored and maintained with changes made occasionally depending on the changing conditions in the reservoir as its pressure decreases over time. The main source of work effort and financial expense occurs during production and is due to the *maintenance* that must be performed on equipment that ages and breaks over the years.

Once the fluid from the reservoir is at the surface, we discover that it is a combination of gaseous hydrocarbons (natural gas), liquid hydrocarbons (crude oil), water, and particulate matter most of which is sand. The composition of the fluid, called *multiphase flow*, into these four groups, and into the many different forms of hydrocarbons, is particular to the individual reservoir and well. This fluid must be separated into these four components and that is often done on site.

All these stages from exploration over drilling, completions, and production are together called *upstream* oil and gas. Often the term oil and gas industry implicitly refers to the upstream side. This is not the end of the journey, however. For an excellent treatment of the upstream industry, see Hyne (2012).

The useful components are transported from the wellhead to processing centers by pipelines. These are pipes that may extend over hundreds of miles of rugged terrain and can transport both oil and gas without mixing them. The pipelines may end at shipping terminals where the substances are loaded onto tankers or they may end at the refinery. The process of transporting crude oil and natural gas from the wellhead to the refinery is the *midstream* oil and gas industry.

As crude oil is a haphazard mix of many different hydrocarbons, it cannot be used for practical purposes as it is. It must be processed and, the various hydrocarbons must be separated into categories of similarity. Particularly long hydrocarbon chains are not practically useful, and they must be split into shorter chains. Processing crude oil into categories of hydrocarbons chains by their length is known as *refining* and occurs in a large processing facility called a *refinery*. These refineries may be located close to the reservoirs or close to the major end users of the products, depending on the cost of transportation.

Some products of a refinery are useful end products in themselves. Particularly, they are gas, gasoline, and heating oil. Gas is generally used either as a fuel for producing heat or electricity. Gasoline comes in various special forms like kerosene as a fuel for airplanes, normal gasoline as a fuel for cars, and diesel as a fuel for some cars and trucks as well as ships. Heating oil is a major fuel for producing district heat in many cold countries.

Other products of a refinery are further processed to make a wide range of products. These products are generally either especially pure hydrocarbons that are used industrially in other processes to make yet other products. Or they are modified and enriched to produce useful end products. Due to the variety of products, the facilities that do this are quite varied and each one specializes in a substance. All such plants together are known as the *petrochemical* industry or *downstream* oil and gas. What follows from there before the substance arrives in a consumer's home is the *chemical* industry that produces an even broader range of products.

It is often underestimated how important crude oil is as a basic source material for our modern civilization. By volume, the main use is of course as a fuel for the propulsion of cars, planes, ships and so on and as a source of heat that may be further transformed into electricity. Via the petrochemical industry however, crude oil finds its way into a vast array of goods that we use every day. It would be difficult to produce most of these goods without starting with crude oil. Here is a list of some selected few products that rely on crude oil:

- Most plastics and all products made from plastics like bottles and toys

- Synthetic fibers used in clothing and other fabric-based products like tents, umbrellas, curtains, carpets
- Wheels, toys, glasses, helmets, paint
- Shampoo and many cosmetics like lipstick, toothpaste, and shaving cream
- Dentures, heart valves, and artificial limbs
- Linoleum and other synthetic surfaces
- Fertilizers and pesticides

This list reveals that we deal with oil-based products multiple times a day in our normal lives. Many of the material comforts in our lives are based on oil as an essential raw material.

1.4 Brief history of oil exploration

There are few places on Earth where hydrocarbons are available freely at the surface. One place is the Dead Sea where one can obtain bitumen, a very thick form of oil, by hand. This was the primary source of bitumen for the Egyptian civilization in antiquity where it was used as sealant, adhesive, incense, pigment, and waterproofing, among others. Already in antiquity, the bitumen was heated and combined with other substances to produce some product, an early form of refining. Thus, the use of oil stretches back to the beginnings of human civilization and perhaps even further.

Oil was a valuable substance in ancient times, virtually unobtainable in most places. Naturally, people strove to find more. As we know today, a successful search depends on knowledge of geoscience that gives insight into the evolution of the Earth's crust. It is not surprising that proper exploration could not occur until modern times.

Modern systematic exploitation of oil can be traced to August 27, 1859 when Colonel Drake successfully drilled a well striking oil in Titusville, Pennsylvania, United States. Prior to this point, the production of oil was haphazard and based more on circumstantial finds rather than a systematic premeditated search. The industry quickly developed in the United States and other countries. In about 1900, the Russian empire was the primary producer while the United States quickly caught up in the early 1900s with some notable advances in Mexico in the 1920s. The countries of central and South America as well as the Caribbean started to play a role in the 1930s.

Significant discoveries were made in the Middle East just prior to the Second World War. It was at this point that the British Admiralty considered oil as a principal strategic element to win the war as oil-based fuel allowed the Navy to stay at sea longer and move faster. The countries of the Middle East, first and foremost Iran, were largely under British influence at this time creating the problematic tension that continues to the present day: Countries that do not have (enough) oil for their strategic needs, meddle in the affairs of other countries that have more oil than they can use. A few hidden agendas and betrayals later, oil is still on center stage where world politics and warfare are concerned, particularly in the Middle East that continues to play the primary role as an oil-producing region (Frankopan, 2015).

More recently, the innovations of drilling horizontally and hydraulic fracturing (artificially breaking the porous rock that contains oil to increase the flow) allowed oil companies to access reservoirs hitherto inaccessible. These measures transformed the United States from a net importer to a net exporter of oil in 2018 with important economic and geopolitical consequences.

In modern times, the primary use of oil was and remains as a propellant for vehicles. Scientific advances, particularly in chemistry, have led to more and more sophisticated products that ultimately are based on oil as a principal raw material. The petrochemical industry started in the 1930s and made major advances during and after the Second World War. Currently, about 5% of global oil production ends up in petrochemical processes and about 40% of all chemicals are made directly from oil.

1.5   Oil and gas as limited resources

Oil and gas are finite resources that eventually will be used up. There are places on Earth where we can observe the creation of future oil deposits such as large swamplands. However, these processes work on geological timescales. It is also clear that many deposits cannot be accessed in a cost-effective manner with present technology. Exploration continues and new finds of oil fields are reported every year. There comes a point, called **peak oil**, when oil production reaches a maximum and thereafter declines. Despite numerous attempts to declare a time for peak oil, this point has not arrived yet.

The Club of Rome published a famous study in 1972 entitled "The Limits to Growth" that analyzed the dependency of the world upon finite resources (Meadows, Meadows, Randers, & Behrens, 1972). While the chronological predictions made in this report have turned out to be overly pessimistic, their fundamental conclusions remain valid, albeit at a future time. A much-neglected condition was made clear in the report: The predictions made were made considering the technology available at that time, that is, the prediction could be extended by technological innovation. This is, of course, exactly what happened. Humanity innovated itself a postponement of the deadline. The problem remains, however. Growth cannot occur indefinitely and especially not based on resources that are finite. The ultimate finite resource is land, but this is quickly followed by oil, considering its place in our world. There is a 30-year update to the study that makes for interesting reading (Meadows, Randers, & Meadows, 2004).

The global climate is changing at a fast rate. This fact is largely related to the burning of fossil fuels by humanity since the start of the industrial revolution. Considering the present-day problem of climate change, the finite nature of oil and gas may be a theoretical problem-it is likely that the consequences of climate change will become significant for the ordinary individual far earlier than the consequences of reduced oil availability.

Both problems, however, have the same solution: We must find a way to live without consuming as many resources. In part because they are finite and in part because their use is harmful. This new way is likely to involve two major elements: Technological innovation and life-style change.

Technological innovations will be necessary to overcome some fundamental challenges such as how wasteful the process is from the reservoir to the point where a barrel of crude is made into a useful product. As this is a scientific challenge, we must figure out how complex processes work and how we can influence them. This is the heart of *data science* with *ML* as its primary toolbox.

Life-style change will be needed from every individual on the planet in that we must use our resources with more care · essentially, we must use far fewer resources. This will involve effort on an individual level, but it will also involve a transformation of societies. Public transportation must supplant individual transportation. Consumerism must be overcome, and products must be of better quality to last longer. Society must rely more on society, rather than material objects. To transform society in a manner that most people are happy with the transition requires careful analysis. In large numbers, humans are predictable and amenable to numeric description and study. It is again the field of data science that is pivotal in enabling the transformation through insight and design of the best systems for the future.

1.6   Challenges of oil and gas

These issues provide a range of challenges to the oil and gas industry worldwide. We must continue to find and produce oil during the transition to a new social norm. The process must become more efficient and less wasteful. There are many dangers along the way that must be mitigated. This starts from injury to individual workers and goes to global threats as illustrated by the Deepwater Horizon disaster in 2010. The *digital transformation* is a journey that will help to solve these challenges. It promises to do this by increasing our understanding and control of the process. At the beginning are sensors and communication software. At the end are physical technologies like drones, 3D printing, autonomous vehicles, or augmented reality.

The heart of it all is ML. ML is the piece of the puzzle that converts a dataset into a formula, also called a model. Once we have a model and we are confident that the model is right, we can do many things with it.

A popular challenge these days is classifying images into various categories like bird, dog, table, house, and so on, see Fig. 1.1. Any individual human being correctly classifies about 95% of images and makes 5% errors. Until 2014, computers performed worse. Starting in 2015 however, computer models based on machine learning started to outperform humans for the first time. This is an

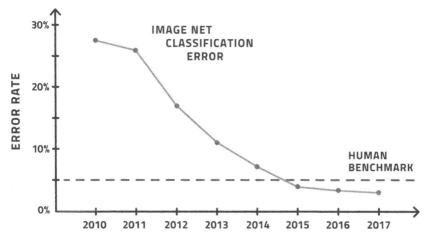

FIGURE 1.1 The evolution of machine learning as compared to human performance relative to a common standard dataset.

example that is consistently repeated time and again with many different datasets and tasks. Machine learning models are now more accurate than humans, calculate their output in a fraction of a second, and can keep doing it with the same accuracy without limit.

As they require little human effort to make and maintain as well as being fast to compute, ML models are much cheaper than any alternative. These two features also enable many novel use cases and business models that simply could not be implemented until now. In relation to the oil and gas industry, there are six main areas that ML can help with, see Fig. 1.2:

1. Is this operation normal or not?
2. What will the value or operational condition be at a future time?
3. Which category does this pattern belong to?
4. Can this complex, expensive, fragile, or laboratory measurement be substituted by a calculation?
5. How shall we adjust set-points in real-time to keep the process stable?
6. How and when shall we change set-points to improve the process according to some measure of success?

One of the biggest topics in industrial ML is *predictive maintenance*, which is the combination of the first three points in this list. Is my equipment performing well right now, how long will that remain and when it fails, what is the damage type? Once we know a damage type and a failure time in the future, we can procure spare parts in advance and schedule a maintenance measure to take place preemptively. This will prevent the actual failure and thus prevent collateral damage and spillage, which is typically 90% of the total financial cost of a failure.

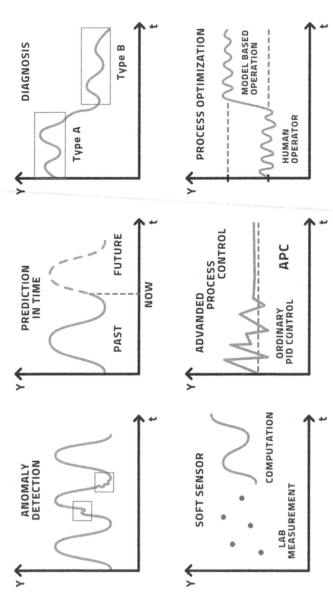

FIGURE 1.2 The six main application areas of machine learning in oil and gas.

References

Frankopan, P. (2015). *The silk roads*. Bloomsbury.

Hyne, N. J. (2012). In *Nontechnical guide to petroleum geology, exploration, drilling and production* (3rd ed.). PennWell.

Meadows, D. H., Meadows, D. L., Randers, J., & Behrens, W. W., III (1972). *The limits to growth: a report for the club of rome's project on the predicament of mankind*. Universe Books.

Meadows, D. H., Randers, J., & Meadows, D. L. (2004). *The limits to growth: the 30-year update*. Chelsea Green Publishing Co.

Chapter 2

Data Science, Statistics, and Time-Series

Patrick Bangert

Artificial Intelligence Team, Samsung SDSA, San Jose, CA, United States; algorithmica technologies GmbH, Küchlerstrasse 7, Bad Nauheim, Germany

In this chapter, we will introduce the concepts, tools, and steps toward generating a clean and representative dataset for later analysis. Simply reading out the last 12 months of control system data will generally not yield good results. Here, we discuss the decisions to be made prior to extracting data, and the treatment steps having extracted the data. After all this is done, the machine learning analysis will be discussed in the next chapter. Following the preparatory steps of this chapter will take time and effort but they will increase the chances of success greatly.

The data that we are likely to encounter in the oil and gas industry is usually numerical data ordered in time. For example, a temperature that is measured every 10 minutes for years. The source of this data is a sensor deployed in the field. This sensor may need to be recalibrated, or repaired, or exchanged every so often but the chain of measurements continues. The collection of these measurements is called a *time-series*. The label in the database that identifies this time-series versus others is usually called a *tag*. For a presentation of more traditional time-series analysis that does not use machine learning (Hamilton, 1994).

In any one plant, or field, we are likely to encounter tens of thousands of tags. Each of these may have a history of years. The frequency of records may range from one measurement a day to several measurements per second, depending on the application.

Much of the data science and machine learning in the popular literature concerns itself with data in different forms. Images, series of words, audio recordings, videos, and the like are not frequently encountered in the industry. These have become popular through a variety of consumer applications of data science and we will not discuss them here.

Apart from the time-series, we encounter another type of data in the industry: the spectrum. A vibration is typically measured as the amount of movement for each of many different frequencies. The collection of these measurements

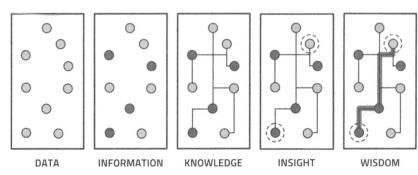

DATA   INFORMATION   KNOWLEDGE   INSIGHT   WISDOM

FIGURE 2.1   The five qualitatively different stages of data science.

is called a *vibration spectrum*. These may then be recorded at regular intervals over time, forming a time-series in their own right. Normal time-series are numbers, but these time-series are vectors as each observation is a string of numbers. There is significant scientific analysis that deals with spectra and this can be brought to bear on these spectra as well, especially the Fourier Transform (Press, Teukolsky, Vetterling, & Flannery, 2007).

As we look at our data, we gradually develop more high-level understanding of what this data tells us. Fig. 2.1 illustrates the five major stages in this journey. First, all we have is a large collection of numbers, the data itself. Second, the data becomes information once all the duplicates, outliers, useless measurements and similar are removed and we are left with relevant and significant data. Third, knowledge is gained once it is clear how the data is connected, which tags are related to other tags, how causes relate to effects and so on. Fourth, insight is generated once we know where are and where we need to be. Fifth, wisdom is when we know how to get to where we need to be. All that is left at this point is to execute the plan.

This chapter will introduce some common techniques of dealing with numerical data in preparation for machine learning (Pyle, 1999).

2.1   Measurement, uncertainty, and record keeping

Sensors measure quantities based on some sort of physical response and usually convert this into an electrical signal. This is then converted into a number using an analog-to-digital converter and sent to the *process control system*. This cycle may happen many times per second. The process control system typically stores this data only for a short while. It is the job of the *historian* to save the data for the long term and make it available for analysis, typically in diagrammatic form.

Some quantities are easy to measure, like temperature. Some are difficult to measure, like heat lost to the environment. Finally, some quantities are impossible to measure directly and must be computed-based on other measurements, like enthalpy. This is illustrated in Fig. 2.2. When measuring something, we

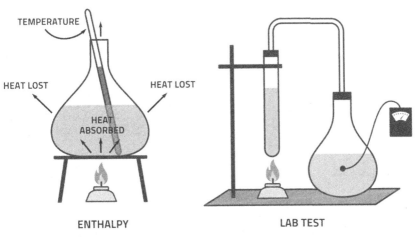

TEMPERATURE

HEAT LOST

HEAT LOST

HEAT ABSORBED

ENTHALPY

LAB TEST

FIGURE 2.2 Measurements both easy and hard.

must choose the right sensor equipment, and this must be placed correctly in order to give sensible outputs.

Suppose that we have measured some temperature to be 1.5°C at time t = 0, 3°C at time t = 1, and 2.5°C at time t = 2. With this data at hand, we ask two crucial questions: (1) How accurate is the measurement, and (2) how shall we record the measurements in the historian?

2.1.1   Uncertainty

Every sensor has a tolerance and so does the analog-to-digital conversion. Some electrical signal loss and distortion is introduced along the way from the sensor to the converter. Sensors drift over time and must be recalibrated at regular intervals. Sensors get damaged and must be repaired. Accumulations of material may cover a sensor and distort its measurement. Numbers are stored in IT systems, such as the control systems, with finite precision and there is loss of such precision with every conversion or calculation. All these factors lead to a de facto uncertainty in any measurement. The total uncertainty in any one measurement may be difficult to assess. In the temperature example, we may decide that the uncertainty is ±1°C. This would make the last two observations—of 3 and 2.5—numerically distinct but actually equal.

Any computation made based on an uncertain input, has an uncertainty of its own as a consequence. Computing this inherited uncertainty may be tricky—depending on the computation itself—and may lead to a larger uncertainty than the human designer anticipated. In general, the uncertainty in a quantity y that has been computed using several uncertain inputs x_i is the total differential,

$$(\Delta y)^2 = \sum_{i=1}^{N} \left(\frac{\partial y}{\partial x_i} \right)^2 (\Delta x_i)^2$$

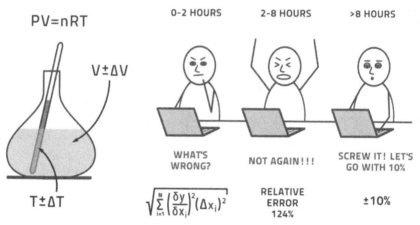

FIGURE 2.3   Determining the uncertainty of a computed result can be difficult and may need to be estimated.

In practice, this might be difficult to use and so we find that it is estimated in many cases, as illustrated in Fig. 2.3.

To anyone who has ever spent time in a laboratory, it is a painful memory that the treatment of measurement errors consumes many hours of work and is competitive in effort to the entire rest of the analysis. When engaged in data science, we must not forget that these data are not precise and therefore any conclusion based on them is not precise either. Some effects that are visible in the data quickly disappear once the uncertainty of the data is considered!

In general, it is advisable never to believe any data science conclusion in the absence of a thorough analysis of uncertainties.

2.1.2   Record keeping

The historian will save the measurements to its database. The IT systems involved were often designed and deployed during a time when hard disk space was expensive and limited. There was a desire to minimize the amount of space that data takes up. An obvious idea is to only record a value if it is different from the last recorded value by more than a certain amount. This amount is known as the *compression factor* and is an essential property of the tag. Someone, usually at the time of commissioning the historian, decided on the compression factor for each tag. Frequently the default value was left in place or some relatively large value was chosen in an effort to save space. Fig. 2.4 illustrates the situation with the temperature measurements alluded to above and a compression factor equal to 1.

The information that is contained in the data points not recorded is lost forever. If the compression factor is lower than the uncertainty, then there is no useful information in the lost data as all we would record are random fluctuations.

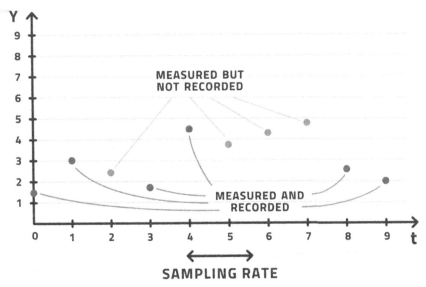

FIGURE 2.4 A sensor outputs its value at every time step but not every value is recorded if the compression factor is 1.

However, if the compression factor is larger than the uncertainty, we lose real information about the system.

It is an easy fix for any company operating a data historian to review the compression factors. In many cases, one will find that they are too large and must be lowered. This change alone will provide the operator with significantly more valuable data.

As the compression factor will prevent the recording of some measurements, what shall we do if the historian is supposed to output the value for a time for which it has no record? See Fig. 2.5 for an illustration. The data for t = 3 is missing. There are several different ways to answer a query for this value:

1. *Staircase*: The value remains constant at the last recording until a new measurement is recorded.
2. *Interpolation*: Some non-linear function is used to interpolate the data. Usually this is a spline curve.
3. *Linear*: A straight line is drawn in between each successive observation to interpolate the data.

While there are arguments both for and against any one of these, it is the general industry standard and agreement to use the first method; the staircase. A tag is equal to the last validly recorded value. This has the distinct advantage of holding in a real-time context, that is, in a context where we need to infer the value at the present moment without knowing the value of the next measurement in the time-series.

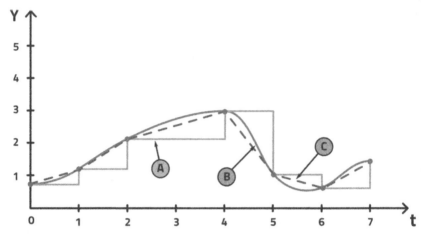

FIGURE 2.5   There are several ways to interpolate missing data. Here the data for t = 3 is missing. Which method shall we choose?

The correct configuration of the data historian forms the basis for the available dataset for any analysis. If we are presented with a ready-made dataset, the uncertainty of each tag is typically dominated completely by the compression factor. In practice therefore, we data scientists may say that the uncertainty *is* the compression factor. Nevertheless, the compression factor should be examined by the process engineers.

2.2   Correlation and timescales

As one tag changes its value, we may be able to say something about the change in another tag's value. For example, as temperature increases, we would expect the pressure to increase also. These two tags are called ***correlated***, see Fig. 2.6. The change in the first tag may provide *some* information about the change in the second tag, but not all. A measure that indicates how much information is provided is the ***correlation coefficient***. This is a number between −1 and 1. Positive numbers indicate that as the first tag increases, so does the second tag. Negative numbers indicate that as the first tag increases, the second decreases. An absolute value of 1 indicates that the change happens in exact lockstep. As far as change is concerned, these two tags are identical. A value of 0 indicates that the two tags are totally unrelated. Change in the first tag provides no information about the change in the second whatsoever.

Datasets from oil and gas facilities concern many interconnected parts. Many tags will be correlated with each other, such as temperature and pressure at one location. Analyzing which values are more and less correlated naturally provides an interesting clustering on the tags.

Generally, it is understood that correlation is measured in the context of linear dependency as illustrated in Fig. 2.6. The correlation coefficient due to

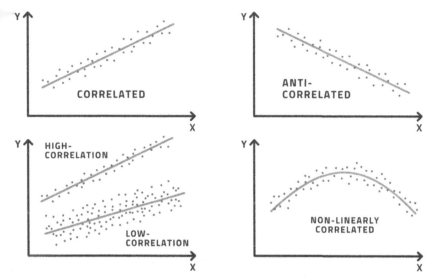

FIGURE 2.6   Correlation between two tags indicates the amount of information that the change in one provides for the other.

Pearson, which is the measure usually used, indicates the strength of a linear relationship. If there is a non-linear relationship, then this may not be picked up by the correlation coefficient in any meaningful way. This is important to take into account, as many physical processes are non-linear. Non-linear correlation analysis presupposes that we know the kind of non-linearity that is present, and this is another challenge. In practice, we find the linear correlation coefficient used exclusively. This is fine if it is understood properly.

As we are dealing with time-series, it is important to say that correlation is typically calculated for values recorded at the same time. If we shift one of the two time-series by a few time-steps, then we get the *correlation function*, that is, the correlation coefficient as a function of the *time lag*. In the special case that we compute the correlation of a tag relative to itself for different time lags, then this is called the *autocorrelation function*. These functions may look like Fig. 2.7.

We see that the correlation is high for a time lag of 0 and then decreases until it reaches a minimum at a time lag of 10 only to rise to a secondary maximum at a lag of 20 and so on. These two time-series are clearly related by some mechanism that takes about 10 time-steps to work. If there is an amount of time that is inherent to the dynamics of the system, we call it the *timescale* of the system. It is very important to know the timescales of the various dynamics in the system under study. Some dynamics may go from cause to effect in seconds while others may take days.

It is important to know the timescales because they determine the frequency at which we must observe the system. For example, in Fig. 2.7 the time scale

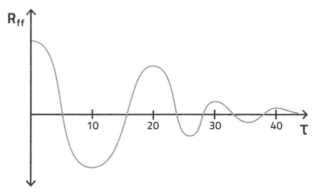

FIGURE 2.7   The correlation function between two tags as a function of the time-lag.

is clearly 10. If we observe the system once every 20 timesteps, then we lose important information about the dynamics of the system. Any modeling attempt on this dataset is doomed to fail.

A famous result in the field of signal processing is known as the ***Nyquist–Shannon sampling theorem*** and it says, roughly speaking: If you want to be able to reconstruct the dynamics of a system based on discrete observations, the frequency of observations must be at least two observations for the shortest timescale present in the system. This is assuming that you know what the shortest timescale is, and that the data is more or less perfect. In industrial reality, this is not true. Therefore, in practical machine learning, there is a rule of thumb that we should try for 10 observations relative to the timescale of the system. In the example of Fig. 2.7, where we measured the system every 1 timestep, the observation frequency was therefore chosen well.

The other reason for choosing to sample more frequently than strictly necessary is an effect known as ***aliasing*** sometimes seen in digital photography. This is illustrated in Fig. 2.8. Based on multiple observations, we desire to reconstruct the dynamics and there are multiple candidates if the sampling frequency is tuned perfectly to the internal timescale. The danger is that the wrong one is chosen, resulting in poor performance. If we measure more frequently, this problem is overcome.

In Fig. 2.8, the dots are the observations and the lines are various possible reconstructions of the system. All reconstructions fit the data perfectly but are quite different from each other. The data is therefore not sufficient to characterize the system properly and we must make an arbitrary choice. This is especially true for transient states and regimes where different time scales come into play. Here we must consider the shortest of the various time scales.

The practical recommendation to a data scientist is to obtain domain knowledge on the system in question in order to determine the best sampling frequency for the task. Sampling the data less frequently than indicated above, results in loss of information. Sampling it more frequently will only cost the analysis in

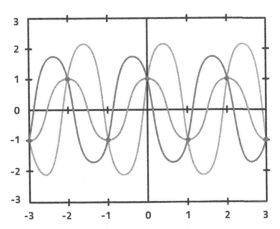

FIGURE 2.8   The aliasing problem in that the observations allow multiple valid reconstructions.

time and effort. On balance, we should err on the side of measuring somewhat more frequently than deemed necessary.

2.3 The idea of a model

Physical phenomena are described by physics using the language of mathematics. The expression of this insight is an equation that allows us to predict the behavior of physical systems and thus establish control over them. Take, as an example, the ideal gas law. It states that an ideal gas of n moles that is contained in a volume V, under pressure P, and at temperature T obeys the relationship

$$PV = nRT$$

where R is a constant. This relationship is simple to understand and allows us to exert control over the gas by changing one of the three variables in order to change the others. For example, if the volume remains the same and we increase the temperature, the pressure will rise.

This equation is an example of a ***model***. A model is an equation that allows us to compute a quantity of interest from other quantities that we can obtain in some way, usually by measurement. The model allows us to calculate this quantity instead of measuring it in the real world. On the one hand, a model saves us the effort and time of measurement and verification. On the other hand, it lets us know in advance what will happen and therefore choose our actions according to what we want to happen. Finally, a model like this also provides understanding.

This model was discovered in 1834 by Émile Clapeyron based on the previous work of several other scientists who had gone through a time-consuming process of performing many physical experiments and carefully measuring all the related quantities. Detailed plots were created by hand and studied until the ideas emerged one by one of the linear relationships between the three state variables and the amount of gas present. Finally, all these partial results were combined into the result, the ideal gas law.

The model was produced by two major factors: patient collection of data and human learning.

Let us say, for the time being, that all we had was a large table of numbers with one column each for volume, pressure, temperature, and amount of gas. We suspect that there is some relationship between these four concepts, but we do not know what it is. What we want is a model that can calculate the pressure from the other three quantities.

Today, what we would do is submit this table of numbers to a computer program that uses techniques called **machine learning**. This computer program would make a model and present some results to us that compare the pressure column from the table and the pressure as obtained from the proposed model. In each experimental case, that is, for each row in the data table, we have these two pressures: the measured and the calculated pressure. If the model is good, these two are always very close to each other. In addition, we usually want the model to be relatively simple in order to overcome some potential problems that we will go into later.

If we are satisfied with this analysis, we may conclude that the model represents the situation sufficiently well to be useful. We have a model that looks like

$$P = f(n, V, T)$$

The function $f(\cdots)$ is not usually something that makes sense to write down on paper, to look at with our eyes, and to try to understand. In most cases, the function $f(\cdots)$ is some large matrix of numbers that is unreadable. We must sacrifice understanding for this model. However, we get a model without having to do the learning ourselves in our head. The more complex and larger the data, the more return we get for our payment in understanding.

Despite being unintelligible—pardon the pun—the model is useful in that it is computable. We can now calculate the pressure from the other quantities. Having performed the analysis of the differences between computed and measured pressures over many known pressures in diverse situations, we are confident that the model produces the correct numerical value and that is all we can expect.

In the modern day, our data sets tend to be so large, the number of variables so many, and the situations so complex, that a manually made model based on human insight is often not feasible, economical, or even possible. In many situations, we already possess the basic understanding of the situation, for example, if this increases, then that increases also, but we need a formula to tell us quantitatively how large the changes are exactly.

This insight is comparable to a long-standing debate in physics: Is it the job of physics to explain why something is happening, or simply to correctly predict what will happen? Where you stand on this idea is the watershed for machine learning as opposed to physical modeling. If you are prepared to sacrifice understanding to some degree, you can gain tremendous performance at the same time as a huge reduction in cost.

2.4 First principles models

Many models, especially those for physical phenomena, can be derived from the laws of physics. Ultimately, this means that we have a set of coupled partial differential equations that probably combine a host of characteristics of the physical system that we must know and provide to the model. We might have to know what material every component is made from, how large every piece is, and how it was put together. Constructing such a model can be a daunting task requiring multiple experts to spend months of their time.

Solving this set of differential equations cannot be done exactly but must be done using numerical approximations that require iterative calculations. These take time. It is not uncommon for models of this nature, for relevant industrial-scale problems, to require a few hours of time on a modern computer to perform a single evaluation of the model.

In situations where we need answers at a speed comparable to the speed of the physical phenomenon, these methods often cannot be employed, and we must resort to approximate small-scale models. In many situations, such models are quite inaccurate. Situations like this are called *real-time* where it is understood that a real-time model does need some time for its calculation, but this time is negligible for the application. Sometimes, real-time can mean minutes and sometimes fractions of seconds; it depends on the application.

In constructing a physical, or so-called *first principles*, model, we provide many details about the construction of the situation; implying that this situation remains as we have specified. However, during the lifetime of most industrial machines, the situation changes. Materials abrade, corrode, and develop cracks. Pipes get dented and material deposits shrink the effective diameter of them. Moving parts move less efficiently or quickly over time. Objects expand with rising temperature. A first-principles model cannot (usually) take such phenomena into account.

We thus have three major problems with first principles models: (1) They take a lot of effort to make and maintain, (2) they often take a long time to compute their result, and (3) they represent an idealized situation that is not the same as reality.

The desire for approximate models made by the computer as opposed to human beings is thus driven mainly by the resource requirements needed to make and to deploy first-principles models. The simplest such model is a straight line.

2.5   The straight line

To illustrate many of these points, we will begin by looking at a straight-line model. Suppose we have an ideal gas in a fixed size container. As the volume and the amount of gas do not change, the ideal gas law of $PV = nRT$ is simple linear relationship between the pressure and the temperature. We could write this in the usual form of a straight line as

$$P = mT + b$$

where the slope of the line is $m = nR/V$ and the y-intercept $b = 0$. Having knowledge of the ideal gas law means we know the answer. Suppose for a moment, that we did not know the answer. We are simply provided with empirical measurements of pressure and temperature. As there are several of each, we will index them with the subscript i and so we have a sequence of pressures P_i and a sequence of temperatures T_i. This is illustrated in Fig. 2.9.

Based on a visual inspection, we believe that a straight-line model is a reasonable functional form for the model. That means that we must find the best slope and y-intercept for the model so that the model fits the data in the best possible way.

Before we can do that however, we must decide what "the best possible way" means, exactly. There is wide-spread agreement that "best" is the lowest sum of squared deviations. This is known as the ***least-squares method***. In other words, we will take the squared difference between model and measurement for each observation and add them all up. That slope and y-intercept are best that makes this sum the lowest possible.

$$\min_{m,b}\left(P_i - mT_i - b\right)^2$$

In this simple case, we can solve this minimization problem explicitly (and we will omit this derivation) to obtain the answer,

$$m = \frac{\left(\sum_{i=1}^{N} P_i T_i\right) - N\overline{T}\,\overline{P}}{\sum_{i=1}^{N} T_i^2 - N\overline{T}^2}, b = \frac{\overline{P}\left(\sum_{i=1}^{N} T_i^2\right) - \overline{T}\left(\sum_{i=1}^{N} P_i T_i\right)}{\sum_{i=1}^{N} T_i^2 - N\overline{T}^2}$$

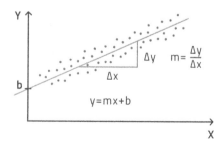

FIGURE 2.9   A collection of temperature and pressure measurement plotted and found to look like a straight line.

where the overbar means taking the average over all values and N is the number of points available. We can thus calculate the best possible values of m and b directly from the empirical data. The way of proceeding from the stated minimization problem to the explicit answer for the model coefficients is the learning or **training algorithm**. It is very rare that the model is simple enough for a theoretical solution like this. In most cases, the minimization must be carried out numerically in each case. These methods are complex, time-consuming calculations that may sometimes end up in suboptimal solutions. Therefore, the bulk of the machine learning literature focuses on the training algorithms.

For the time being, we note that we had to make two important decisions ourselves before executing the learning. We had to decide to use a straight-line model and we had to decide on the least-squares manner of looking for the best set of model coefficients. Those decisions are outside the realm of machine learning as they are made by a human expert.

The empirical data (T_i, P_i) had to be carefully collected. It is understood that if the experimenter made a mistake in the experiment, the value would be deleted from the data set. Mistakes can occur in any number of forms and if they are in the data set, then the model coefficients will contain that information, producing a somewhat false model. It is essential to clean the data set prior to learning, so that we learn only legitimate behavior.

Finally, we must recognize that every measurement made in the real world has a measurement error. Every instrument has a measurement tolerance. The place where the instrument is put may not be the ideal place to take the measurement. Sensors drift away from their calibration, get damaged, or get covered in dirt. Environmental factors may falsify results such as the sun shining on a temperature sensor and thus temporarily inflating the measured temperature.

The dataset is therefore not just (T_i, P_i) but rather $(T_i \pm \Delta T, P_i \pm \Delta P)$. In turn, this measurement uncertainty leads to an uncertainty in the calculated model coefficients and therefore in the model itself, see Fig. 2.10. Any calculation based on an uncertain value produces an uncertain value in return. If we expect models to deliver numerical answers to our questions, it is crucial that we

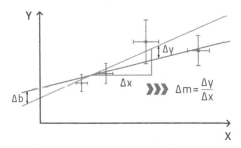

FIGURE 2.10 When each data point has a measurement uncertainty in both axes (indicated by the *cross*), there is an implied uncertainty in the slope and intercept of the best straight-line fit.

know how accurate we can expect those answers to be. This depends directly on the quality of the data used to make the model and the data used to evaluate the model whenever it is used. More on linear models can be found in (Hastie, Tibshirani, & Friedman, 2008).

2.6 Representation and significance

Before we analyze data, we must select or obtain the data to be analyzed. This must be done carefully with the desired outcome in mind. Suppose that you want to be able to tell the difference between circles and squares. What you would do is to collect a set of images of circles and squares and feed them into some modeling tool with the information of which images are circles and which images are squares. The modeling tool is then supposed to come up with a model that can tell the difference with some high accuracy. Suppose you give it the dataset in Fig. 2.11. That dataset contains 37 circles and 1 square.

The model that *always* says "it is a circle" would be correct in 37 out of 38 cases, using this dataset. That is an accuracy of 97%. If you went to a board meeting reporting that you have a model with a 97% accuracy for some industrial effect, you would most likely get congratulated on a job well done and the model might go live. In this case, the accuracy is not real because the dataset is not representative of the problem.

A dataset is ***representative*** of the problem if the dataset contains a similar number of examples of every differentiating factor of the problem. In our case that would mean a similar number of examples of circles and squares. Please note carefully that being representative of the *problem* is not the same as being representative of the *situation*. In the case of industrial maintenance, for example, we may want to distinguish various failure modes of equipment. That is the problem. However, most of the time, there is no failure mode at all because the equipment is functioning normally. That is the situation. In collecting a dataset representative of the failure modes therefore, we must de-emphasize the normal condition even though it is the typical condition.

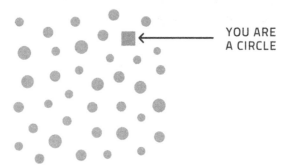

FIGURE 2.11 Sample dataset for distinguishing between circles and squares. It is composed of 37 circles and 1 square.

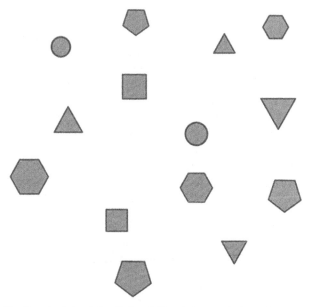

FIGURE 2.12 Sample dataset for distinguishing between *circles* and *squares* containing other shapes that make the problem harder.

Now consider the same task of distinguishing between circles and squares but with the dataset given in Fig. 2.12. In addition to some circles and squares, we now also have some other shapes. This is just distracting because these are not part of the problem statement. This is not just useless data. It is going to cause harm to the model because the model will try to incorporate this into the model and thus distort the internal representation of circles and squares. These disturbance variables should be excluded because they are not significant for the problem.

A dataset is *significant* for the problem if it contains only those factors that are important for the problem. In this case, this means excluding all other shapes from the dataset.

In industrial applications, the dataset should be both representative and significant for the problem at hand. This means that we must gather some domain knowledge to ask which tags are important for the effect that we want to model, either directly or indirectly. This will make the dataset more significant. We must also ask when certain effects were observed so that we can include and exclude parts of the entire timeline into our dataset to make it more representative of the problem. For instance, we will usually want to exclude any downtime of the facility, as this is usually not representative of any conditions we want to model. Conditions that are not at steady state may not be wanted either. These decisions must, however, be taken with the desired outcome in mind.

2.7 Outlier detection

An empirical dataset usually includes points that we do not want to show to a machine-learning algorithm. These are atypical situations or the result of faulty sensors. All such points are called outliers, that is, they lie outside of the phenomena we want to consider. They must be removed from the dataset. Of course, we can do that with domain knowledge, but this takes effort. There are automated methods to recognize outliers.

One way to do this is to cluster the data in an unsupervised manner. One of the most popular methods is the ***k-means clustering*** method. We must specify the number of clusters that we want to find *a priori* and that is the major liability in this method. It creates that many points in the dataspace, called the centers, and associates points to each one by distance. We judge the goodness of a cluster by measuring the internal similarity and the difference between different clusters. Based on this, the centers are moved around until the method converges. Individual points that do not fit easily into this scheme are outliers, see Fig. 2.13. For more on data preparation, see Aggarwal (2015).

2.8 Residuals and statistical distributions

When we have a model $\hat{y}_i = f(\underline{x}_i)$ relative to some dataset $(\underline{x}_i, y_i)$, we ask how good it is. There are many ways to quantify an answer to this question. Remember that we created the model by choosing model coefficients so that the squared difference between measurement and model is a minimum. Because of

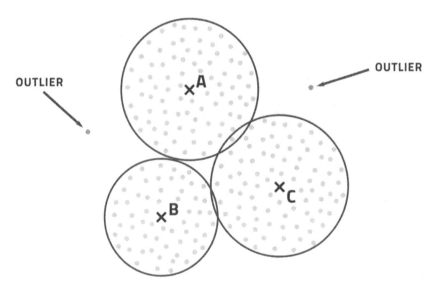

FIGURE 2.13　One method to detect outliers by creating clusters.

this, it makes sense to judge the quality of this model versus another model on the grounds of ***root-mean-square-error*** or ***RMSE***,

$$RMSE = \sqrt{\frac{\sum_{i=1}^{N}(\hat{y}_i - y_i)^2}{N}}$$

This measure takes its name from first computing the error e_i, also known as the ***residual***, between the measurement y_i and the model output $\hat{y}_i$, that is, $e_i = \hat{y}_i - y_i$. The error is then squared so that we do not distinguish between deviations above and below the measurement. We then average these values over the whole dataset and take the square root so that the final answer is in the same units of measurement as the original value.

A related measure is the ***mean-absolute-error*** or ***MAE***,

$$MAE = \frac{\sum_{i=1}^{N}|\hat{y}_i - y_i|}{N}$$

which is often numerically quite similar to the RMSE but can be considered a slightly more robust measure.

As the model and measurement are supposed to be equal in an ideal world, it makes sense to compute the linear ***correlation coefficient*** or R^2 between them, which should be very close to 1 for a good model.

$$R^2 = \frac{\sum_{i=1}^{N}(y_i - \bar{y})(\hat{y}_i - \bar{\hat{y}})}{\sqrt{\sum_{i=1}^{N}(y_i - \bar{y})^2}\sqrt{\sum_{i=1}^{N}(\hat{y}_i - \bar{\hat{y}})^2}}$$

Apart from these numerical measures of goodness, it is best to examine the ***probability distribution function*** of the residuals, see Fig. 2.14 for an

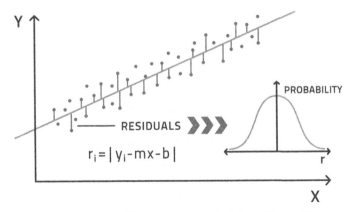

FIGURE 2.14 **Residuals are the differences between the data and the model.**

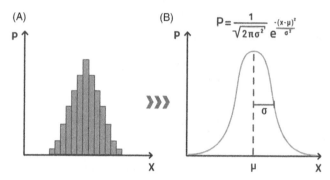

FIGURE 2.15 As the number of bins in the histogram (A) get large, it becomes the probability distribution function. The best distribution is the so-called normal distribution (B).

illustration. The residuals are the vertical lines between the data points and the model that has been fitted to the data. A residual can be a positive number if the data point is above the model and a negative number of the data point is below the model.

To analyze the residuals, find the smallest and largest residual. Break up the range between smallest and largest residual into several bins; let's say 100 bins which is usually sufficient for practical purposes. Go through all residuals and put them into the appropriate bin. So, each bin has a count for the number of residuals that fell into this bin. Divide the number in each bin by the total number of residuals. The result is a histogram that has been normalized to a total sum equal to 1. As the number of bins get larger and larger, this histogram becomes the theoretical probability distribution function of residuals, see Fig. 2.15. For practical purposes, the histogram is fine.

The best possible distribution that we can observe in the distribution of residuals is called the ***normal distribution*** or ***Gaussian distribution***, see Fig. 2.15. This distribution has several important characteristics:

1. It is centered around zero, that is, the most common residual is vanishingly small.
2. Its width is small, that is, the residuals are all quite close to the (small) mean value.
3. It is symmetric, that is, there is no systematic difference in the deviations of the model above and below the data.
4. It falls off exponentially on either side of the mean so that the probability of finding a residual far from the mean is very small. This also implies that there should only be one peak in the distribution. If the distribution looks like a mountain range instead of a solitary peak, there is a fundamental problem with the model.

While the distribution can be drawn and looked at graphically, these four important characteristics can also be quantified in numerical terms and they are known as the first four moments of the distribution. In the following definitions, recall that the residual error is $e_i = \hat{y}_i - y_i$.

The first moment μ_0 is the **mean** of the distribution. This should be very close to zero.

$$\mu_0 = \overline{e} = \frac{1}{N} \sum_{i=1}^{N} e_i$$

The second moment μ_1 is the **variance** of the distribution. Most commonly, we look at the **standard deviation** $\sigma = \sqrt{\mu_1}$ that is the square root of the variance. This is easier to interpret because it has the same units of measurement as the model. The standard deviation is a measure for the width of the distribution. Ideally this value is comparable to the measurement uncertainty of the empirical data. If this is much larger than that, the model should be reassessed.

$$\mu_1 = \frac{1}{N} \sum_{i=1}^{N} (e_i - \overline{e})^2$$

The third moment μ_2 is the **skewness** of the distribution. If the distribution is symmetric, this value will be zero. If the left side of the distribution contains more data points, the skewness will be negative, and vice-versa. Ideally, this number is very small.

$$\mu_2 = \frac{\frac{1}{N} \sum_{i=1}^{N} (e_i - \overline{e})^3}{\left(\frac{1}{N-1} \sum_{i=1}^{N} (e_i - \overline{e})^2 \right)^{3/2}}$$

The fourth moment μ_3 is the **kurtosis** of the distribution. The kurtosis of the normal distribution is 3 and so what we define below is known as the **excess kurtosis**. If your distribution has an excess kurtosis close to zero, it is like the normal distribution and that is a good thing. If it has a positive excess kurtosis, then there are more data points in the tails than the normal distribution has. This is a common observation in industrial practice. If it has negative excess kurtosis, then there are fewer data points in the tails than the normal distribution has.

$$\mu_3 = \frac{\frac{1}{N} \sum_{i=1}^{N} (e_i - \overline{e})^4}{\left(\frac{1}{N} \sum_{i=1}^{N} (e_i - \overline{e})^2 \right)^2} - 3$$

Because of outliers and the fact that exceptional circumstances are not quite as rare as we would hope, real datasets typically have heavier tails (positive excess kurtosis) than the normal distribution. This is the central burden on the model to overcome. If you get a model with low kurtosis, then the atypical situations are well taken care of.

2.9 Feature engineering

The basic task of modeling is to find some function that will take in a vector of tags and compute what we want to know. The independent variables are assumed to be sensor measurements in our industrial context, that is, they are numbers resulting from some sort of direct measurement; they are tags. Sometimes these sources of information are called *features* of the dataset (Guyon & Elisseeff, 2003).

In many cases, we possess some domain knowledge that allows us to create further inputs to the function that are not new variables but rather derived variables. For example, a distillation column generally has a pressure measurement at the bottom and the top of the column. The reason is that the differential pressure over the column is an interesting variable. This is a piece of domain knowledge that we have. The differential pressure is itself not a measurement, but a calculation based on two pressure measurements. It is an additional feature of the dataset that we create based on our understanding of the situation. Practically, we create a new column in our dataset equal to the difference in the two pressures. This may even be a tag in the control system.

We could provide the modeling procedure with both pressure measurements and expect it to learn that it is their difference that has an important impact on the outcome. Otherwise, we could also create a derived variable that is equal to this difference and feed it into the modeling procedure ourselves. The difference is twofold.

First, providing raw data only, means that the modeling procedure must figure out basic variable transformations and this requires information. Information is contained in the dataset, but it is finite. As some information is used up to learn these transformations, less information is available to learn other dependencies meaning that the final model will most likely perform less well.

Second, these basic transformations must be represented in the model and this requires parameters. For a certain limited number of parameters in the model, this restricts the expressiveness of the model unnecessarily.

It is therefore to our advantage to include in the dataset all derived quantities that are known to be of interest and this requires domain knowledge and some thought. It is well worth doing because this creates better and more robust models. This process is known as *feature engineering*. The amount of effort that goes into this must not be underestimated and should be regarded as an essential step in the treatment of industrial datasets.

The essential point to understand is that the information present in a dataset is limited and that there is a *principle of diminishing returns* in the size of the dataset. When the dataset is small, each new data point adds information. At some point, the dataset is rich enough that new data points do not add information. The dataset is saturated in information. Despite being able to grow the dataset with effort and money, you cannot provide new information. Recognizing that information is a limited resource, we must be careful how we spend it. Spending it on learning to establish features already known to human experts is wasteful.

Having looked at adding some features, it is also necessary and instructive to remove some features as they do not necessary encapsulate any information.

2.10 Principal component analysis

Suppose that we have a dataset with three independent variables as displayed in Fig. 2.16. We can see that this dataset has three rough clusters of data points in it. As this is the main visible structure in the dataset, we want to retain this, but we ask if we are able to lower the dimensionality of the dataset somewhat.

If we project the dataset onto a single one of the three coordinate axes, we get the result displayed in Fig. 2.17 on the top right, that is, all three such projections are just blobs of data in which the clustered structure is lost.

However, if we were to first rotate the dataset in three-dimensional space to align it with the axes indicated by the two arrows in Fig. 2.16 and then project them onto the axes, we get the result displayed in Fig. 2.17 on the bottom right. The rotated dataset projected onto the two-dimensional plane defined by the two axes is displayed in Fig. 2.17 on the left. We easily see that the two-dimensional representation captures most of the information and the one-dimensional representation indicated by "pc1" also retains the clustering structure. We are thus able to reduce the dimensionality of the dataset to either two or even one dimension by rotating it appropriately.

The rotation we are looking for is such that the new axes are along the directions of greatest variance in the dataset. This is displayed easily in Fig. 2.18

FIGURE 2.16   A three-dimensional dataset to illustrate principal component analysis.

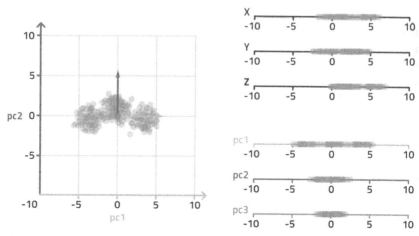

FIGURE 2.17   The dataset of Fig. 2.16 is projected onto its coordinate axes *(top right)* and its principal components *(bottom right)* as well as the plane of the first two principal components *(left)*.

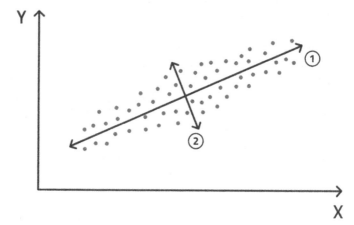

FIGURE 2.18   An indication of the principal components of a simple linear dataset.

where a dataset that we can fit with a straight-line is displayed. The direction of that straight-line model is the direction of greatest variance in the dataset and could be used to represent most of the information. The perpendicular direction then represents the direction of second greatest variance. These directions in order from the most significant to the least significant in the sense of variance are called the principal components and the method that computes them is called **principal component analysis** or **PCA**. A discussion of the calculations involved are beyond this book but many software packages include this procedure as an essential data science feature and so it can be readily carried out in practice.

The variances can be quantified precisely, and they can be added up. We can then ask: How many principal components must we keep in order to explain a certain amount, 99%, say, of the total variance in the original dataset? With industrial datasets where many of the variables are related to each other in various ways, we often find that the dimensionality can be reduced by as much as one-half, without significantly reducing the information content of the dataset. The reason for this is simple: As the tags originate from a single coordinated system, they duplicate a lot of information. More information can be found in (James, Witten, Hastie, & Tibshirani, 2015).

Dimensionality reduction is an important pre-processing step for machine learning and data science in general. It reduces the complexity of the dataset and the model, which generally leads to more robust models. Principal component analysis is recommended in virtually all modeling situations.

2.11   Practical advice

When faced with a data analysis challenge, what should we do in practice?

First, define very clearly what the desired outcome is. Do you need to forecast something at a future time? Do you need to characterize the state of something into various categories? Do you need decision support in ranking options? Do you want a graphical display of various pieces of information? How accurate does this analysis have to be in order to achieve its purpose? How often does the analysis need to get done—once a day or in real-time?

Second, decide which data sources you need to tap into to get the data for this outcome. This is going to be the control system or historian and so the real question is which tags do you want to look at? It will almost invariably be a small selection of all available tags and so you need to make a careful choice based on domain knowledge. Tools like correlation analysis can help in this selection. Having got this far, you can extract historical data from your database.

Third, decide how you are going to measure success. Such metrics are available from a data analytics point of view as discussed in this chapter, but success metrics are also of a practical nature such as (1) how much money is this earning, (2) how many predictions were correct, (3) how many proposals made by the analysis were adopted, (4) how many of the people who need to use/implement/follow this system accept it willingly.

Fourth, this dataset is now raw and must be cleaned before analysis can take place. Data that is irrelevant should be removed, such as data when the plant was offline or being maintained. If a sensor was temporarily broken, those data points might need to be interpolated or removed as well. Unusual operating conditions may need to be removed, depending on the use case. Most data will be quite similar, and some data will represent rare conditions. For the dataset to be representative and significant relative to the use case, you may need to resample the dataset so that analysis can pick up the dynamics you want to focus on.

Fifth, based on the clean dataset, think if there are some combinations of these data that would provide useful information to the analysis. If you know, for instance, that the ratio of two quantities is crucial for the outcome, then you should introduce a further column in the data matrix with that ratio. Particularly in downstream plants, material balance is an important aspect and so calculating material in and material out are examples of important features that can guide the analysis.

Sixth, analysis will benefit from dealing with as few data columns as possible that nonetheless contain all the information needed. Dimensionality reduction methods such as principal component analysis can do this elegantly.

At this point, you have a clean, streamlined dataset that includes all the information and domain knowledge you can give it. You are now ready to perform the analysis of this data relative to your use case and you know how to assess the success of this analysis so that if you try more than one method, you can compare them.

Please note that we have not talked about machine learning yet. In practice, getting to this point requires 80% of the person-hours for the whole project. Getting to a good dataset is the main work effort. Recognizing this alone will increase your chances of success greatly. Let's talk about machine learning next.

References

Aggarwal, C. C. (2015). *Data mining*. Springer.

Guyon, I., & Elisseeff, A. (2003). An introduction to variable and feature selection. *Journal of Machine Learning Research, 3*, 1157–1182.

Hamilton, J. D. (1994). *Time series analysis*. Princeton University Press.

Hastie, R., Tibshirani, R., & Friedman, J. (2008). In *The elements of statistical learning* (2nd ed.). Springer.

James, G., Witten, D., Hastie, T., & Tibshirani, R. (2015). *An introduction to statistical learning*. Springer.

Press, W. H., Teukolsky, S. A., Vetterling, W. T., & Flannery, B. P. (2007). In *Numerical recipes* (3rd ed.). Cambridge University Press.

Pyle, D. (1999). *Data preparation for data mining*. Morgan Kaufmann.

Chapter 3

Machine Learning

Patrick Bangert

Artificial Intelligence Team, Samsung SDS America, San Jose, CA, United States; algorithmica technologies GmbH, Küchlerstrasse 7, Bad Nauheim, Germany

In this chapter, we give an introduction to machine learning, its pivotal ideas, and practical methods as applied to industrial datasets. The dataset was discussed in the previous chapter and so here we assume that a clean dataset already exists alongside a clear problem description. Some principal ideas are introduced first, after which we discuss some different model types, and then discuss how to assess the quality and fitness-for-purpose of the model.

3.1 Basic ideas of machine learning

Machine learning is the name given to a large collection of diverse methods that aim to produce models given enough empirical data only. They do not require the use of physical laws or the specification of machine characteristics. They determine the dependency of the variables among each other by using the data, and only the data.

That is not to say that there is no more need for a human expert. The human expert is essential but the way the expertise is supplied is very different to the first principles model. In machine learning, domain expertise is supplied principally in these four ways:

1. Providing all relevant variables and excluding all irrelevant variables. This may include some elementary processed variables via feature engineering, for example, when we know that the ration between two variables is very relevant, we may want to explicitly supply that ratio as a column in the data table.
2. Providing empirical data that is significant and representative of the situation.
3. Assessing the results of candidate models to make sure that they output what is expected.
4. Explicitly adding any essential restrictions that must be obeyed.

Machine Learning and Data Science in the Oil and Gas Industry.
http://dx.doi.org/10.1016/B978-0-12-820714-7.00003-0

These inputs are important, but they are easily supplied by an expert who knows the situation well and who knows a little about the demands of data science.

The subject of *machine learning* has three main parts. First, it consists of many prototypical models that could be applied to the data at hand. These are known by names such as neural networks, decision trees, or k-means clustering. Second, each of these comes with several prescriptions, so called *algorithms* that tell us how to calculate the model coefficients from a data set. This calculation is also called *training* the model. After training, the initial prototype has been turned into a model for the specific data set that we provided. Third, the finished model must be deployed so that it can be used. It is generally far easier and quicker to evaluate a model than to train a model. In fact, this is one of the primary features of machine learning that make it so attractive: Once trained, the model can be used in real-time. However, it needs to be embedded in the right infrastructure to unfold its potential.

Associated to machine learning are the two essential topics that are at the heart of *data science*. First, the data must be suitably prepared for learning, which we discussed in Chapter 2. Second, the resultant model must be adequately tested, and its performance must be demonstrated using rigorous mathematical means. This pre-processing and post-processing before and after machine learning is applied to round out the scientific part of a data science project. In addition to these scientific parts, there are managerial and organizational parts that concern collecting the data and dealing with the stakeholders of the application.

In this book, we will treat all these topics, except the training algorithms. These form the bulk of the literature on machine learning and are technically challenging. This book aims to provide an overview to the industrial practitioner and not a university course on machine learning. The practitioner has access to various computer programs that include these methods and so a detailed understanding of how a model is trained is not necessary. On the other hand, it is essential that a practitioner understand what must be put into such a training exercise and how to determine whether the result is any good. So, this will be our focus.

Methods of machine learning are often divided into two groups based on two different attributes. One grouping is into *supervised* and *unsupervised* methods. The other grouping is into *classification* and *regression* methods.

Supervised methods deal with datasets for which we possess empirical data for the inputs to the model as well as the desired outputs of the model. Unsupervised methods deal with datasets for which we only have the inputs. Imagine teaching a child to add two numbers together. Supervised learning consists of examples like $1 + 2 = 3$ and $5 + 4 = 9$ whereas unsupervised learning consists of examples like $1 + 2$ and $5 + 4$. It is clear from this difference that we can expect supervised methods to be much more accurate in reproducing the outcome. Unsupervised methods are expected to learn the structure of the input data and recognize some patterns in them.

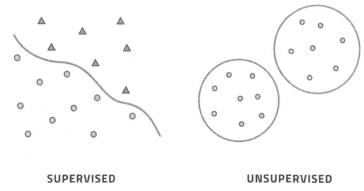

SUPERVISED   UNSUPERVISED

FIGURE 3.1   An illustration of the difference between supervised and unsupervised learning.

Fig. 3.1 illustrates the difference in the context of learning the difference between two collections of data points. In the first example, we know that circles and triangles are different and must learn where to divide the dataspace between them. In the second example, we only have points and must learn that they can be meaningfully divided into two clusters such that points in one cluster are maximally similar and points in different clusters are maximally different from each other—for some measure of similarity that makes sense in the context of the problem domain.

The difference between *classification* and *regression* methods is similarly fundamental. Classification methods aim to place a datapoint into one of several groups while regression methods aim to compute some numerical value. Fig. 3.2 illustrates the difference. In the first example, we only wish to draw a dividing line between the circles and the triangles, differentiating one category from the other. In the second example, we want to reproduce a continuous numeric value as accurately as possible.

This distinction is closely related to the difference between a *categorical* variable and a *continuous* variable. A categorical variable is equal to one of several

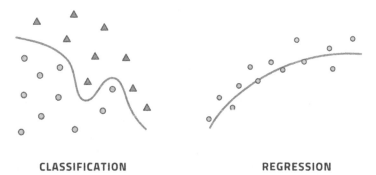

CLASSIFICATION   REGRESSION

FIGURE 3.2   An illustration of the difference between classification and regression methods.

values, like 1, 2, or 3, where the values indicate membership in some group. The difference between two values is therefore meaningless. If datapoint A has value 3 and datapoint B has value 2, all this means is that A belongs to group 3 and B belongs to group 2. Taking the difference of the values, $3-2=1$, has no meaning, that is, the difference between the two datapoints does not necessarily belong to group 1 nor does group membership necessarily even make sense for the difference.

A continuous variable is different in that it measures a quantity and taking the difference does mean something real. For instance, if datapoint A has a flow rate of 1 ton per hour and datapoint B has a flow rate of 2 tons per hour, we can take the difference $2-1=1$ and the difference is 1 ton per hour of flow, which is a meaningful quantity that we can understand.

Machine learning has many methods and all methods are either supervised or unsupervised, and either classification or regression. Any task that uses machine learning can also be divided into these categories. These two groupings are the first point of departure in selecting the right method to solve the problem at hand. So, ask yourself:

1. Do you have data for the result you want to compute, or only for the factors that go into the computation?
2. Is the result a membership in some category, or a numerical value with intrinsic meaning?

We will present methods for all these possibilities guided by the state-of-the-art and industrial experience. This book will not present an exhaustive list of all possible methods as this would go well beyond our scope.

As we need to be brief on the technical aspects of machine learning, here are some great books for further reading on machine learning in and of itself. A great book on the ideas of machine learning without diving into its technical depths is Domingos (2015). If you are more interested in the general economic ramifications of the field, an excellent presentation is in Agrawal, Gans, and Goldfarb (2018). More mathematical books that present a great overview are Mitchell (2018), Bishop (2006), MacKay (2003), and Goodfellow, Bengio, and Courville (2016).

3.2 Bias-variance complexity trade-off

One of the most important aspects in machine learning concerns the quality of the model that you can expect, given the quality of the data you have to make the model. Three aspects of the model are important here.

The *bias* of the model is an assessment of how far off the average output of the model is from the average expected output. The *variance* of the model is the extent to which inputs that are very close to each other result in outputs that are far from each other. The *complexity* of the model is generally measured by the number of parameters that must be determined using a machine learning algorithm from the empirical data provided for training.

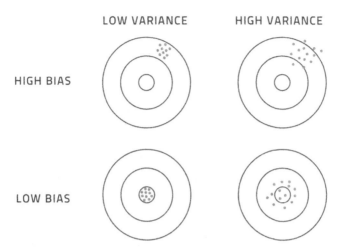

LOW VARIANCE   HIGH VARIANCE

HIGH BIAS

LOW BIAS

FIGURE 3.3   **The concepts of bias and variance illustrated by throwing darts at a target.**

The concepts of bias and variance are depicted in Fig. 3.3. Imagine throwing darts at a target. If the darts are all close together, the variance is low. If the darts are, on average, close to the center of the target, the bias is low. Of course, we want a model with low bias and low variance.

In order to get such a model, we first choose a type of model, that is, a formula with unknown parameters for which we believe that it can capture the full dynamics of our dataset if only the right values for the parameters are found. An example of this is the straight-line model that naturally has two parameters for one independent variable. We shall meet several other models later on that are more expressive because they have more parameters. Having thought about the straight-line, we can easily understand that a quadratic polynomial (three parameters for one independent variable) can model a more complex phenomenon than a straight-line.

Choosing a model that has very few parameters is not good because it will not be able to express the complexity present in the dataset. The model performance in bias and variance will be poor because the model is too simple. This is known as ***underfitting***.

However, choosing a very complex model with a great many parameters is not necessarily the solution because it may get so expressive that it can essentially memorize the dataset without learning its structure. This is an interesting distinction. Learning something suggests that we have understood some underlying mechanism that allows us to do well on all the examples we saw while learning, *and* any other similar tasks that we have *not* seen while learning. Memorization will not do this for us and so it is generally thought of as a failed attempt to model something. It is known as ***overfitting***. See Fig. 3.4 for an illustration of underfitting and overfitting.

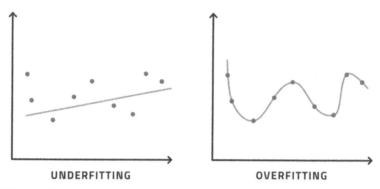

UNDERFITTING   OVERFITTING

FIGURE 3.4 **If the model is too simple, we will underfit.** If the model is very complex, it will overfit.

There must exist some optimal number of parameters that is able to capture the underlying dynamics without being able to simply memorize the dataset. This is some medium number of parameters.

If the model reproduces the data used in its construction poorly, then the model is definitely bad, and this is usually either underfitting or a lack of important data. The fix is to either use a more complex model, to get more data points, or to search for additional independent variables (additional sources of information). The data used in the construction of the model is the *training data* and the difference between the model output and the expected output for the training data is the *training error*.

To properly assess model performance, we cannot use the training data, however. We must use a second dataset that was not used to construct the model. This is the *testing data* that results in a *testing error*. The testing data is often known as *validation data*. More complex training algorithms need two datasets, one to train on and another to assess when to stop training. In situations like this, we may need to create three datasets and these are then usually called training, testing, and validation datasets where the testing dataset is used to assess whether training is done and the validation dataset is used to assess the quality of the model.

Fig. 3.5 displays the typical model performance as a function of the model complexity. As the model gets more complex, the training error decreases. The testing error decreases at first and then increases again as the model starts to overfit. There is a point, at minimum testing error, when the model achieves its best performance. This is the performance as measured by the bias, that is, the deviation between computed and expected values. This has not yet taken the variance into account. Often, we find that to get better variance, we must sacrifice some bias and vice-versa. This is the nature of the bias-variance-complexity trade-off.

While this picture is typical, we usually cannot draw it in practice as each point on this image represents a full training of the model. As training is usually

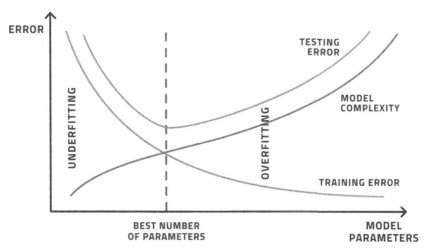

FIGURE 3.5   As the model gets more complex, the training and testing errors change, and this indicates an optimal number of parameters.

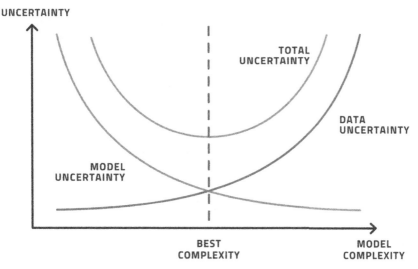

FIGURE 3.6   As model complexity rises, model uncertainty decreases but data uncertainty increases. There is an optimal point in the middle.

not a deterministic process, the training must be repeated several times for each sample complexity to get a representative answer. This is also needed to measure the variance. It is, in most practical situations, an investment of time and effort no one is prepared to make.

A similar diagram, see Fig. 3.6, can be drawn in terms of variance. There are two major sources of uncertainty in machine learning. First, the uncertainty inherent in the dataset and the manner in which it was obtained. This includes both the method of measurement and the managerial process of selecting which variables to include in the first place. Second, the uncertainty contained in the model and the basic ability of it to express the underlying dynamics of the data. There is a similar compromise that must happen.

In the end, we must choose the right model type and the right number of parameters for this model to achieve a reasonable compromise between bias and variance, avoiding both underfitting and overfitting. To prove to ourselves and other that all this has been done, we need to examine the performance on both training and testing data.

3.3 Model types

There are many types of model that we can choose from. This section will explore some popular options briefly with some hints as to when it makes sense to use them. To learn more about diverse model types and machine learning algorithms, we refer to Goodfellow et al. (2016).

3.3.1 Deep neural network

The *neural network* is perhaps the most famous and most used technique in the arsenal of machine learning (Hagan, Demuth, & Beale, 1996). The recent advent of *deep learning* has given new color to this model type through novel methods of training its parameters. Training methods are beyond the scope of this book, however.

While it took its inspiration from the network of neurons in the human brain, the neural network is merely a mathematical device. A schematic diagram like the one shown in Fig. 3.7 is commonly displayed for neural networks. On the left side, the input vector $\underline{x}$ enters the network. This is called the *input layer* and

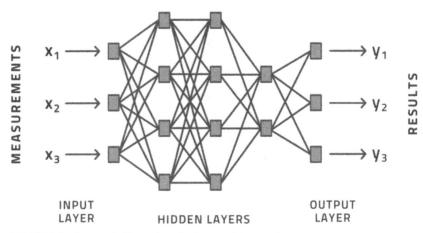

FIGURE 3.7   Schematic illustration of a deep neural network.

has several neurons equal to the number of elements in the input vector. On the right side, the output vector $\underline{y}$ exits from the network. This is called the **output layer** and has several neurons equal to the number of elements in the output vector; this is the result of the calculation. In between them are several **hidden layers**. The number of hidden layers and the number of neurons in each are up to us to choose and is referred to as the **topology** of the neural network.

In bringing the data from one layer to the next, we multiply it by a matrix, add a vector, and apply a so-called **activation function** to it. These are the parameters for each layer that the machine learning algorithm must choose from the data so that the neural network fits to the data in the best possible way. A neural network with two hidden layers, therefore, looks like

$$\underline{y} = A_2 \cdot f\left(\underline{A_1} \cdot f\left(\underline{A_0} \cdot x + \underline{b_0}\right) + \underline{b_1}\right) + \underline{b_2}$$

The activation function $f(\dots)$ is not learnt but decided on beforehand. Considerable research has been put into this choice. While older literature prefers $\tanh(\dots)$, we now have evidence that $ReLu(\dots)$ performs significantly better in virtually all tasks for all layers except the outermost for which we usually choose the identity function. They are both displayed in Fig. 3.8.

$$ReLu(x) = \begin{cases} 0 & x \le 0 \\ x & x > 0 \end{cases}$$

The neural network is not some magical entity. It is the above equation. After we choose values for the parameter matrices $\underline{A_i}$ and vectors $\underline{b_i}$, the neural network becomes a specific model. The main reason it is so famous is that it has the property of **universal approximation**. This property states that the formula

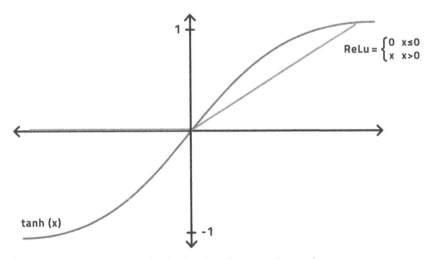

FIGURE 3.8   Two popular activation functions for a neural network.

can represent any continuous function with arbitrary accuracy. There are other formulae that have this property, but neural networks were the first ones that were practically used with this feature in mind. In practice, this means that a neural network can represent your data. What the property does not do is tell us how many layers and how many neurons per layer to choose. It also does not tell us how to find the correct parameter values so that the formula does in fact represent the data. All we know is that the neural network *can* represent the data.

Machine learning as a research field is largely about designing the right learning algorithms, that is, how to find the best parameter values for any given dataset. There are multiple desiderata for it, not just the quality of the parameter values. We also want the training to be fast and to use little processing memory so that training does not consume too many computer resources. Also, we want it to be able to learn from as few examples as possible. Beyond the hype of "big data," there is now significant research into "small data."

Many practical applications provide a certain amount of data. Getting more data is either very difficult, expensive, or virtually impossible. In those cases, we cannot get out of machine learning difficulties by the age-old remedy of collecting more data. Rather, we must be smarter in getting the information out of the data we have. That is the challenge of small data and it is not yet solved. We mention it here to put the property of universal approximation into proper context. It is a nice property to be sure, but it has little practical value.

The neural network is the right model to use if the empirical observations x are independent of each other. For example, if we take a manual fluid sample from a well once per day and perform a laboratory analysis on it, we can assume that today's observation is largely independent of yesterday's observation but may well depend on the pressure and temperature right now. In this context, we may use pressure and temperature as inputs and expect the neural network to represent the laboratory measurement.

If, however, time delays do play a role, then we must incorporate time into the model itself. Let's say we change the choke setting on a well and measure its wellhead pressure, we will see that the wellhead pressure responds to our action a few minutes later. If we measure these quantities every minute, then the observations are not independent of each other because the observation a few minutes ago causally brought about the current observation. Applying the neural network to such datasets is not a good idea because the neural network cannot represent this type of correlation. For this, we need a recurrent neural network.

3.3.2 Recurrent neural network or long short-term memory network

A *recurrent neural network* is very similar in spirit to the regular, the so-called feed-forward, neural network. Instead of having connections that move strictly from left to right in the diagram, it also include connections that amount to

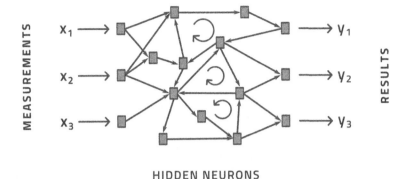

MEASUREMENTS

$x_1 \longrightarrow$

$x_2 \longrightarrow$

$x_3 \longrightarrow$

$\longrightarrow y_1$

$\longrightarrow y_2$

$\longrightarrow y_3$

RESULTS

HIDDEN NEURONS

FIGURE 3.9 Schematic illustration of a recurrent neural network.

cycles, see Fig. 3.9. These cycles act as a memory in the formula by retaining—in some transformed way—the inputs made at prior times.

As mentioned in the previous section, the purpose of recurrence is to model the interdependence between successive observations. In practice such interdependence is virtually always due to time and represents some mechanism of causation or control. If my foot touches the brake pedal in my car, the car slows down a short time later. Modeling the dependence of car speed on the position of the brake pedal can lead to important models, that is, the advanced process control behind the automatic braking system in your car. The time delay involved is important as the car will have moved some distance in that time and so braking must begin sufficiently early so that the car stops before it hits something. It is cases such as this for which the recurrent neural network has been developed.

There are many different forms of recurrent neural networks differing mainly in how the cycles are represented in the formula. The current state-of-the-art is the *long short-term memory* network, or *LSTM* (Hochreiter and Schmidhuber, 1997; Yu, Si, Hu, & Zhang, 2019; Gers, 2001). This network is made up of cells. A schematic of one cell is shown in Fig. 3.10. These cells can be stacked both horizontally and vertically to make up an arbitrarily large network of cells. Once again, it is up to the human data scientist to choose the topology of this network.

The inner working of one cell is defined by the following equations that refer to Fig. 3.10.

$$\underline{f}(t) = \sigma\left(\underline{W}_{fh} \cdot \underline{h}(t-1) + \underline{W}_{fx} \cdot \underline{x}(t) + \underline{b}_f\right)$$
$$\underline{i}(t) = \sigma\left(\underline{W}_{ih} \cdot \underline{h}(t-1) + \underline{W}_{ix} \cdot \underline{x}(t) + \underline{b}_i\right)$$
$$\underline{\tilde{c}}(t) = \tanh\left(\underline{W}_{\tilde{c}h} \cdot \underline{h}(t-1) + \underline{W}_{cx} \cdot \underline{x}(t) + \underline{b}_{\tilde{c}}\right)$$
$$\underline{c}(t) = \underline{f}(t) \cdot \underline{c}(t-1) + \underline{i}(t) \cdot \underline{\tilde{c}}(t)$$
$$\underline{o}(t) = \sigma\left(\underline{W}_{oh} \cdot \underline{h}(t-1) + \underline{W}_{ox} \cdot \underline{x}(t) + \underline{b}_o\right)$$
$$\underline{h}(t) = \underline{o}(t) \cdot \tanh \underline{c}(t)$$

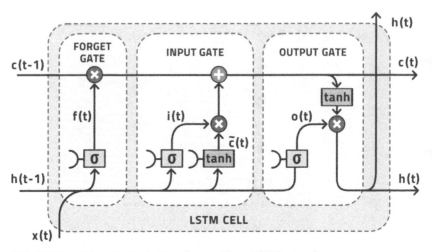

FIGURE 3.10 Schematic illustration of one cell in an LSTM network.

The various $\underline{W}$ and $\underline{b}$ are the weight and bias parameters of the LSTM cell. All weights and biases of the entire network must be tuned by the learning algorithm. The evolution of the internal state of the network $\underline{h}(t)$ is the memory of the network that can encapsulate the time dynamics of the data.

The network is trained by providing it with a time-series $\underline{x}(t)$ for many successive values of time t. The model outputs the same time-series but at a later time $t+T$, where T is the forecast horizon. This is the essence of how LSTM can forecast a time-series,

$$\underline{x}(t+T) = \mathscr{A}_t\big(\underline{x}(t)\big)$$

It is generally not a good idea to model the next timestep and then to chain the model

$$\underline{x}(t+3) = \mathscr{A}_{t+2}\big(\underline{x}(t+2)\big) = \mathscr{A}_{t+2}\Big(\mathscr{A}_{t+1}\big(\underline{x}(t+1)\big)\Big) = \mathscr{A}_{t+2}\Big(\mathscr{A}_{t+1}\big(\mathscr{A}_t\big(\underline{x}(t)\big)\big)\Big)$$

because this compounds errors and makes for a very unreliable model if we want to predict more than one or two timesteps into the future. It is much better to choose the forecast horizon to be some larger number of timesteps. Having done that, the intermediate values can always be computed using the same model with input vectors from earlier in time.

3.3.3 Support vector machines

The *support vector machine* (SVM) is a technique mainly used for classification both in a supervised and unsupervised manner and can be extended to regression as well (Cortes & Vapnik, 1995). The idea, as illustrated in Fig. 3.11 is

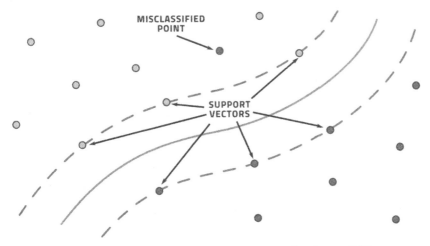

FIGURE 3.11   A support vector machine draws a line through a dataset dividing one category from another as efficiently as possible.

that we have a set of points that belong to one of two categories and we draw a non-linear plane through the space dividing the data into two pieces. Everything above the plane is classified as one category and everything below the plane is classified as the other. The plane itself is an interpolated spline-like object that is based on a few selected data points—the support vectors—that must be found.

The plane chosen is that which has the maximum distance from all points possible. The points at the boundary of this space form the support vectors. In practice, the classification problem may require multiple planes for a good result. As the space of all points gets divided into sections by a collection of planes, we may find that a few data points are isolated in one area. This is another way of identifying outliers.

3.3.4   Random forest or decision trees

The *decision tree* is familiar perhaps from management classes where it is used to structure decision making. At the base of the tree is some decision to be made. At each branching, we ask some question that is relatively easy to answer. Depending on the answer, we go down one branch as opposed to another and eventually reach a point on the tree that has no more branches, a so-called leaf node. This leaf node represents the right answer to the original decision.

Fig. 3.12 illustrates this using a toy example. The question is whether the equipment needs maintenance or not. The left-hand tree starts by analyzing the equipment age in years. If this is larger than 10 years, we go down the left-hand path and otherwise the right-hand path. The left-hand path terminates at a leaf node with the value 1. The right-hand path leads to a secondary question where

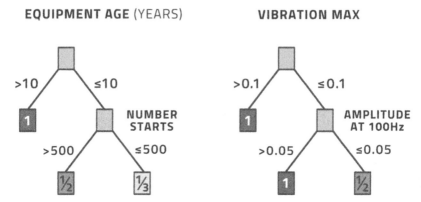

FIGURE 3.12   An example of two decision trees that can be combined into a random forest.

we ask for the number of starts of the equipment. If this is larger than 500, we go down that left-hand path leading us to the number 1/2 and otherwise we go down the right-hand path leading to the number 1/3. The number that we end up with is just a numerical score for now.

The numerical score now needs to be interpreted, which is usually done by thresholding, that is, if the score is greater than some fixed threshold, we decide that the equipment needs preventative maintenance and otherwise we decide that it can operate a little longer without inspection. This is an example of a decision tree in the industrial context (Breiman, Friedman, Olshen, & Stone, 1984).

We can combine several decision trees into a single decision-making exercise. Fig. 3.12 displays a second tree that operates in the same way. Having gone through all the trees, we add up the scores and threshold the final result. If the left tree leads to ½ and the right tree leads to 1, then the total is a score of 3/2. This combination of several decision trees is called a *random forest*.

Constructing the trees, their structures, the decision factors at each branching, and the values for each branching are the model parameters that must be learnt from data. There are sophisticated algorithms for this that have seen significant evolution in recent years making random forest one of the most successful techniques for classification problems. Incidentally, the forest is far from random, of course. The forest is carefully designed by the learning method to give the best possible classification result possible. Random forests can also be used for regression tasks, but they seem to be better suited to classification tasks.

3.3.5   Self-organizing maps (SOM)

A *self-organizing map (SOM)* is an early machine learning technique that was invented as a visualization aid for the human expert (Kohonen, 2001). Roughly, it belongs to the classification kind of techniques as it groups or clusters data points by similarity. The idea is to represent the dataset with special points that are constructed by the learning method in an iterative manner. These special

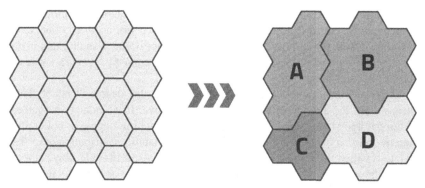

FIGURE 3.13   An example of a self-organizing map.

points are mapped on a two-dimensional grid that usually has the topology of a honeycomb. The topology is important because the learning, that is, the moving around of the special points, occurs over a certain neighborhood of the special point under consideration.

When this process has converged, the special points are arrayed in the topology in a certain way. Any point in the dataset is associated with the special point that it is closest to in the sense of some metric function that must be defined for the problem at hand. This generates a two-dimensional distribution of the entire data. It is now usually a task for the human expert to look at this distribution and determine to what extent the cells in the topology (the honeycomb cells) are similar to each other. Frequently it turns out that several cells are very similar and can be associated with the same macroscopic phenomenon.

For example, in Fig. 3.13 we begin with a topology of 22 honeycomb cells. Having done the learning and the human examination, it is determined that there are only four macroscopically relevant clusters present. Each cluster is represented by several cells. These cells can either have a more subtle meaning within the larger group, or they can simply be a more complex representation of a single cluster than one special point; as one would do using algorithms like k-means clustering (Seiffert & Jain, 2002).

Ultimately, when one encounters a new data point, one would compute the metric distance between it and all the special points learnt. The special point closest to the new point is its representative and we know which cluster it belongs to. The SOM method has turned out to be a very good classification method in its own right. In addition to this however, the current position of a system can be plotted on the graphic and one can see the temporal evolution on the state chart, which is an elegant visual aid for any human working with the data source.

3.3.6   Bayesian network and ontology

Bayesian statistics follows a different philosophy than standard statistics and this has some technical consequences (Gelman, Carlin, Stern, & Rubin, 2004).

Without going into too much detail, standard statistics assesses the probability of an event by the so-called frequentist approach. We will need to collect data as to how frequently this event occurs and does not occur. Usually, this is no problem but sometimes the empirical data is hard to come by, events are rare or very costly, and so on.

Bayesian statistics concerns itself with the so-called degree of belief. This encapsulates the amount of knowledge that we have. If we manage to gather additional knowledge, then we may be able to update our degree of belief and thereby sharpen our probability estimate. The probability distribution after the update is called the posterior distribution. There is a specific rule, Bayes theorem, on how to perform this updating and there is no doubt or difficulty in its application. The tricky part is establishing the probability distribution at the very beginning, the prior distribution. Usually we assume that all events are equally likely if we have no knowledge at all or we might prime the probabilities using empirical frequencies if we do have some knowledge.

Supposing that we have gathered quite a bit of knowledge about the general situation, we can start to create multiple conditional probability distributions. These are conditional on knowing or not knowing certain information.

Let's take the case of a root-cause analysis for a mechanical defect in a gas turbine, see Fig. 3.14. We know from past experience that certain measurements provide important information relating to the root cause of this problem, for example, the axial deviation, which is the amount of movement of the rotor away from its engineered position. This, in turn, is influenced by the bearing temperature and the maximum vibration. These two factors are influenced by other factors in turn. We can build up a network of mutually influencing events and factors. At one end of this matrix are the events we are concerned about such as the mechanical defect. At the other end are the events we can directly cause to happen such as the addition of lubricant into the system.

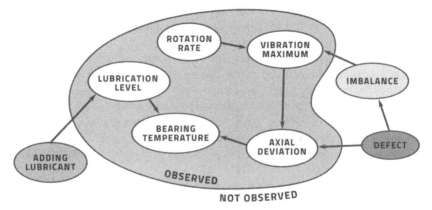

FIGURE 3.14 A very simple example of a Bayesian network for deciding on the root cause of a mechanical defect in a turbine.

Based on our past experience we can establish that if the maximum vibration is above a certain limit, the chance of a rotor imbalance increases by a certain amount, and so on for all the other factors. We can then use statistics to obtain a conditional probability distribution for the mechanical defect as a function of all the factors. This cascading matrix of conditional probabilities is called a ***Bayesian network*** (Scutari and Denis, 2015). Once we have this network, it can be used like so: Initially, the chance of a mechanical defect is very small; we assess this from the frequency of observations of such defects. Then we encounter a high bearing temperature and a high vibration for a low rotation rate. Combining this information using the various conditional probabilities updates our probability of defect. If this is sufficiently high, we may release an alarm and give a specific probability for a problem. We can also compute what effect it would have on the probability if we added lubricant and, based on the result, may issue the recommendation for this specific action.

Please note that this is very different from an ***expert system***. An expert system looks similar at first glance, but it is a system of rules written by human experts as opposed to the Bayesian network that is generated by an algorithm based on data. The expert system has the advantage that all components are designed by people who deeply understand the system. However, combinations of rules often have logical consequences that were not intended by those experts and these are difficult to detect and prevent. Due to this complexity, expert systems have usually failed to work, or at least be commercially relevant given all the effort that flows into them. At present, expert systems are considered an outdated technology. In comparison, Bayesian networks are constructed automatically and therefore can be updated automatically as new data becomes available. An understanding of each individual link in the chain is possible, just like an expert system, but the totality of a network can usually not be understood, as a practical network is usually quite large.

Bayesian networks are great as decision support systems. How much will this factor contribute to an outcome? If we perform this activity, how much will the risk decrease? In our asset intensive industry, the main use cases are: (1) root-cause analysis for some problem, (2) deciding what to do and to avoid based on a concrete numerical target value, (3) scenario planning as a reaction to changes in consumer demand, market prices, and other events in the larger world.

A related method is called an ***ontology*** (Ebrahimipour and Yacout, 2015). This is not a method of machine learning but rather a way to organize human understanding of a system. We may draw a tree-like structure where a branch going from one node to another has a specific meaning such as "is a." For example, the root node could be "rotating equipment." The node "turbine" then has a link to the root node because a turbine is an example of a rotating equipment. The nodes "gas turbine," "steam turbine," or "wind turbine" are all examples of a turbine. This tree can be extended both by adding more nodes as well as adding other types of branches such as "has a" or "is a part of" and so on. Having such a structure allows us to draw certain inferences on the fly. For example, being told that something is a gas turbine, immediately tells us that this object is a rotating equipment that has blades, requires fuel, and generates electricity and so on. This can be used practically in many asset management tasks. Combined with the Bayesian network approach, it reveals causal links between parts of an asset.

3.4 Training and assessing a model

In training any model, we use empirical data to tune the model parameters such that the model fits best to the data. That data is called the *training data*. Having made the model, we want to know how good the model is. While it is interesting how well it performs on the training data, this is not the final answer. We are usually much more concerned by how the model will perform on data it has not seen during its own construction, the *testing data*.

Some training algorithms not only the training data to tune the model parameters but, additionally, use a dataset to determine when the training should stop because no significant model performance improvement can be expected. It is common to use the testing data for this purpose. It then may become necessary to generate a third dataset, the *validation data*, for testing the model on data that was not used at all during training.

It is common practice therefore to split the original dataset into two parts. We generally use 70%–85% of the data for training and the remainder for testing. Fig. 3.15 illustrates this process.

The choice of which data goes into the training or testing datasets needs to be made carefully because both datasets should be representative and significant for the problem at hand. Often, these are chosen randomly, and so chances are that some unintended bias enters the choice. This gave birth to the idea of

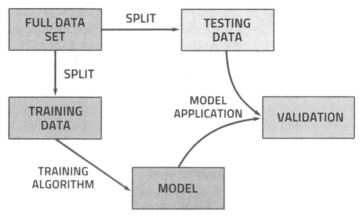

FIGURE 3.15 The full dataset must be split into data to be used for model training and data to be used for testing the model's performance.

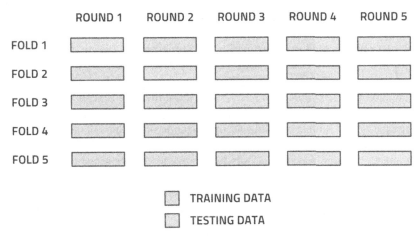

FIGURE 3.16   An illustration of how to divide the full dataset into folds for cross-validation.

cross-validation, which is the current accepted standard for demonstrating a model's performance.

The original dataset is divided into several, roughly equally sized, parts. It is typical to use 5 or 10 such parts and they are called *folds*. Let's say that we are using N folds. We now construct N different training datasets and N different testing datasets. The testing datasets consists of one of the folds and the training datasets consist of the rest of the data. Fig. 3.16 illustrates this division.

We now go through N distinct trainings, resulting in N distinct models. Each model is tested using its own testing dataset. The performance is now estimated to be the average performance over all the trained models. Generally, only the best model is kept but its performance has been estimated in this averaged fashion. This way of assessing the performance is more independent of the dataset bias than assessing it using any one choice of testing dataset and so it is the expected performance estimation method.

Having done all this, we could get the idea that we could use all N models, evaluate them all in each real case and take the average of them all to be the final output of an overarching model. This is a legitimate idea that is called an *ensemble model*. The ensemble model consists of several fully independent models that may or may not have different architectures and that are averaged to yield the output of the full model. These methods are known to perform better on many tasks as compared to individual models, at least in the sense of variance. However, it comes at the expense of resources as several models must not only be trained but maintained and executed in each case. Especially for industrial applications, that often require real-time model execution, this may be prohibitive.

It is important to mention that the model and the training algorithm usually have some fixed parameters that a data scientist must choose prior to training. An example is the number of layers of a neural network and the number of neurons in each layer. Such parameters are called *hyperparameters*. It is rare indeed that we know what values will work best in advance. In order to achieve the best overall result, we must perform *hyperparameter tuning*. Often this is done in a haphazard trial-and-error method by hand because we do not want to try too many combinations. After all, we should ideally perform the full cross-validation for each choice and each one of those involves several model trainings. All this training takes time. It is possible however to perform automated hyperparameter tuning by using an optimization algorithm. If model performance is very important and we have both time and resources to spare, hyperparameter tuning is a very good way to improve model performance. Once this has been exhausted, there is little more one can do.

3.5   How good is my model?

On a weekly basis, we read of some new record of accuracy of some algorithm in a machine learning task. Sometimes it's image classification, then it's regression, or a recommendation engine. The difference in accuracy of the best model or algorithm to date from its predecessor is shrinking with every new advance.

A decade ago, accuracies of 80% and more were already considered good on many problems. Nowadays, we are used to seeing 95% already in the early days of a project. Getting (1) more sophisticated algorithms, (2) performing laborious hyperparameter tuning, (3) making models far more complex, (4) having access to vastly larger data sets, and last—but certainly not least—(5) making use of incomparably larger computer resources has driven accuracies to 99.9% and thereabouts for many problems.

Instinctively, we feel that greater accuracy is better and all else should be subjected to this overriding goal. This is not so. While there are a few tasks for which a change in the second decimal place in accuracy might actually matter, for most tasks this improvement will be irrelevant—especially given that this improvement usually comes at a heavy cost in at least one of the above five dimensions of effort.

Furthermore, many of the very sophisticated models that achieve extremely high accuracies are quite brittle and thus vulnerable to unexpected data inputs. These data inputs may be rare or forbidden in the clean data sets used to produce the model, but strange inputs do occur in real life all the time. The ability for a model to produce a reasonable output even for unusual inputs is its *robustness* or *graceful degradation*. Simpler models are usually more robust and models that want to survive the real world must be robust.

A real-life task of machine learning sits in between two great sources of uncertainty: Life and the user. The data from life is inaccurate and incomplete. This source of uncertainty is usually so great that a model accuracy difference of tens of a percent may not even be measurable in a meaningful way. The user who sees the result of the model makes some decision on its basis and we know from a large body of human-computer-interaction research that the user cares much more about how the result is presented, than the result itself. The usability of the interface, the beauty of the graphics, the ease of understanding and interpretability count more than the sheer numerical value on the screen.

In most use cases, the human user will not be able to distinguish a model accuracy of 95% from 99%. Both models will be considered "good" meaning that they "solve" the underlying problem that the model is supposed to solve. The extra 4% in accuracy are never seen but might have to be bought by many more resources both initially in the model-building phase as well as in the on-going model execution phase. This is the reason we see so many prize-winning algorithms from competitions never being used in a practical application. They have high accuracy but either this high accuracy does not matter in practice or it is too expensive (complexity, project duration, financial cost, execution time, computing resources, etc.) in real operations.

We must not compare models based on the simplistic criterion of accuracy alone but measure them in several dimensions. We will then achieve a balanced understanding of what is "good enough" for the practical purpose of the underlying task. The outcome in many practical projects is that we are done much faster and with less resources. Machine learning should not be perfectionism but pragmatism.

3.6   Role of domain knowledge

Data science aims to take data from some domain and come to a high-level description or model of this data for practical use in solving some challenge in that domain. How much knowledge about the domain does the data scientist have to have to do a good job?

Before starting on a data science project, someone must define (1) the precise domain to focus on, (2) the particular challenge to be solved, (3) the data to be used, and (4) the manner in which the answer must be delivered to the beneficiary. All four of these aspects are not data science in themselves but have significant impact on both the data science and the usefulness of the entire effort. Let's call these aspects the *framework* of the project.

While doing the data science, the data must be assessed for its quality: precision, accuracy, representativeness, and significance.

- Precision: How much uncertainty is in a value?
- Accuracy: How much deviation from reality is there?
- Representativeness: Does the dataset reflect all relevant aspects of the domain?
- Significance: Does the dataset reflect every important behavior/dynamic in the domain?

In seeking a high-level description of the data, be it as a formulaic model or some other form, it is practically expedient to be guided by existing descriptions

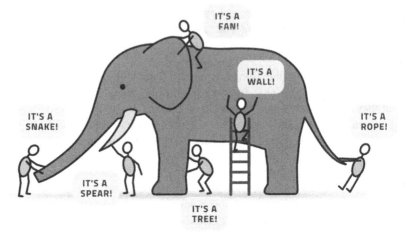

FIGURE 3.17   The parable of the six blind people trying to describe an elephant by touching part of it.

that may only exist in textual, experiential, or social forms, that is, in forms inaccessible to structured analysis. In real projects we find that data science often finds (only) conclusions that are trivial to domain experts or does not find a significant conclusion at all. Incorporating existing descriptions will prevent the first and make the second apparent a lot earlier in the process.

It thus becomes obvious that domain knowledge is important both in the framework as well as the body of a data science project. It will make the project faster, cheaper, and more likely to yield a useful answer.

This situation is beautifully illustrated by the famous elephant parable. Several blind persons, who have never encountered an elephant before, are asked to touch one and describe it. The descriptions are all good descriptions, given the experience of each person, but they are all far from the actual truth, because each person was missing other important data, see Fig. 3.17.

This problem could have been avoided with more data or with some contextual information derived from existing elephantine descriptions. Moreover, the effort might be better guided if it is clear what the description will be used for.

A domain expert typically became an expert by both education and experience in that domain. Both imply a significant amount of time spent in the domain. As most domains in the commercial world are not freely accessible to the public, this usually entails a professional career in the domain. This is a person who could define the framework for a data science project, as they would know what the current challenges are and how they must be answered to be practically useful, given the state of the domain as it is today. The expert can judge what data is available and how good it is. The expert can use and apply the deliverables of a data science project in the real world. Most importantly, this person can communicate with the intended users of the project's outcome. This

Domain Expert & Data Scientist as same Person?

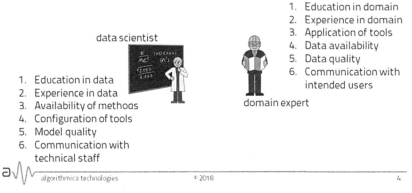

data scientist

1. Education in data
2. Experience in data
3. Availability of methods
4. Configuration of tools
5. Model quality
6. Communication with
   technical staff

1. Education in domain
2. Experience in domain
3. Application of tools
4. Data availability
5. Data quality
6. Communication with
   intended users

domain expert

algorithmica technologies   © 2018   4

FIGURE 3.18   The difference in roles between a data scientist and a domain expert.

is crucial, as many projects end up being shelved because the conclusions are either not actionable or not acted upon.

A data scientist is an expert in the analysis of data. Becoming such an expert also requires a significant amount of time spent being educated and gaining experience. Additionally, the field of data science is developing rapidly, so a data scientist must spend considerable time keeping up with innovations. This person decides which of the many available analysis methods should be used in this project and how these methods are to be parametrized. The tools of the trade (usually software) are familiar to this person, and he or she can use them effectively. Model quality and goodness-of-fit are evaluated by the data scientist. Communication with technical persons, such as mathematicians, computer scientists, and software developers, can be handled by the data scientist (Fig. 3.18).

The expectation that a single individual would be capable of both roles is unrealistic, in most practical cases. Just the requirement of time spent, both in the past as well as regular upkeep of competence, prohibits dual expertise. In some areas it might be possible for a data scientist to learn enough about the domain to make a good model, but assistance would still be needed in defining the challenge and communicating with users, both of which are highly non-trivial. It may also be possible for a domain expert to learn enough data science to make a reasonable model, but probably only when standardized tools are good enough for the job (Fig. 3.19).

If there are two individuals, they can get excellent results quickly by good communication. While the domain expert (DE) defines the problem, the data scientist (DS) chooses and configures the right toolset to solve it. The representative, significant, and available data are chosen by the DE and processed by the DS. The DS builds the tool (that might involve programming) for the

The Role of Domain Expertise in Data Science

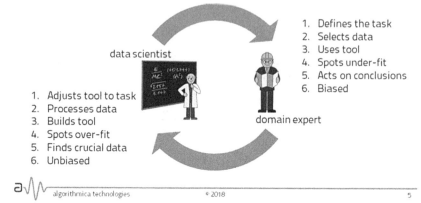

data scientist

1. Defines the task
2. Selects data
3. Uses tool
4. Spots under-fit
5. Acts on conclusions
6. Biased

1. Adjusts tool to task
2. Processes data
3. Builds tool
4. Spots over-fit
5. Finds crucial data
6. Unbiased

domain expert

algorithmica technologies   © 2018   5

FIGURE 3.19   **A project is much more likely to succeed if a domain expert and data scientist collaborate.**

task given the data, and the DE uses the tool to address the challenge. Having obtained a model, the DS can spot over-fitting, where the model has too many parameters so that it effectively memorizes the data, leading to excellent reproduction of training data, but poor ability to generalize. The DE can spot under-fitting, where the model provides too little accuracy or precision to be useful or applied in the real world. The DS can isolate the crucial data in the dataset needed to make a good model; frequently this is a small subset of all the available data. The DE then acts on the conclusions by communicating with the users of the project and makes appropriate changes. The DS approaches the project in an unbiased way, looking at data just as data. The DE approaches the project with substantial bias, as the data has significant meaning to the DE, who has pre-formed hypotheses about what the model should look like. It is important to note that bias, in this context, is not necessarily a detriment to the effort.

It is instructive to think what the outcome can be if we combine a certain amount of domain knowledge with a certain level of data science capability. The vision below is the author's personal opinion but is probably a reasonable reflection of what is possible today.

First, for the sake of this discussion, let's divide domain knowledge into four levels: (1) Awareness is the basic level where we are aware of the nature of the domain. (2) Foundation is knowing what the elements in the domain do, equivalent to a theoretical education. (3) Skill is having practical experience in the domain. (4) Advanced is the level where there is little left to learn and where skill and knowledge can be provided to other people, that is, this person is a domain expert.

Similarly, we can divide up data science into five levels: (1) Data is where we have a table or database of numbers. We can do little more than draw diagrams with this. (2) Information is where we have descriptive statistics about the available data, such as correlations and clusters. (3) Knowledge is when we have some static models. (4) Understanding is when we have dynamic models. The distinction between static and dynamic models is whether the model incorporates the all-important variable of time. A static model makes a statement about how one part of the process affects another, whereas a dynamic model makes additional statements about how the past affects the future. (5) Wisdom is when we have both dynamic models and pattern recognition, for now we can tell what will happen when.

In Fig. 3.20 are several technologies that exist today ordered by the level of data science that they represent and the amount of domain knowledge that was necessary to create them. There are no technologies in the upper left of the diagram because one cannot make such advanced data science with so little domain knowledge.

In conclusion, data science needs domain knowledge. As it is unreasonable to expect any one person to fulfill both roles, we are necessarily looking at a team effort.

3.7   Optimization using a model

We train a machine learning model by adjusting its parameters such that its performance—the least-squares difference between model output and expected output—is a minimum. Such a task is called *optimization* and methods that do this are called optimization algorithms. Every training algorithm for a model is an example of an optimization algorithm. Usually these algorithms are highly customized to the kind of machine learning model that they deal with. We will not treat these methods as that level of detail is beyond the scope of the book.

There are also some general-purpose optimization algorithms that work on any model. These methods can be used on the machine learning model that was just made. If we have a model for the efficiency of a gas turbine, for example, as a function of a multitude of input variables, then we can ask: How should the input variables be modified in order to achieve the maximum possible efficiency? This is a question of *process optimization* or *advanced process control*, which is also sometimes—strangely—called *statistical process control* (Coughanowr, 1991; Qin and Badgwell, 2003).

For the sake of illustration, let's say that we have only two manipulated variables that we can change in order to affect the efficiency. Fig. 3.21 is a contour plot that illustrates the situation. The horizontal and vertical directions are the two manipulated variables. The efficiency resulting from these is displayed in contour lines just like on a topographical map. The points labeled A, B, and C are the peaks in the efficiency.

Any practical problem has boundary conditions, that is, conditions that prevent us from a particular combination of manipulated variables. These usually arise from safety concerns, physical constraints, or engineering limitations and

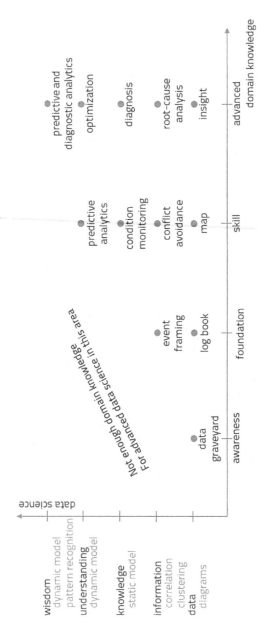

FIGURE 3.20   More is possible if sophisticated domain knowledge is combined with deep data science expertise.

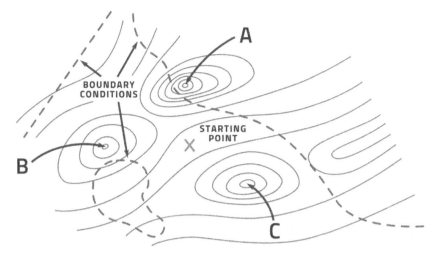

FIGURE 3.21   A contour plot for a controlled variable like efficiency in terms of two manipulated variables.

are indicated by the dashed lines. The peak labeled A for instance is not allowed
due the boundary conditions.

The starting point in the map is the current combination of manipulated
variables. Whatever we do is a change away from this point. What is the best
point? It is clearly either B or C. The resolution of the map does not tell us but
in practice these two efficiencies may be quite close to one another. This is like
a car's GSP choosing one route over another because it computes one path to be
a few seconds shorter. Whichever point we choose to head for, the next task is
to find the right path. In terms of efficiency this is perpendicular to the contour
lines, that is, the steepest ascent.

Computing these actions in advance, with the appropriate amount of time
that is needed in between variable changes, is the purpose of the optimization. Then the actions can be realized in the physical process and the efficiency
should rise as a result.

There is a special case of optimization where the objective is a linear function of all the measurable variables. This is known as *linear programming* and
is displayed in Fig. 3.22. The boundary conditions, or *constraints*, define a region that contains all the points allowed. We can prove that the best solution is
on the boundary of the region and solving such problems is relatively simple
and fast. Linear optimization problems are routinely solved in the oil and gas
industry, particularly in financial optimization, for example, how much of each
refinery end product to make during any one week in a refinery.

So, we have a model $y = f(\underline{x})$ where the function $f(\cdots)$ is potentially a
highly complex machine learning model, $\underline{x}$ is potentially a long vector of various measurements, and y is the target of the optimization, that is, the quantity
that we want to maximize. Of the elements of $\underline{x}$, there will be some that the

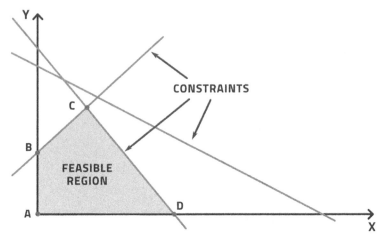

FIGURE 3.22   Linear optimization illustrated as a region bounded by constraints with the optimal points necessarily on the border.

operators of the facility can change at will. These are called *set-points* because we can set them. There will also be values that are fixed by powers that are completely beyond our control, for example, the weather, market prices, raw material qualities and other factors determined outside our facility. These are fixed by the outside world and so correspond to *constraints*.

Now we can ask: What are the values of the set-points, given all the boundary conditions and constrained values, that lead to the largest possible value of y? This is the central question of optimization. Let us call any setting of the set-points that satisfies all boundary conditions, a *solution* to the problem. We then seek the solution with the maximum outcome in y.

One of the most general ways of answering this question is the method of *simulated annealing*. This method starts from the current setting of the set-points. We then consider a random change in the vector of set-points such that this second point is still a solution. If the new point is better than the old, we accept the change. If it is worse, we accept it with a probability that starts off being high and that gradually decreases over the course of the iterations. We keep on making random changes and going through this probabilistic acceptance scheme. Once the value of the optimization target does not change anymore over some number of iterations, we stop the process and choose the best point encountered over the entire journey. See Fig. 3.23 for an illustration of a possible sequence of points in search for the global minimum.

This method provably converges to the best possible point (the global optimum). Even if we cut off the method after a certain finite amount of resources (such as computation time or number of transitions) has been spent on it, we can rely on the method having produced a reasonable improvement relative to the resources spent.

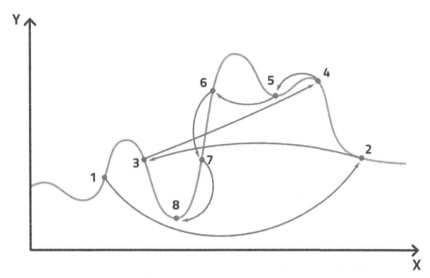

FIGURE 3.23   Simulated annealing illustrated as a sequence of jumps, sometimes for the better and sometimes for the worse, eventually converging on the global minimum.

There are a few fine points of simulated annealing and there is much discussion in the literature on how to calculate the probability of accepting a temporary change for the worse, how to judge if the method has converged, how to generate the new trial point and so on (Bangert, 2012).

3.8   Practical advice

If you want to perform a machine learning project in real-life, here is some advice on how to proceed.

First, all the steps outlined in Section 2.10 should be followed so that you have a clean, representative, significant dataset and you are clear on what you want as the final outcome and how measure whether it is good enough for the practical purpose.

Second, as mentioned in Section 3.6 it is a good idea to conduct such a project with more than one person. Without going into too much detail here, you may need a small team consisting of one or several of these persons: an equipment operator, a maintenance engineer, a process engineer, a process control specialist, a person from the information technology (IT) department, a sponsor from higher management, a data scientist, and a machine learner. From practical experience, it is virtually impossible to do such a project by oneself.

Third, select the right kind of mathematical modeling approach to solve your problem. Several popular ones were outlined earlier. There are more but these represent most models applied in practice. You may ask yourself the following questions:

1. Do you need to calculate values at a future time? You will probably get the best results from a recurrent neural network such as an LSTM as it is designed to model time dependencies.
2. Do you need to calculate a continuous number in terms of others? This is the standard regression problem and can be reliably solved using deep neural networks.
3. Do you need to identify the category that something belongs to? This is classification and the best general classification method is random forest.
4. Do you need to compute the likelihood that something will happen (risk analysis) or has happened (root-cause analysis)? Bayesian networks will do this very nicely.
5. Do you need to do several of these things? Then you will need to make more than one model.

Fourth, every model type and training method has settings. These are the hyperparameters as the model coefficients are usually called the parameters. Having selected the model type, you will now have to set the hyperparameters. This can be done by throwing resources at the issue and doing an automated hyperparameter tuning in which a computer tries out many of them and selects the best. If you have the time to do this, do it. If not, you will need to think the choices through carefully. The main consideration is the bias-variance-complexity trade-off that mainly concerns those hyperparameters that influence how many parameters the model has.

Fifth, perform cross-validation as described in Section 3.4 to calculate all the performance metrics you have chosen correctly. This will get you an objective assessment of how good your model is as described in Section 3.5.

Sixth, if your goal was the model itself, then you now have it. However, if your goal is to derive some action from the model, you now need to compute that action using the model and this usually requires an optimization algorithm as described in Section 3.7. Several such are available but simulated annealing has several good properties that make it an ideal candidate for practical purposes.

Seventh, communicate your results to the rest of the team and to the wider audience in a way that they can understand it. You may need the domain experts to help you do this properly. This audience may need to change their long-established ways based on your results, and they may not wish to do so. The implied change management in this communication is the most difficult challenge of the entire project. In my personal experience this is the cause of failure in almost all failed data science projects in the industry, which is why we will address this at length in other chapters of this book.

References

Agrawal, A., Gans, J., & Goldfarb, A. (2018). *Prediction machines*. Harvard Business Review Press.

Bangert, P. D. (2012). *Optimization for industrial problems*. Springer.

Bishop, C. M. (2006). *Pattern recognition and machine learning*. Springer.

Breiman, L., Friedman, J. H., Olshen, R. A., & Stone, C. J. (1984). *Classification and regression trees*. Chapman & Hall.

Cortes, C., & Vapnik, V. N. (1995). Support-vector networks. *Machine Learning, 20*, 273–297.

Coughanowr, D. R. (1991). In *Process systems analysis and control* (2nd ed.). McGraw-Hill.

Domingos, P. (2015). *The master algorithm*. Basic Books.

Ebrahimipour, V., & Yacout, S. (2015). *Ontology modeling in physical asset integrity management*. Springer.

Gelman, A., Carlin, J. B., Stern, H. S., & Rubin, D. B. (2004). *Bayesian data analysis*. Chapman & Hall.

Gers, F. (2001): Long Short-Term Memory in Recurrent Neural Networks. PhD thesis at Ecole Polytechnique Federale de Lausanne.

Goodfellow, I., Bengio, Y., & Courville, A. (2016). *Deep learning*. MIT Press.

Hagan, M. T., Demuth, H. B., & Beale, M. (1996). *Neural network design*. PWS.

Hochreiter, S., & Schmidhuber, J. (1997). Long short-term memory. *Neural Computation, 9*, 1735–1780.

Kohonen, T. (2001). In *Self-organizing maps* (3rd ed.). Springer.

MacKay, D. J. C. (2003). *Information theory, inference, and learning algorithms*. Cambridge University Press.

Mitchell, T. M. (2018). *Machine learning*. McGraw Hill.

Qin, S. J., & Badgwell, T. A. (2003). A survey of industrial model predictive control technology. *Control Engineering Practice, 11*, 733–764.

Scutari, M., & Denis, J. B. (2015). *Bayesian networks*. Chapman & Hall.

Seiffert, U., & Jain, L. C. (2002). *Self-organizing neural networks*. Physica Verlag.

Yu, Y., Si, X., Hu, C., & Zhang, J. (2019). A review of recurrent neural networks: LSTM cells and network architectures. *Neural Computation, 31*, 1235–1270.

Chapter 4

Introduction to Machine Learning in the Oil and Gas Industry

Patrick Bangert

Artificial Intelligence Team, Samsung SDSA, San Jose, CA, United States; algorithmica technologies GmbH, Küchlerstrasse 7, Bad Nauheim, Germany

This chapter will attempt to provide an overview over some of the practical applications that machine learning has found in oil and gas. The aim of the chapter is twofold: First, it is to show that there are many applications that are realistic and have been carried out on real-world assets, that is, machine learning is not a dream. Second, the status of machine learning in oil and gas is in its early days as the applications are specialized and localized. It must be stated clearly that most of the studies done, have been done at universities and that the applications fully deployed in commercial companies are the exception. This chapter makes no attempt at being complete or even representative of the work done. It just provides many starting points for research on use cases and presents an overview. There are some use cases that attract a vast number of papers and this chapter will present such use cases with just one or a few exemplary papers chosen at random.

This chapter can be used as a starting point to searching for literature on a specific application case and as an overview over the fact that there are a great many different use cases in oil and gas for machine learning both in terms of the type of use as well as the area of application. Some review papers have already appeared on this topic that also provide a good overview (Mohammadpoor & Torabi, 2018; Balaji et al., 2018; Hanga and Kovalchuk, 2019; Li, Yu, Cao, Tian, & Cheng, 2020; Nguyen, Gosine, & Warrian, 2020).

4.1 Forecasting

Most data in the oil and gas industry is in the form of a time-series. That is a series of values for the same measurement obtained at regular intervals of time. For such data, we are often interested in future values and this is called

Machine Learning and Data Science in the Oil and Gas Industry.
http://dx.doi.org/10.1016/B978-0-12-820714-7.00004-2

69

forecasting. We may, for instance, want to forecast production in unconventional wells (Cao et al., 2016), in wet gas shale (Anderson et al., 2016), or in general (Sheremetov, González-Sánchez, López-Yáñez, & Ponomarev, 2013; Bikmukhametov and Jäschke, 2019). More long-term, we may want to forecast the ultimate recovery (EUR) for horizontal shale wells (Gaurav, 2017).

Forecasting can lead to immediate action in a kind of process control application. An example is the modeling of lost circulation while drilling. This is a major expense, can be forecasted, and thereby mitigated (Al-Hameedi et al., 2018).

Financially speaking, the oil price itself can be forecasted for better financial planning and investment (Dimitriadou, Gogas, Papadimitriou, & Plakandaras, 2019).

It is possible to forecast the need for flaring (Giuliani et al., 2019), which leads to an advanced process control method to mitigate and avoid flaring. This is significant as flaring is one of the most wasteful processes of the modern age accounting for as much as 25% of the natural gas consumption in the United States. An effective flare-reducing method would contribute tremendously toward an effort to curb climate change.

4.2 Predictive maintenance

The general topic of *predictive maintenance* features largely in the popular literature and discussion in the industry. In that phrase, the word "predictive" means different things to different people.

On the one hand, it means forecasting an event. That is, we calculate at what time in the future something will happen and alert the operator to that future fact. In that sense, prediction is a synonym for forecast.

On the other hand, it sometimes means recognizing an event. That is, we calculate that there is something worrisome going on right now. In that sense, prediction is a synonym for computation. This use case is often called *anomaly detection*. For marketing reasons, many products offered under the label predictive maintenance, deliver anomaly detection or some form of advanced *condition monitoring*. Anomaly detection can be done in an unsupervised way as discussed in a whole book (Aldrich and Auret, 2013).

It is worthwhile to be aware of these different usages of the term so that one can be careful when reading texts about some technologies. In the context of this book, predictive maintenance will be used in the sense of forecasting something in the future whereas anomaly detection will be used in the context of detecting something going on now. The two are clearly very distinct use cases because predictive maintenance allows you to prepare, plan, and execute an action before the forecasted moment comes to pass, that is, it allows you to prevent the failure event. For this reason, the forecast must be made a macroscopic amount of time into the future—a forecast by several seconds is usually useless.

General methods for forecasting problems in refineries are studied (Su-ursalu, 2017; Helmiriawan, 2018). Problems with electric submersible pumps can be forecasted (Kandziora, 2019). Generally, one may trace and forecast the remaining useful life of a physical asset (Fauzi, Aziz, & Amiruddin, 2019; Animah and Shafiee, 2018). Forecasting can be combined with the diagnosis of what failure mechanism will be present as illustrated by the case of sucker rod pumps (Bangert and Sharaf, 2019).

Anomaly detection has led to many more publications and applications in the industry. Here is a short list of a few use cases:

- Pipeline defects (Mohamed, Hamdi, & Tahar, 2015a; Layouni, Hamdi, & Tahar, 2017) and using long-range ultrasound (Akrama, Isa, Rajkumar, & Lee, 2014).
- Leak detection (Ahn, Kim, & Choi, 2019).
- General drilling operations (Kejela, Esteves, & Rong, 2014; Noshi & Schubert, 2018).
- Monitoring the bottom-hole pressure during drilling (Sui, Nybø, Gola, Ro-verso, & Hoffmann, 2011; Ignatov, Sinkov, Spesivtsev, Vrabie, & Zyuz-in, 2018).
- Offshore facility architecture (Tygesen, Worden, Rogers, Manson, & Cross, 2019).
- Structural health (Farrar and Worden, 2013).
- Artificial lift problems (Pennel, Hsiung, & Putcha, 2018).
- Corrosion detection (Chern-Tong and Aziz, 2016; Zukhrufany, 2018).
- Gas turbines (Yan and Yu, 2019; Bangert, 2020).
- Exhaust fan (Amruthnath and Gupta, 2018).
- Multiphase flow meters (Barbariol, Feltresi, & Susto, 2020).
- Rotating equipment in general (Bangert, 2017a,b).

Once the problem is recognized, we can employ recommended systems to decide what to do about it (Madrid and Min, 2020).

In modeling some mechanisms, it is sometimes necessary to choose which feature (input variables) to choose among many that are available. One inter-esting study chooses to use self-organizing maps to help humans make the se-lection, and neural networks to do the actual modeling (Mohamed, Hamdi, & Tahar, 2015b). Clever uses of multiple machine learning methods in sequence to approach a single problem are an excellent way to approach complex situations.

With some insight as to which equipment is in what state and can be ex-pected run, a plant may want to *schedule planned outages* in the best possible way so that it can conduct all the necessary maintenance measures in that same outage. It is undesirable to maintain something too early. On the other hand, one also does not want something to fail a few weeks after a major outage. Good scheduling as well as good planning of measures is critical in keeping the plant running in between major planned outages. These events can be scheduled and planned by models (Dalal, Gilboa, Mannor, & Wehenkel, 2019). Dynamic decision making in the context of predictive maintenance can also be supported by models (Susto, Schirru, Pampuri, McLoone, & Beghi, 2015).

In studying predictive maintenance in all its forms, it is important to distinguish a *drifting sensor* from a real change in signal. All physical sensors will drift over time and need to occasionally be recalibrated. This effect may lead either to seemingly abnormal conditions or happen so gradually as to redefine what a model might consider normal. In any case, telling the difference is important and this too can be done with analytics (Zenisek, Holzinger, & Affenzeller, 2019).

After anomaly detection is done, we naturally ask what type of anomaly it is. This is asking for the failure mode, the physical location, the type of problem or generally for some sort of description or *diagnosis* of what is going on. Ultimately, this would lead to a maintenance measure plan in which certain individuals will be sent on site with tools and spare parts to perform a task. We must know the task to plan for the people, tools, and parts to be on hand at the right time. The issue of diagnosis is important (Jahnke, 2015).

The diagnosis of a physical fault is challenging because training a model to recognize fault A as opposed to fault B requires a sufficient number of examples of all the possible faults. As equipment, fortunately, does not fail very often, most operators have a limited library of dataset for known faults—at least relative to the same make and model of equipment. In the cases where operators do have this data, one can use this to diagnose the problem automatically. An example is rod pumps for which many operators have a large fleet of hundreds or thousands. As these pumps also fail once or twice a year, collecting the required dataset is no problem and diagnosis can be done accurately (Bangert, 2019; Sharaf et al., 2019).

4.3   Production

Machine learning is often used in decision support systems to assist a human decision maker to make the decision faster, more objectively, and utilizing more information. An example is deciding which enhanced oil recovery (EOR) method to employ (Alvarado et al., 2002; Muñoz Vélez, Romero Consuegra, & Berdugo Arias, 2020). Having chosen the method, EOR needs to be used in the best way and its benefits predicted (Krasnov, Glavnov, & Sitnikov, 2017).

Waterflooding is a method by which water in injected into a reservoir to increase the pressure on the oil and thus increase production. The interaction between reservoir, water injection, and resultant oil production can be modeled well using generative adversarial networks (Zhong, Sun, Wang, & Ren, 2020).

A productive well produces a mixture of gas, oil, water, and, often, particulate matter such as sand. This mixture is usually called *multiphase flow* and we are interested in how much of this is in the form of oil, water, and gas. This is the job of the multiphase flow meter, which is a device installed on the wellhead. Such devices are expensive and fragile. There is a growing body of research on substituting this device with a soft sensor, that is, a calculation based on more readily available measurements on the well. Some examples include a gas flow calculation (Ricardo, Jiménez, Ferreira, & Meirelles, 2018) and full virtual multiphase flow meters (Andrianov, 2018; Shoeibi Omrani, Dobrovolschi, Belfroid, Kronberger, & Munoz, 2018; Bikmukhametov and Jäschke, 2020) and classification of the fluid regimes (Vahabi and Selviah, 2019). In pipelines, we may be interested in which phase (water, oil, gas) is present at any moment and place (Vahabi, 2019).

During its productive life, a well may leak some gas into the atmosphere circumventing the well itself. This gas is lost to production and contributes to climate change. This effect, known as gas migration, can be modeled, and thus mitigated (Montague, Pinder, & Watson, 2018).

As well ages over time, it generally produces less and less volume each year. This is the famous decline curve. There are some standard models of decline curves that are quite simple in their assumptions. Decline can also be modeled using more variables to achieve a better fit and thus ability to plan and support decision making (Aliyuda, Howell, & Humphrey, 2020).

Keeping a field in production requires generating electricity and this releases pollutants into the atmosphere. Modeling the amount of NOx released is important for regulatory approval as well as for the environment and can be done using machine learning methods (Si, Tarnoczi, Wiens, & Du, 2019).

4.4 Modeling physical relationships

Crude oil extracted from the ground is not a homogenous fluid but rather varies greatly depending on the geographical location. Many of the crude's idiosyncrasies are characterized by the so-called PVT properties. The acronym PVT stands for pressure-volume-temperature and is a reference to the ideal gas law but these properties include more items such as gravity and molecular weight, bubble-point pressure, bubble-point oil formation volume factor, isothermal compressibility, undersaturated oil formation volume factor, oil density, dead oil viscosity, bubble-point oil viscosity, undersaturated oil viscosity, gas/oil interfacial tension, and water/oil interfacial tension. All these properties can be computed using machine learning (El-Sebakhy et al., 2007; Ramirez, Valle, Romero, & Jaimes, 2017). Similarly, the dew-point pressure of a gas condensate can be computed (Majidi, Shokrollahi, Arabloo, Mahdikhani-Soleymanloo, & Masihia, 2014). The equivalent alkane carbon number of oil can be computed (Creton, Lévêque, & Oukhemanou, 2019).

For reservoir geologists, characterizing the reservoir is a major challenge. Machine learning can help by modeling a variety of sub-surface properties (Anifowose, 2009). Examples include the rock porosity and permeability (Helmy, Fatai, & Faisal, 2010; Berezovsky, Belozerov, Bai, & Gubaydullin, 2019). Interpreting sub-surface data from a petrophysical point of view in order to determine if a region is worth drilling or not is a major task for exploration and production companies and requires significant labor, which can largely be automated (Brown et al., 2008). After that, the location for the individual wells needs to be determined (Nwachukwu, Jeong, Sun, Pyrcz, & Lake, 2018). Various risks in exploration must be combined and weighed to make decisions and these can be objectified (Silva et al., 2019). Fractures in shales can be detected using deep learning methods from image processing (Tian and Daigle, 2018). Rock facies can be classified (Zhang and Zhan, 2017).

Dead oil viscosity has been difficult to determine but can be successfully modeled using machine learning (Sinha, Dindoruk, & Soliman, 2020). Sodium glycinate solutions are often used to extract CO_2 from flue gas emissions and thus cleaning the exhaust released to the atmosphere. We can model the solubility of CO_2 in these solutions and thus optimize the extraction efficiency (Baghban and Bahadori, 2016).

4.5 Optimization and advanced process control

When drilling a well, we want to get to the bottom as fast as possible. The rate of penetration depends on many variables some of which can be controlled on the rig and some of which are inherent to the rock. Optimizing the rate is a complex task for real-time control (Wallace, Hegde, & Gray, 2015).

Deployment of advanced process control applications can be very work intensive over larger fields. Machine learning can be used to cluster the wells into groups and then each group needs only one manual configuration. In this way, manual effort can be reduced significantly (Patel and Patwardhan, 2019).

Enhanced oil recovery (EOR) relies on a delicate hydrophilic-lipophilic balance that can be maintained using modeling systems (Baghban and Bahadori, 2016; Baghban et al., 2016). A modern popular method of EOR is to inject CO_2 into the well to extract oil. This results in long-term storage of CO_2 and is thus beneficial relative to climate change. The relationship between the injected and extracted volumes relative to the rock structure must be modeled to do this optimally (You et al., 2019).

Another important environmental task is removing sulfur from crude oil and this process can be made more efficient by modeling (Al-Jamimi, Al-Azani, & Saleh, 2018).

One of the most important aspects of advanced process control is to deal with variations in the process that occur due to outside influences that are not under the direct control of the operator. Such events need to be detected and corrected for by using the influence that the operator has. This happens for instance when there are oscillations in the process (Dambros, Trierweiler, Farenzena, & Kloft, 2019) or swings in process pressure (Subraveti, Li, Prasad, & Rajendran, 2019).

The profitable running of a refinery requires careful scheduling of when it should produce which products. This is often done based on linear programming optimization, but this relies on assumptions made by analysts. One can optimize the whole situation using more complex modeling (Gao, Shang, Jiang, Huang, & Chen, 2014).

Optimization based on modeling is discussed in more detail and generality in Bangert (2012). The model that describes the process is made separately using machine learning. Having gotten this model, one can use an optimization method to derive the best action for accomplishing the goal at any one time. Simulated annealing is the best general-purpose optimization method for this purpose that is capable of running in resource constrained real-time applications.

4.6   Other applications

For an asset-heavy industry, assessing risks as a decision-support system and planning tool is important. Even risk assessment can be automated using machine learning. This can be done holistically by investigating incidence reports culminating is a risk matrix analysis (Kurian, Ma, Lefsrud, & Sattari, 2020). It can also be done in a target way by focusing on individual risk factors such as drive-off for a drilling rig (Paltrinieri, Comfort, & Reniers, 2019).

Most data in the oil and gas industry is in the form of time-series and so is structured data. Some data is also available in text documents and can be analyzed by natural language processing systems. An example is the extraction of risk event from text descriptions (Antoniak, Dalgliesh, Verkruyse, & Lo, 2016).

The oil and gas industry is severely influenced by the *price of oil and gas*, which are commodity prices available on public marketplaces and known to all. These prices are influenced heavily by changes in production brought about by engineering and maintenance events as well as by exploration and drilling operations that uncover new sources. Perhaps an even larger influence is international politics and warfare. It is therefore a difficult proposition to forecast the economic value of a region's oil and gas exports for any length of time. Nonetheless this has been attempted with some success (Windarto, Dewi, & Hartama, 2017). One may also attempt to model the oil price directly (Gao and Lei, 2017; Naderi, Khamehchi, & Karimi, 2019). In contrast, one may wish to model the crash events in oil prices (Zhang and Hamori, 2020).

The boom-and-bust nature of the oil price market causes large shifts in demand by exploration and production (EP) companies for services and equipment. The vast majority of the people in the industry work for the equipment manufacturers or service companies, however. A small upset in oil demand, causes a larger upset with the EP companies, which causes an even larger upset with the services and equipment companies and this results in a significant shift in employment levels and spending on products and services even further down the supply chain. This increasing cascade effect is known as the *Bullwhip Effect* and can be studied using machine learning to try to cope with future swings (Sousa, Ribeiro, Relvas, & Barbosa-Póvoa, 2019).

Monitoring the production of a well is an important task for production, maintenance, and planning. Installing instrumentation is challenging for various reasons. Sensors are expensive to buy, maintain and connect at scale over large fields in inaccessible areas that have inhospitable weather. Even if they work, they have inherent uncertainties and cannot be totally relied upon. Calculating the oil rate, for example, is possible without having to measure it (Khan, Alnuaim, Tariq, & Abdulraheem, 2019).

Especially bottom-hole information is hard to come by, as these sensors are hard to place and replace. One can however infer the bottom-hole pressure by modeling (Spesivtsev et al., 2018).

Well logs need to be interpreted to ascertain oil and gas identification. This is particularly complex in low-porosity and low-permeability reservoirs but can be automated (Shi, 2009).

Deciding which artificial lift method is best of any individual well is an engineering task fraught with details that can take up a significant amount of time and effort. Decision-support systems are available that suggest the best method based on many variables (Ounsakul, Sirirattanachatchawan, Pattarachupong, Yokrat, & Ekkawong, 2019).

Satellite data is useful for many things and correctly interpreting it for particular use cases is a major task for data-driven systems. For example, one can detect oil spills from satellite data and thus recognize problems sooner (Pelta, Carmon, & Ben-Dor, 2019).

References

Ahn, B., Kim, J., & Choi, B. (2019). Artificial intelligence-based machine learning considering flow and temperature of the pipeline for leak early detection using acoustic emission. *Engineering Fracture Mechanics, 210*, 381–392.

Aldrich, C., & Auret, L. (2013). *Unsupervised process monitoring and fault diagnosis with machine learning methods*. Springer.

Al-Hameedi, A. T. T., Alkinani, H. H., Dunn-Norman, S., Flori, R. E., Hilgedick, S. A., Amer, A. S., & Alsaba, M. T. (2018). *Using machine learning to predict lost circulation in the Rumaila field*. Iraq: Society of Petroleum Engineersdoi: 10.2118/191933-MS.

Aliyuda, K., Howell, J., & Humphrey, E. (2020). *Impact of geological variables in controlling oil-reservoir performance: an insight from a machine-learning technique*. Society of Petroleum Engineersdoi: 10.2118/201196-PA.

Al-Jamimi, H. A., Al-Azani, S., & Saleh, T. A. (2018). Supervised machine learning techniques in the desulfurization of oil products for environmental protection: A review. *Process Safety and Environmental Protection, 120*, 57–71.

Akrama, N. A., Isa, D., Rajkumar, R., & Lee, L. H. (2014). Active incremental support vector machine for oil and gas pipeline defects prediction system using long range ultrasonic transducers. *Ultrasonics, 54*(6), 1534–1544.

Alvarado, V., Ranson, A., Hernandez, K., Manrique, E., Matheus, J., Liscano, T., & Prosperi, N. (2002). *Selection of EOR/IOR opportunities based on machine learning*. Society of Petroleum Engineersdoi: 10.2118/78332-MS.

Amruthnath, N., & Gupta, T. (2018). In *A research study on unsupervised machine learning algorithms for early fault detection in predictive maintenance* (pp. 355–361.). Singapore: 2018 5th International Conference on Industrial Engineering and Applications (ICIEA)doi: 10.1109/IEA.2018.8387124. 2018.

Anderson, R. N., Xie, B., Wu, L., Kressner, A. A., Frantz, J. H., Ockree, M. A., & Brown, K. G. (2016). *Petroleum analytics learning machine to forecast production in the wet gas marcellus shale*. San Antonio, Texas: Unconventional Resources Technology Conference. 1–3 August 2016.

Andrianov, N. (2018). A machine learning approach for virtual flow metering and forecasting. *IFAC-PapersOnLine, 51*(8), 191–196. 2018.

Anifowose, F. (2009). *Hybrid AI models for the characterization of oil and gas reservoirs: concept, design and implementation*. VDM Verlag.

Animah, I., & Shafiee, M. (2018). Condition assessment, remaining useful life prediction and life extension decision making for offshore oil and gas assets. *Journal of Loss Prevention in the Process Industries, 53*, 17–28.

Antoniak, M., Dalgliesh, J., Verkruyse, M., & Lo, J. (2016). *Natural language processing techniques on oil and gas drilling data*. Society of Petroleum Engineersdoi: 10.2118/181015-MS.

Baghban, A., Bahadori, M., Lee, M., Bahadori, A., & Kashiwao, T. (2016). Modelling of CO_2 separation from gas streams emissions in the oil and gas industries. *Petroleum Science and Technology, 34*(14), 1291–1299.

Baghban, A., & Bahadori, A. (2016). Determination of efficient surfactants in the oil and gas production units using the SVM approach. *Petroleum Science and Technology, 34*(20), 1691–1697.

Balaji, K., Rabiei, M., Suicmez, V., Canbaz, C. H., Agharzeyva, Z., Tek, S., Bulut, U., & Temizel, C. (2018). *Status of data-driven methods and their applications in oil and gas industry*. Society of Petroleum Engineersdoi: 10.2118/190812-MS.

Bangert, P. (2017a). *Predicting and detecting equipment malfunctions using machine learning*. Offshore Technology Conferencedoi: 10.4043/28109-MS.

Bangert, P. (2017b). *Smart condition monitoring using machine learning*. Society of Petroleum Engineers. doi: 10.2118/187936-MS.

Bangert, P. (2012). *Optimization for industrial problems*. Springer.

Bangert, P. (2019). *Diagnosing and predicting problems with rod pumps using machine learning*. Society of Petroleum Engineersdoi: 10.2118/194993-MS.

Bangert, P., & Sharaf, S. (2019). *Predictive maintenance for rod pumps*. Society of Petroleum Engineersdoi: 10.2118/195295-MS.

Bangert, P. (2020). *Predictive maintenance for gas turbines*. International Petroleum Technology Conferencedoi: 10.2523/IPTC-19864-Abstract.

Barbariol, T., Feltresi, E., & Susto, G.A. (2020). A machine learning-based system for self-diagnosis multiphase flow meters. International Petroleum Technology Conference. doi:10.2523/IPTC-19865-MS.

Berezovsky, V., Belozerov, I., Bai, Y., & Gubaydullin, M. (2019). Digital rock modeling of a terrigenous oil and gas reservoirs for predicting rock permeability with its fitting using machine learning. In V. Voevodin, S. Sobolev (Eds.), *Supercomputing. 5th Russian Supercomputing Days, Moscow, Sept 23–24*. Heidelberg: Springer.

Bikmukhametov, T., & Jäschke, J. (2019). Oil production monitoring using gradient boosting machine learning algorithm. *IFAC-PapersOnLine, 52*(1), 514–519. 2019.

Bikmukhametov, T., & Jäschke, J. (2020). First principles and machine learning virtual flow metering: a literature review. *Journal of Petroleum Science and Engineering, 184*, 106487.

Brown, N., Roubíčková, A., Lampaki, I., MacGregor, L., Ellis, M., & de Newton, P V. (2008). Machine learning on Crays to optimize petrophysical workflows in oil and gas exploration. *Concurrency and Computation: Practice and Experience, 20*, 1721–1723.

Cao, Q., Banerjee, R., Gupta, S., Li, J., Zhou, W., & Jeyachandra, B. (2016). *Data driven production forecasting using machine learning*. Society of Petroleum Engineersdoi: 10.2118/180984-MS.

Chern-Tong, H., & Aziz, I. B. A. (2016). A corrosion prediction model for oil and gas pipeline using CMARPGA. 2016 3rd International Conference on Computer and Information Sciences (ICCOINS), Kuala Lumpur, pp. 403–407. doi: 10.1109/ICCOINS. 2016.7783249.

Creton, B., Lévêque, I., & Oukhemanou, F. (2019). Equivalent alkane carbon number of crude oils: A predictive model based on machine learning. *Oil & Gas Science and Technology - Revue d'IFP Energies nouvelles.*, *74*, 30.

Dalal, G., Gilboa, E., Mannor, S., & Wehenkel, L. (2019). Chance-constrained outage scheduling using a machine learning proxy. *IEEE Transactions on Power Systems*, *34*(4), 2528–2540. doi: 10.1109/TPWRS. 2018.2889237.

Dambros, J. W. V., Trierweiler, J. O., Farenzena, M., & Kloft, M. (2019). Oscillation detection in process industries by a machine learning-based approach. *Industrial & Engineering Chemistry ResearchV 58*(31), 14180–14192.

Dimitriadou, A., Gogas, P., Papadimitriou, T., & Plakandaras, V. (2019). Oil market efficiency under a machine learning perspective. *Forecasting*, *1*(1), 157–168. 2019.

El-Sebakhy, E. A., Sheltami, T., Al-Bokhitan, S. Y., Shaaban, Y., Raharja, P. D., & Khaeruzzaman, Y. (2007). *Support vector machines framework for predicting the PVT properties of crude oil systems*. Society of Petroleum Engineers. doi: 10.2118/105698-MS.

Farrar, C. R., & Worden, K. (2013). *Structural health monitoring: a machine learning perspective*. Wiley.

Fauzi, M. F. A. M., Aziz, I. A., & Amiruddin, A. (2019): The prediction of remaining useful life (RUL) in oil and gas industry using artificial neural network (ANN) algorithm. 2019 IEEE Conference on Big Data and Analytics (ICBDA), Pulau Pinang, Malaysia, pp. 7–11. doi: 10.1109/ICBDA47563.2019.8987015.

Gaurav, A. (2017). *Horizontal shale well EUR determination integrating geology, machine learning, pattern recognition and multivariate statistics focused on the Permian basin*. Society of Petroleum Engineersdoi: 10.2118/187494-MS.

Gao, S., & Lei, Y. (2017). A new approach for crude oil price prediction based on stream learning. *Geoscience Frontiers*, *8*(1), 183–187. January 2017.

Gao, X., Shang, C., Jiang, Y., Huang, D., & Chen, T. (2014). Refinery scheduling with varying crude: A deep belief network classification and multimodel approach. *AIChE Journal*, *60*(7), 2525–2532.

Giuliani, M., Camarda, G., Montini, M., Cadei, L., Bianco, A., Shokry, A., Baraldi, P., & Zio, E. (2019). Flaring events prediction and prevention through advanced big data analytics and machine learning algorithms. Offshore Mediterranean Conference.

Hanga, K. M., & Kovalchuk, Y. (2019). Machine learning and multi-agent systems in oil and gas industry applications: A survey. *Computer Science Review*, *34*, 100191.

Helmiriawan, H. (2018): Scalability analysis of predictive maintenance using machine learning in oil refineries. Master's Thesis. TU Delft Electrical Engineering.

Helmy, T., Fatai, A., & Faisal, K. (2010). Hybrid computational models for the characterization of oil and gas reservoirs. *Expert Systems with Applications*, *37*(7), 5353–5363.

Ignatov, D. I., Sinkov, K., Spesivtsev, P., Vrabie, I., & Zyuzin, V. (2018). Tree-based ensembles for predicting the bottomhole pressure of oil and gas well flows. In W. van der Aalst et al.,et al. (Ed.), *Analysis of Images, Social Networks and Texts. AIST 2018. Lecture Notes in Computer Science* (vol. 11179). Cham: Springer.

Jahnke, P. (2015): Machine learning approaches for failure type detection and predictive maintenance. Technische Universität Darmstadt. Master's Thesis.

Kandziora, C. (2019). Applying artificial intelligence to optimize oil and gas production. Offshore Technology Conference. doi:10.4043/29384-MS.

Khan, M. R., Alnuaim, S., Tariq, Z., & Abdulraheem, A. (2019). *Machine learning application for oil rate prediction in artificial gas lift wells*. Society of Petroleum Engineersdoi: 10.2118/194713-MS.

Kejela, G., Esteves, R. M., & Rong, C. (2014). In *Predictive analytics of sensor data using distributed machine learning techniques* (pp. 626–631.). Singapore: 2014 IEEE 6th International Conference on Cloud Computing Technology and Sciencedoi: 10.1109/CloudCom.2014.44.

Krasnov, F., Glavnov, N., & Sitnikov, A. (2017). A machine learning approach to enhanced oil recovery prediction. International Conference on Analysis of Images, Social Networks and Texts AIST 2017: Analysis of Images, Social Networks and Texts, pp. 164–171.

Kurian, D., Ma, Y., Lefsrud, L., & Sattari, F. (2020). Seeing the forest and the trees: Using machine learning to categorize and analyze incident reports for Alberta oil sands operators. *Journal of Loss Prevention in the Process Industries*, *64*, 104069.

Layouni, M., Hamdi, M. S., & Tahar, S. (2017). Detection and sizing of metal-loss defects in oil and gas pipelines using pattern-adapted wavelets and machine learning. *Applied Soft Computing*, *52*, 247–261.

Li, H., Yu, H., Cao, N., Tian, H., & Cheng, S. (2020). Applications of Artificial Intelligence in Oil and Gas Development. *Archives of Computational Methods in Engineering*. https://doi.org/10.1007/s11831-020-09402-8.

Madrid, J., Min, A. (2020). Reducing oil well downtime with a machine learning recommender system. Massachusetts Institute of Technology. Supply Chain Management Capstone Projects.

Majidi, S. M. J., Shokrollahi, A., Arabloo, M., Mahdikhani-Soleymanloo, R., & Masihia, M. (2014). Evolving an accurate model based on machine learning approach for prediction of dew-point pressure in gas condensate reservoirs. *Chemical Engineering Research and Design*, *92*(5), 891–902.

Mohamed, A., Hamdi M.S., & Tahar, S. (2015). Self-organizing map-based feature visualization and selection for defect depth estimation in oil and gas pipelines. 2015 19th International Conference on Information Visualisation, Barcelona, pp. 235–240. doi: 10.1109/iV. 2015.50.

Mohamed, A., Hamdi, M. S., & Tahar, S. (2015b). A machine learning approach for big data in oil and gas pipelines. 2015 3rd International Conference on Future Internet of Things and Cloud, Rome, pp. 585–590. doi: 10.1109/FiCloud.2015.54.

Mohammadpoor, M., & Torabi, F. (2018). Big data analytics in oil and gas industry: An emerging trend. Petroleum. In Press. Available from: https://www.sciencedirect.com/science/article/pii/S2405656118301421.

Montague, J. A., Pinder, G. F., & Watson, T. L. (2018). Predicting gas migration through existing oil and gas wells. *Environmental Geosciences*, *25*(4), 121–132.

Muñoz Vélez, E. A., Romero Consuegra, F., & Berdugo Arias, C. A. (2020). *EOR screening and early production forecasting in heavy oil fields: a machine learning approach*. Society of Petroleum Engineersdoi: 10.2118/199047-MS.

Naderi, M., Khamehchi, E., & Karimi, B. (2019). Novel statistical forecasting models for crude oil price, gas price, and interest rate based on meta-heuristic bat algorithm. *Journal of Petroleum Science and Engineering*, *172*, 13–22.

Nguyen, T., Gosine, R. G., & Warrian, P. (2020). A systematic review of big data analytics for oil and gas industry 4.0. *IEEE Access*, *8*, 61183–61201. doi: 10.1109/ACCESS.2020.2979678.

Noshi, C. I., & Schubert, J. J. (2018). *The role of machine learning in drilling operations: a review*. Society of Petroleum Engineersdoi: 10.2118/191823-18ERM-MS.

Nwachukwu, A., Jeong, H., Sun, A., Pyrcz, M., & Lake, L. W. (2018). *Machine learning-based optimization of well locations and WAG parameters under geologic uncertainty*. Society of Petroleum Engineersdoi: 10.2118/190239-MS.

Ounsakul, T., Sirirattanachatchawan, T., Pattarachupong, W., Yokrat, Y., & Ekkawong, P. (2019). *Artificial lift selection using machine learning.* International Petroleum Technology Conferencedoi: 10.2523/IPTC-19423-MS.

Paltrinieri, N., Comfort, L., & Reniers, G. (2019). Learning about risk: Machine learning for risk assessment. *Safety Science, 118,* 475–486.

Patel, K., & Patwardhan, R. (2019). *Machine learning in oil & gas industry: a novel application of clustering for oilfield advanced process control.* Society of Petroleum Engineersdoi: 10.2118/194827-MS.

Pelta, R., Carmon, N., & Ben-Dor, E. (2019). A machine learning approach to detect crude oil contamination in a real scenario using hyperspectral remote sensing. *International Journal of Applied Earth Observation and Geoinformation, 82,* 101901.

Pennel, M., Hsiung, J., & Putcha, V. B. (2018). *Detecting failures and optimizing performance in artificial lift using machine learning models.* Society of Petroleum Engineersdoi: 10.2118/190090-MS.

Ramirez, A. M., Valle, G. A., Romero, F., & Jaimes, M. (2017). *Prediction of PVT properties in crude oil using machine learning techniques MLT.* Society of Petroleum Engineersdoi: 10.2118/185536-MS.

Ricardo, D. M. M., Jiménez, G. E. C., Ferreira, J. V., & Meirelles, P. S. (2018). Multiphase gas-flow model of an electrical submersible pump. *Oil & Gas Science and Technology - Rev IFP Energies nouvelles, 73,* 29–36.

Sharaf, S. A., Bangert, P., Fardan, M., Alqassab, K., Abubakr, M., & Ahmed, M. (2019). *Beam pump dynamometer card classification using machine learning.* Society of Petroleum Engineersdoi: 10.2118/194949-MS.

Sheremetov, L. B., González-Sánchez, A., López-Yáñez, I., & Ponomarev, A. V. (2013). Time series forecasting: applications to the upstream oil and gas supply chain. *IFAC Proceedings Volumes, 46*(9), 957–962.

Shi, G. R. (2009). The use of support vector machine for oil and gas identification in low-porosity and low-permeability reservoirs. *International Journal of Mathematical Modelling and Numerical Optimisation, 1*(1), 75–87.

Shoeibi Omrani, P., Dobrovolschi, I., Belfroid, S., Kronberger, P., & Munoz, E. (2018). *Improving the accuracy of virtual flow metering and back-allocation through machine learning.* Society of Petroleum Engineersdoi: 10.2118/192819-MS.

Si, M., Tarnoczi, T. J., Wiens, B. M., & Du, K. (2019). Development of predictive emissions monitoring system using open source machine learning library – Keras: a case study on a cogeneration unit. *IEEE Access, 7,* 113463–113475. doi: 10.1109/ACCESS.2019.2930555.

Silva, F., Fernandes, S., Casacão, J., Libório, C., Almeida, J., Cersósimo, S., Mendes, C.R., Brandão, R., & Cerqueira, R. (2019): Machine-learning in oil and gas exploration: a new approach to geological risk assessment. Conference Proceedings, 81st EAGE Conference and Exhibition 2019, vol. 2019, pp. 1–5.

Sinha, U., Dindoruk, B., & Soliman, M. (2020). Machine learning augmented dead oil viscosity model for all oil types. *Journal of Petroleum Science and Engineering, 195,* 107603.

Sousa, A. L., Ribeiro, T. P., Relvas, S., & Barbosa-Póvoa, A. (2019). Using machine learning for enhancing the understanding of bullwhip effect in the oil and gas industry. *Machine Learning and Knowledge Extraction, 1*(3), 994–1012.

Spesivtsev, P., Sinkov, N., Sofronov, I., Zimina, A., Umnov, A., Yarullin, R., & Vetrov, D. (2018). Predictive model for bottomhole pressure based on machine learning. *Journal of Petroleum Science and Engineering, 166,* 825–841.

Subraveti, S. G., Li, Z., Prasad, V., & Rajendran, A. (2019). Machine learning-based multiobjective optimization of pressure swing adsorption. *Industrial & Engineering Chemistry Research, 58*(44), 20412–20422.

Sui, D., Nybø, R., Gola, G., Roverso, D., & Hoffmann, M. (2011). Ensemble methods for process monitoring in oil and gas industry operations. *Journal of Natural Gas Science and Engineering, 3*(6), 748–753.

Susto, G. A., Schirru, A., Pampuri, S., McLoone, S., & Beghi, A. (2015). Machine learning for predictive maintenance: a multiple classifier approach. *IEEE Transactions on Industrial Informatics, 11*(3), 812–820. doi: 10.1109/TII. 2014.2349359.

Suursalu, S. (2017): Predictive maintenance using machine learning methods in petrochemical refineries. Master's Thesis. TU Delft Electrical Engineering.

Tian, X., & Daigle, H. (2018). Machine-learning-based object detection in images for reservoir characterization: A case study of fracture detection in shales. *The Leading Edge, 37*(6), 435–442.

Tygesen, U. T., Worden, K., Rogers, T., Manson, G., & Cross, E. J. (2019). State-of-the-art and future directions for predictive modelling of offshore structure dynamics using machine learning. In S. Pakzad (Ed.), *Dynamics of Civil Structures, Volume 2. Conference Proceedings of the Society for Experimental Mechanics Series*Cham: Springer.

Vahabi, N. (2019): Machine learning algorithms for analysis of oil, gas and water well acoustic datasets. University College London, PhD Thesis.

Vahabi, N., & Selviah, D.R. (2019): Convolutional neural networks to classify oil, water and gas wells fluid using acoustic signals. IEEE International Symposium on Signal Processing and Information Technology (ISSPIT), Ajman, United Arab Emirates, pp. 1–6. doi: 10.1109/IS-SPIT47144.2019.9001845.

Wallace, S. P., Hegde, C. M., & Gray, K. E. (2015). *A system for real-time drilling performance optimization and automation based on statistical learning methods.* Society of Petroleum Engineersdoi: 10.2118/176804-MS.

Windarto, A. P., Dewi, L. S., & Hartama, D. (2017). Implementation of artificial intelligence in predicting the value of indonesian oil and gas exports with BP algorithm. *International Journal of Recent Trends in Engineering & Research, 03*(10), .

Yan, W., & Yu, L. (2019). On accurate and reliable anomaly detection for gas turbine combustors: a deep learning approach. PHM 2015 Conference. arXiv:1908.09238.

You, J., Ampomah, W., Kutsienyo, E. J., Sun, Q., Balch, R. S., Aggrey, W. N., & Cather, M. (2019). *Assessment of enhanced oil recovery and CO_2 storage capacity using machine learning and optimization framework.* Society of Petroleum Engineersdoi: 10.2118/195490-MS.

Zenisek, J., Holzinger, F., & Affenzeller, M. (2019). Machine learning based concept drift detection for predictive maintenance. *Computers & Industrial Engineering, 137*.

Zhang, L., & Zhan, C. (2017). Machine learning in rock facies classification: an application of XG-Boost. International Geophysical Conference, Qingdao, China, 17–20 April 2017.

Zhang, Y., & Hamori, S. (2020). Forecasting crude oil market crashes using machine learning technologies. *Energies, 13*(10), 2440.

Zhong, Z., Sun, A. Y., Wang, Y., & Ren, B. (2020). Predicting field production rates for waterflooding using a machine learning-based proxy model. *Journal of Petroleum Science and Engineering, 194*, 107574.

Zukhrutany, S. (2018): The utilization of supervised machine learning in predicting corrosion to support preventing pipelines leakage in oil and gas industry. University of Stavanger. Master's Thesis.

Chapter 5

Data Management from the DCS to the Historian

Jim Crompton

Reflections Data Consulting, Colorado Springs, CO, United States

5.1 Introduction

Many operators in the oil and gas industry are increasingly looking to move toward a data-driven and technology-enabled way of working. Often this is called the Digital Oilfield 2.0. Although in principle, this may sound relatively straightforward, the reality is that things have become a lot more complicated since companies deployed their initial real-time information technology solutions. Organizations face many challenges of scale and performance, data quality, data silos, integration, and cybersecurity. There are also an increasing number of confusing and conflicting messages from within the industry as to how best to support digital operations in order to enable further automation and advanced "Big Data" predictive analysis.

Most companies are trying to see how the new "digital" approach fits and how techniques like machine learning, artificial intelligence, and advanced statistics can improve the performance of field operations and reservoir performance. It is not that modeling; simulation and automation projects are something new to the oilfield; the industry has been implementing these techniques and technologies for many years. What is different is that the Digital Oilfield 2.0 is more instrumented and more connected than ever before. With this degree of digitization, operators can develop data-driven models that combine physics, statistics, and best practice engineering principles in order to gain greater insight on what is going on in the field, and how they can improve performance of the total asset lifecycle.

Seismic acquisition, processing, and interpretation have been "Big Data" programs (at least lots-of-data programs) since the 1970s. The industry began the Digital Oilfield (i.e., integrated operations), smart field, integrated field, intelligent field, or whatever you want to call it, programs nearly 2 decades ago. The oil and gas industry has learned quite a bit about workflow optimization, remote decision support, and real-time operations. Some projects have gone well and created a lot of value. Others we do not talk about as much.

Machine Learning and Data Science in the Oil and Gas Industry.
http://dx.doi.org/10.1016/B978-0-12-820714-7.00005-4

The barriers are usually not the technology. The wise advice is to think about people, process, and technology when starting a Digital Oilfield program. This chapter adds some much-needed consideration around the data foundation as well. The technology is the "bright shiny object" that gets too much attention. The trouble spots are often difficult access to relevant data, uncertain data quality (especially when used outside our infamous functional and asset silos), a resistant organizational culture that still values experience over new statistical models and often a lack of digital literacy within the organization. All of this is especially true when a company starts adding in data from sensors in the field and down the wellbore, to the volumes of data they have from transactional and engineering systems.

5.1.1 Convergence of OT and IT

This chapter is about data management. It starts with the oilfield instrumentation and control systems perspective from the ground up; which brings a new data type to traditional analytical projects. This area of Operations Technology, or OT, has developed independently from the Information Technology, or IT, world that companies are probably more familiar with. OT systems have largely been in the domain of electrical engineering, instrumentation, and control systems technology. Recently, this operations world is converging with consumer information technology, whereas previous proprietary systems have become more open. The desired learning outcome of this chapter is to better understand and be able to describe this convergence of OT and IT and understand the vocabulary used.

Why is this important? This convergence brings the physical world of instrumented equipment and meters into the digital world where predictive models can:

- Identify and monitor field performance
- Recognize key ways that the field system can failure
- Develop models to predict those failure points
- Build safe, secure, reliable, and performant systems to optimize all of the elements of the system to increase production, while lowering operating costs and minimizing the environmental footprint of the oilfield

Wearables and mobile devices connect the human operator to the process systems. Sensor measure physical behavior and actuators allow the human to control specific process steps. Data collection and aggregation bridge the gap to the Internet of Things (IoT), or edge computing devices, that bring the field data into the cloud where it can be sent anywhere, processed by large computing resources, remotely monitored, and modeled to create the digital twin of the physical field system.

The Digital Oilfield 1.0 was about installing more sensors on field equipment, increasing information intensity, and improving the surveillance of what was going on. The Digital Oilfield 2.0 is about connectivity and building predictive models to be able to respond, and predictively intervene in order to improve process results.

5.1.2 A maturity model for OT/IT convergence

One place one can go to see this convergence taking place is a modern control room. A visitor's eyes will be drawn to all of the screens on the front wall of the operations room. All of the data captured from the field is visualized in a variety of different ways, and process alarms will alert operations staff when equipment or process behavior start to vary from normal conditions, allowing emergency situations to be managed with minimal impact.

These control rooms are getting to the state where they have too much data for the human to comfortably comprehend. So, behind the screen operators build predictive models (or AI assistants) that help the human operator understand all the big events, and even many of the small events that are going on in the field. When processes are well-known and appropriately instrumented, the control systems behind the screens can be set to automatic and the human is alerted only when necessary.

The convergence process is not happening overnight, the journey is taking several decades. One way to understand this convergence is through a maturity model. The following describes how OT and IT began to grow toward a common integrated perspective. One key observation is that the industry does not swap out its operations technology as often as you change your smart phone, or as quickly as agile development methods update software and websites. An engineer in the control room will probably come across facilities that are in each of the three phases. Dealing with legacy technology and infrastructures is a challenge that all petroleum data analytics professionals have to deal with. In one situation, the engineer has more data than they need, yet in another situation, they are lacking critical information and are uncertain of the data quality that they have on hand.

Generation One:

- Focus on keeping the plant running
- Simple Data Processing: Data In-Data Out to trending reports (Process Book)
- Plant operations centers are local

Generation Two:

- Data is Contextualized (Asset Framework and Asset Analytics)
- Custom workflows as fit-for-purpose solutions
- Remote decision support environments begin to lower manning levels
- Analytics and links to specialized advanced algorithms and modeling are beginning to be used

Generation Three:
- A Digital Twin concept is developed
- Predictive capabilities (process alarms and optimization, (event frame) emerge
- Autonomous (normally unmanned) operations complement field operations

The key message from the convergence of OT and IT is about building the bridge between the technology used in the field, to the processing power, and engineering expertise in the office. The technology is not the hard part of this convergence; it is the organization change and access to trusted data. The operations groups need to think beyond reacting to what is happening at this moment in the field, to planning and responding to a "manage-by-exception" mentality where field surveillance is combined with the predictive power of a digital twin model, and the algorithms that are developed from field data, and engineering and physics principles.

For their part, the IT group needs to expand their scope beyond the data center and the desktop environment at headquarters, to the field environment. The digital ecosystem now includes the world of sensors and control systems, of telecommunications all the way to the well head, the operations reliability, and security of a full lifecycle view of the system.

5.1.3 Digital Oilfield 2.0 headed to the edge

The Digital Oilfield 2.0 is now headed for the edge, edge computing that is. Advances are now allowing companies to place their intelligent algorithms in instrumentation and control systems. While the information intensity of the oilfield has grown by orders of magnitude over the last several decades, most of that data has been trapped in legacy SCADA and engineering systems or by limited communications solutions, and not available to build holistic asset lifecycle models. The twin dangers are that the previous generation of operators did not use or trust the data they had; and that the new digital generation may believe data-driven models too much, without a good understanding of the data that is driving these models.

Whether one is talking about predictive models for maintenance, or optimization, or full digital twins of their assets, understanding of data is still very important. How can an operator implement a "manage-by-exception" process when the data used in building the model behind the prediction engine is biased, or poor quality, missing key attributes, or is out of date? How can automation replace experienced workers when the rules and logic of the models are based upon flawed data?

Let's take a deeper dive into the operational data world and find out just what an operator is dealing with before the data scientists start programming the robots to do the heavy lifting.

5.2   Sensor data

Sensor data is the output of a device that detects and responds to some type of input from the physical environment. The output may be used to provide information, or input, to another system or to guide a process. Sensors can be used to detect just about any physical element. Here are a few examples of sensors, just to give an idea of the number and diversity of their applications:

- An accelerometer detects changes in gravitational acceleration in a device it is installed in; such as a smart phone or a game controller, to determine acceleration, tilt, and vibration.
- A photosensor detects the presence of visible light, infrared transmission (IR) and/or ultraviolet (UV) energy.
- Lidar, a laser-based method of detection, range finding, and mapping; typically uses a low-power, eye-safe pulsing laser working in conjunction with a camera.
- A charge-coupled device (CCD) stores and displays the data for an image in such a way that each pixel is converted into an electrical charge, the intensity of which is related to a color in the color spectrum.
- Smart grid sensors can provide real-time data about grid conditions, detecting outages, faults and load, and triggering alarms.

Wireless sensor networks combine specialized transducers with a communications infrastructure for monitoring and recording conditions at diverse locations. Commonly monitored parameters include: temperature, humidity, pressure, wind direction and speed, illumination intensity, vibration intensity, sound intensity, power-line voltage, chemical concentrations, pollutant levels, and vital body functions.

Sensor data is in integral component of the increasing reality of the Internet of Things (IoT) environment. In the IoT scenario, almost any entity imaginable can be outfitted with a unique identifier (UID) and the capacity to transfer data over a network. Much of the data transmitted is sensor data. The huge volume of data produced and transmitted from sensing devices can provide a lot of information but is often considered the next big data challenge for businesses. To deal with that challenge, sensor data analytics is a growing field of endeavor (Rouse et al., 2015).

According to Talend, a Big Data Software company, sensor data and sensor analytics are poised to become the next Big Thing in information technology; with experts predicting that the volume of sensor data will soon dwarf the amount of data that social media is currently producing. Gathered from cell phones, vehicles, appliances, buildings, meters, machinery, medical equipment, and many other machines, sensor data will likely completely transform the way organizations collect information and process business intelligence.

Working with sensor data is problematic for most organizations today. Most enterprises are not equipped to integrate data from such a wide range of sensor sources—their legacy integration technology simply is not up to the task. On top of infrastructure problems, few organizations have developers with the skills, like MapReduce programming, required to manipulate massive sets of sensor data, and training existing developers is extremely costly and time-consuming (Pearlman, 2019).

5.2.1 There are problems with data from sensors: data quality challenges

In today's data-driven world, it has become essential for companies using IoT technologies to address the challenges posed by exploding sensor data volumes; however, the sheer scale of data being produced by tens of thousands of sensors on individual machines is outpacing the capabilities of most industrial companies to keep up.

According to a survey by McKinsey, companies are using a fraction of the sensor data they collect. For example, in one case, managers at a gas rig interviewed for the survey said they only used one percent of data generated by their ship's 30,000 sensors when making decisions about maintenance planning. At the same time, McKinsey found serious capability gaps that could limit an enterprise's IoT potential. In particular, many companies in the IoT space are struggling with data extraction, management, and analysis.

Timely, and accurate, data is critical to provide the right information at the right time for business operations to detect anomalies, make predictions, and learn from the past. Without good quality data, companies hurt their bottom line. Faulty operational data can have negative implications throughout a business, hurting performance on a range of activities from plant safety, to product quality, to order fulfillment. Bad data has also been responsible for major production and/or service disruptions in some industries (Moreno, 2017).

Sensor data quality issues have been a long-standing problem in many industries. For utilities who collect sensor data from electric, water, gas, and smart meters, the process for maintaining sensor data quality is called Validation, Estimation and Editing, or VEE. It is an approach that can be used as a model for other industries, as well when looking to ensure data quality from high volume, high velocity sensor data streams.

Today, business processes and operations are increasingly dependent on data from sensors, but the traditional approach of periodic sampling for inspecting data quality is no longer sufficient. Conditions can change so rapidly that anomalies and deviations may not be detected in time with traditional techniques.

Causes of bad sensor data quality include:

- Environmental conditions—vibration, temperature, pressure or moisture—that can impact the accuracy of measurements and operations of asset/sensors.

- Misconfigurations, miscalibrations, or other types of malfunctions of asset/ sensors.
- Different manufacturers and configurations of sensors deliver different measurements.
- Loss of connectivity interrupts the transmission of measurements for processing and analysis.
- Tampering of sensor/device and data in transit, leading to incorrect or missing measurements.
- Loss of accurate time measurement because of use of different clocks, for example.
- Out-of-order, or delayed, data capture and receipt.

5.2.2 Validation, estimation, and editing (VEE)

The goal of VEE is to detect and correct anomalies in data before it is used for processing, analysis, reporting, or decision-making. As sensor measurement frequencies increase, and are automated, VEE is expected to be performed on a real-time basis in order to support millions of sensor readings per second, from millions of sensors. Fig. 5.1 displays examples of real time data within the industry.

The challenge for companies seeking actionable operational and business knowledge from their large-scale sensor installations is that they are not able to keep up with the VEE processes needed to support data analytics systems because of the high volume and velocity of data. This is why so many companies today are using less than 10% of the sensor data they collect (Tomczak, 2019).

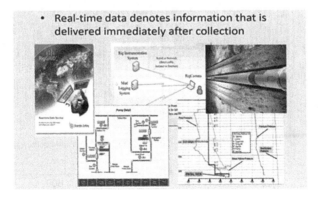

FIGURE 5.1   **What is real time data**. (*Crompton*).

5.3 Time series data

A time series is a series of data points indexed, listed, or graphed in time order. Most commonly, a time series is a sequence taken at successive equally spaced points in time. Thus, it is a sequence of discrete-time data. Examples of time series are; heights of ocean tides, counts of sunspots, well head pressure and temperature readings, and the daily closing value of the Dow Jones Industrial Average.

Time series are very frequently plotted via line charts. As displayed in Fig. 5.2, they are used in statistics, signal processing, pattern recognition, econometrics, mathematical finance, weather forecasting, earthquake prediction, electroencephalography (heart monitoring), control engineering, astronomy, communications engineering, and largely in any domain of applied science and engineering which involves temporal measurements.

Time series analysis comprises methods for analyzing time series data in order to extract meaningful statistics and other characteristics of the data. Time series forecasting is the use of a model to predict future values based on previously observed values. While regression analysis is often employed in such a way as to test theories that the current values of one or more independent time series affect the current value of another time series, this type of analysis of time series is not called "time series analysis;" which focuses on comparing values of a single time series, or multiple dependent time series at different points in time. Interrupted time series analysis is the analysis of interventions on a single time series.

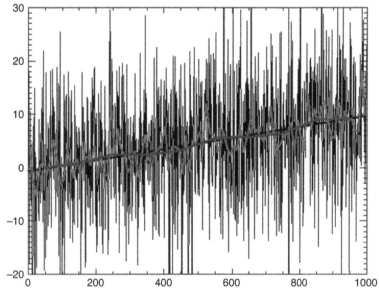

FIGURE 5.2   Time series data. (*Maksim., 2006*).

Time series data have a natural temporal ordering. This makes time series analysis distinct from cross-sectional studies in which there is no natural ordering of the observations (e.g., explaining people's wages by reference to their respective education levels, where the individuals' data could be entered in any order). Time series analysis is also distinct from spatial data analysis where the observations typically relate to geographical locations (i.e., a leasing or drilling program in a basin). A stochastic model for a time series will generally reflect the fact that observations close together in time will be more closely related than observations further apart. In addition, time series models will often make use of the natural one-way ordering of time so that values for a given period will be expressed as deriving in some way from past values, rather than from future values. Transactional data is recorded in a tabular format with values associated by columns in each row. Real-time data is recorded with only time context (i.e., value and timestamp) (Ullah et al., 2013).

Time series analysis can be applied to real-valued, continuous data, discrete numeric data, or discrete symbolic data (i.e., sequences of characters, such as letters and words in the English language). Methods for time series analysis may be divided into two classes: frequency-domain methods and time-domain methods. The former includes: spectral analysis and wavelet analysis; the latter include auto-correlation and cross-correlation analysis. In the time domain, correlation and analysis can be made in a filter-like manner using scaled correlation, thereby mitigating the need to operate in the frequency domain.

Additionally, time series analysis techniques may be divided into parametric and non-parametric methods. The parametric approaches assume that the underlying stationary stochastic process has a certain structure, which can be described using a small number of parameters (e.g., using an autoregressive or moving average model). In these approaches, the task is to estimate the parameters of the model that describes the stochastic process. By contrast, non-parametric approaches explicitly estimate the covariance or the spectrum of the process without assuming that the process has any particular structure. Methods of time series analysis may also be divided into linear and non-linear, and univariate and multivariate (Lin et al., 2003).

5.4   How sensor data is transmitted by field networks

The goal of all information technology implementations, regardless of industry, should be to improve productivity. The bottleneck for oilfield data flow until now has been the transfer of real-time data to the engineers' desktop in an accurate, timely, and useful fashion. Engineers typically have seen only a subset of the field data available (i.e., daily production volumes and rates, along with a few gauge pressures and temperature settings). With databases updated only periodically from real-time historians, engineers have lacked sufficient insight into the dynamics of platform or field operations. What's needed, *"is an alarm system to inform engineers of under-performing or critical conditions of a well or reservoir,"* before it begins to degrade production and the revenue stream. Oilfield operations need to move beyond the familiar data management mantra of the *"right data to the right person at the right time"* and adopt the 21st century goal of *"validated data, to the decision maker, before the critical event,"* (Piovesan and Jess, 2009).

As field automation moved from plants, where wired connections are possible, to the field, the evolution of radio transmission networks began. There are countless SCADA radio networks monitoring and controlling municipal water, waste-water, drainage, oil and gas production, power grids, automatic meter reading, and much more. These SCADA radio technology-based communications systems have ranges from yards, to hundreds of miles.

Selecting the right technology, defining the projects requirements, and creating a robust SCADA communications network is crucial. Having an understanding of the advantages and disadvantages of each technology can help the engineer, or operator, to make the best business decisions for a system that will have a ten to twenty-year service lifetime and are often mission critical or high importance SCADA systems.

5.4.1 From Plant to Field: Communications Protocols (HART, Fieldbus, OPC, OPC-UA and Wireless Hart)

Here are just a few examples of field, or plant, to office communications protocols.

The **HART Communications Protocol** (Highway Addressable Remote Transducer) is an early implementation of Fieldbus, a digital industrial automation protocol. Its most notable advantage is that it can communicate over legacy 4–20 mA analog instrumentation wiring, sharing the pair of wires used by the older system. An example of this protocol is shown in Fig. 5.3.

Fieldbus is the name of a family of industrial computer network protocols used for real-time distributed control, standardized as IEC 61158. A complex automated industrial system—such as manufacturing assembly line—usually needs a distributed control system—an organized hierarchy of controller systems—to function. In this hierarchy, there is usually a Human Machine Interface at the top, where an operator can monitor or operate the system. This is typically linked to a middle layer of programmable logic controllers via a non-time-critical communications system. At the bottom of the control chain is the fieldbus that links the PLCs to the components that actually do the work, such as; sensors, actuators, electric motors, console lights, switches, valves, and contactors.

Requirements of Fieldbus networks for process automation applications (i.e., flowmeters, pressure transmitters, and other measurement devices and control valves in industries such as hydrocarbon processing and power generation) are different from the requirements of Fieldbus networks found in discrete manufacturing applications and is shown in Fig. 5.4. Examples of implementation of

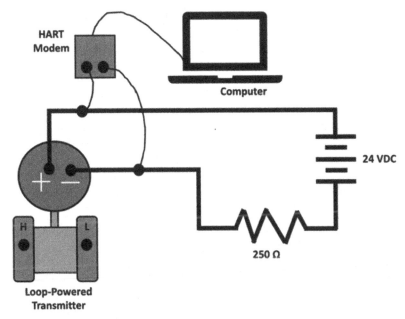

FIGURE 5.3   Hart communications protocol. (*Crompton*).

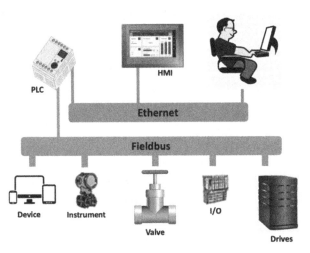

FIGURE 5.4   Fieldbus architecture. (*Crompton*).

these Fieldbus networks are; automotive manufacturing, where large numbers of discrete sensors are used including motion sensors, position sensors, and so on. Discrete Fieldbus networks are often referred to as "device networks," (Anderson, 2009).

OPC (OLE for Process Control) was first defined by a number of players in automation together with Microsoft in 1995. Over the following ten years it became the most used versatile way to communicate in the automation layer in all types of industry. Over the years it has evolved from the start with simple Data access (DA), over Alarm and Events (AE), to the more advanced Historical Data Access (HDA) to have quite extensive functionality and reach. Though there were always some gaps where it did not cover the needs and requirements from the more advanced control systems, it was out of those needs for model-based data and getting more platform independent that resulted in the creation of the OPC UA standard (Marffy, 2019).

Building on the success of OPC Classic, **OPC UA** was designed to enhance and surpass the capabilities of the OPC Classic specifications. OPC UA is functionally equivalent to OPC Classic, yet capable of much more.

1. Discovery: find the availability of OPC servers on local PCs and/or networks
2. Address space: all data is represented hierarchically (e.g., files and folders) allowing for simple and complex structures to be discovered and utilized by OPC clients
3. On-demand: read and write data/information based on access-permissions
4. Subscriptions: monitor data/information and report-by-exception when values change based on a client's criteria
5. Events: notify important information based on client's criteria
6. Methods: clients can execute programs, etc., based on methods defined on the server

Integration between OPC UA products and OPC Classic products is easily accomplished with COM/Proxy wrappers (Unified, 2019). The differences are visualized in Fig. 5.5.

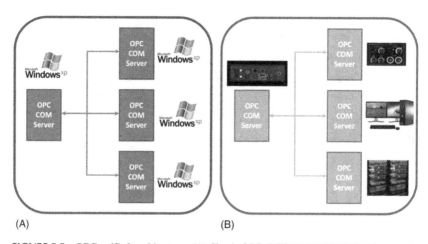

(A) (B)

FIGURE 5.5 OPC unified architecture. (A) Classic OPC (DCOM), (B) OPC UA. (*Crompton*).

The most significant difference between classical OPC and **OPC UA (OPC Unified Architecture)** is that it does not rely on OLE or DCOM technology from Microsoft that makes implementation possible on any platform (i.e., Apple, Linux [JAVA] or Windows). The other very important part of UA is the possibility to use structures or models. This means that the data tags, or points, can be grouped and be given context, making governance and maintenance much easier. These models can be identified in runtime, which makes it possible for a client to explore connection possible by asking the server.

The information modeling is very modern in OPC UA. These models can be defined by manufactures, or protocols like BACNet; but it can also contain more of a MESH structure where very complex relations and connections between points and nodes can be defined. The possibility also exists to have data structures so that certain data always is grouped and handled as one piece. This is important in many applications where you want to be sure that the data set is taken at the same time.

OPC UA, as said before, is built to be platform independent and the communication is built into layers on top of the standard TCP/IP stack. Above the standard transport layers there are two layers, one that handles the session, and one to establish a secure channel between the client and server. The transport layer is made up of TCP/IP and on top of that SSL, HTTP, or HTTPS. The Communication layer secures the communication channel, not just that the data is corrupted but also it secures the authentication so that the end points cannot be infiltrated and changed. This is based on X.509 certificates that have three parts to it and the first peer to peer trust needs to be manually done; but after that the rest is taken care of securely.

With approximately 30 million HART devices installed and in service worldwide, HART technology is the most widely used field communication protocol for intelligent process instrumentation. With the additional capability of wireless communication, the legacy of benefits this powerful technology provides continues to deliver the operational insight users need to remain competitive.

Even though millions of HART devices are installed worldwide, in most cases the valuable information they can provide is stranded in the devices. An estimated 85% of all installed HART devices are not being accessed to deliver device diagnostics information, with only the Process Variable data communicated via the 4–20 mA analog signal. This is often due to the cost and the difficulty of accessing the HART information.

5.4.2 Wireless SCADA radio

Wireless SCADA is required in those applications when wircline communications to the remote site is prohibitively expensive, or it is too time-consuming to construct. In particular types of industry, such as Oil and Gas, or Water and Wastewater, wireless SCADA is often the only solution due to the remoteness of the sites. Wireless SCADA systems can be built on a private radio, licensed or unlicensed, cellular or satellite communications.

One major difference between private radio and cellular or satellite is that private radio has no associated monthly fees. Once an operator builds their hardware infrastructure, they own it. With cellular, or satellite service providers, there is an associated monthly fee.

The list of communications protocols that have been developed goes on from Foundation FieldBus, ProfiBus, HART, wireless HART, ProfiNet, Modbus, and DeviceNet Ethernet.

5.4.3   Which protocol is best?

The right answer is "it depends on the application and what is already installed." An optimized solution will probably use more than one communication type, for example, new "analog" installations benefit greatly from FOUNDATION Fieldbus, while new "discrete" installations can benefit from PROFIBUS, PROFINET, DeviceNet, and Ethernet/IP. Electrical Integration depends on the equipment-supported protocol; but IEC 61850, Ethernet/IP and PROFIBUS could all prove useful. The most common view is to look at the current installed base, HART, PROFIBUS, DeviceNet, MODBUS, or others. Why replace when one can integrate?

For the decision maker, it comes down to deciding between a large capital investment and no monthly service fees, OR a smaller capital investment with monthly service fees. If the client's remote assets are well within the Service Providers coverage area, they have high speed or TCP/IP data requirements, or their remote sites are there for a short term (lasting no more than 1 or 2 years) all of these factors contribute to making Cellular communications a more attractive option.

If the assets are extremely remote (hundreds or thousands of miles away from civilization) and/or they have TCP/IP LAN speed requirements, then Satellite communications is obviously the only solution. Often it is more typical for companies to have a mixture of solutions; using Point-Multipoint Private Radio communications where there is a high density of remote sites served by the one Master site, using Cellular for sites for from Private radio clusters, and/or satellite for the very remote sites.

Another difference between Private Radio networks and Cellular or Satellite networks is expansion capabilities. With Private Radio systems, if distances between communicating sites are too great for a single hop, then repeaters can always be installed to further the range. This certainly adds to the overall project cost, but it is at least an option if required.

With Cellular or Satellite, one is leveraging the vast existing infrastructure of the Service Provider. The advantage of Cellular or Satellite is that the client has access to very wide coverage but the disadvantage is that if there is a site that happens to be outside of the Service Providers coverage, there is no way for the client to increase the range; only the Service Provider can do that (Bentek, 2019).

5.5   How control systems manage data

Supervisory Control and Data Acquisition (SCADA) is a control system architecture that uses computers, networked data communications, and graphical user interfaces for high-level process supervisory management; but uses other peripheral devices, such as programmable logic controllers and discrete PID controllers to interface to the process plant, oilfield operations or critical equipment (seen in Fig. 5.6). The operator interfaces which enables monitoring and the issuing of process commands, such as controller set point changes, are handled through the SCADA supervisory computer system. The real-time control logic, or controller calculations, are performed by networked modules which connects to the field sensors and actuators. The main purpose of a SCADA system is to collect data from sensors on well heads, tanks, and critical equipment (i.e., temperature, pressure, vibrations, flow rates, fill levels, and provide a view to the human operator the condition of field operations).

5.5.1   Cloud-based SCADA and web-based SCADA

Cloud computing is a hot topic. As people become increasingly reliant on accessing important information through the Internet, the idea of storing, or displaying, vital real-time data in the cloud has become more commonplace. With tech giants like Apple, Microsoft, and Google pushing forward the cloud computing concept, it seems to be more than just a passing trend.

While many cloud services are specifically meant as storehouses for data, some cloud-based SCADA systems are offered as a "service;" which is referred to as SaaS (Software as a Service). Instead of having the SCADA system

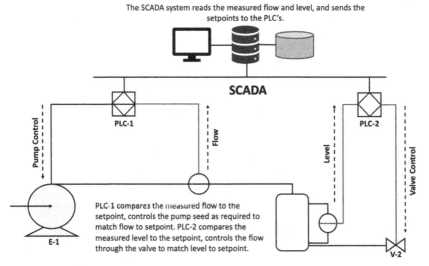

FIGURE 5.6   **Supervisory control and data acquisition.** (*Crompton*).

software installed on local computers, the entire system and its data is stored and maintained in the cloud. SaaS companies offer their customers the power of software applications, off-site IT support, and scalable server space all through the cloud.

The manufacturing industry is big, encompassing a wide variety of companies; likewise, the type of information each company tracks can vary greatly. This fact should be taken into account when determining what information, if any, should be stored in the cloud. Information such as reports, analytics, and configurations are ideal candidates for the cloud; however, information that is vital to safety and control functions—and that which relies on bandwidth availability and reliability—is particularly important to the operation of a manufacturer. It is essential to weigh the risks involved with putting this type of information in the cloud because it can directly affect the functionality and productivity of your company.

5.6 Historians and information servers as a data source

A Data Historian, also known as a Process Historian or Operational Historian, is a software program that records and retrieves production and process data by time. It stores the information in a time series database that can efficiently store data with minimal disk space and fast retrieval. Time series information is often displayed in a trend or as tabular data over a time range (i.e., the last day, last 8 hours, last year).

5.6.1 What can you record in a data historian?

A historian will record data over time from one or more locations for the user to analyze. Whether one chooses to analyze a valve, tank level, fan temperature, or even a network bandwidth, the user can evaluate its operation, efficiency, profitability, and setbacks of production. It can record integers (whole numbers), real numbers (floating point with a fraction), bits (on or off), strings (e.g., product name), or a selected item from a finite list of values (e.g., Off, Low, High).

Some examples of what might be recorded in a data historian include:

- Analog Readings: temperature, pressure, flowrates, levels, weights, CPU temperature, mixer speed, fan speed
- Digital Readings: valves, limit switches, motors on/off, discrete level sensors
- Product Info: product ID, batch ID, material ID, raw material lot ID
- Quality Info: process and product limits, custom limits
- Alarm Info: out of limits signals, return to normal signals
- Aggregate Data: average, standard deviation, moving average

A Data Historian could be applied independently in one or more areas; but can be more valuable when applied across an entire facility, many facilities in a department, and across departments within an organization. An operator can discover that a production problem's root cause is insufficient power supply to the production equipment, or they could discover the two similar units produce significantly different results over time.

A Data Historian is not designed to efficiently handle relational data, and relational databases are not designed to handle time series data. If an operator can have a software package that offers both with the ability to integrate data from each different storage type, then they have a much more powerful solution.

There are many uses for a Data Historian in different industries:

- Manufacturing site to record instrument readings
- Process (e.g., flow rate, valve position, vessel level, temperature, pressure)
- Production Status (e.g., machine up/down, downtime reason tracking)
- Performance Monitoring (e.g., units/hour, machine utilization vs. machine capacity, scheduled vs. unscheduled outages)
- Product Genealogy (e.g., start/end times, material consumption quantity, lot number tracking, product setpoints, and actual values)
- Quality Control (e.g., quality readings inline or offline in a lab for compliance to specifications)
- Manufacturing Costing (e.g., machine and material costs assignable to a production)
- Utilities (e.g., Coal, Hydro, Nuclear, and Wind power plants, transmission, and distribution)
- Data Center to record device performance about the server environment (e.g., resource utilization, temperatures, fan speeds), the network infrastructure (e.g., router throughput, port status, bandwidth accounting), and applications (e.g., health, execution statistics, resource consumption).
- Heavy Equipment Monitoring (e.g., recording of run hours, instrument and equipment readings for predictive maintenance)
- Racing (e.g., environmental and equipment readings for sail boats, race cars)
- Environmental Monitoring (e.g., weather, sea level, atmospheric conditions, ground water contamination)

Information collected within a facility can come from many different types of sources including:

- PLCs (Programmable Logic Controllers) that control a finite part of the process (e.g., one machine or one processing unit)
- DCS (Distributed Control System) that could control an entire facility
- Proprietary Instrument Interface (e.g., Intelligent Electronic Devices): data delivered directly from an instrument instead of a control system (e.g., Weighing system, clean-in-place skid)
- Lab Instrument (e.g., Spectrophotometer, TOC Analyzer, Resonance Mass Measurement)
- Manual Data Entry (e.g., an operator periodically walks the production line and records readings off manual gauges) (Winslow, 2019).

5.7   Data visualization of time series data—HMI (human machine interface)

Human-Machine Interface (HMI) is a component of certain devices that are capable of handling human-machine interactions. The interface consists of hardware and software that allow user inputs to be translated as signals for machines that, in turn, provide the required result to the user. Human-machine interface technology has been used in different industries, like; electronics, entertainment, military, medical, etc. Human-machine interfaces help in integrating humans into complex technological systems. Fig. 5.7 displays how the information can be displayed to the human in different ways (i.e., touch, vision, and sound). Human-machine interface is also known as Man-Machine Interface (MMI), computer-human interface, or human-computer interface (Techopedia, 2019).

In an HMI system, the interactions are basically of two types (i.e., human to machine and machine to human). Since HMI technology is ubiquitous, the interfaces involved can include motion sensors, keyboards and similar peripheral devices, speech-recognition interfaces, and any other interaction in which information is exchanged using sight, sound, heat, and other cognitive and physical modes are also considered to be part of HMIs.

Although considered as a standalone technological area, HMI technology can be used as an adapter for other technologies. The basis of building HMIs largely depends on the understanding of human physical, behavioral, and mental capabilities. In other words, ergonomics forms the principles behind HMIs. Apart from enhancing the user experience and efficiency, HMIs can provide unique opportunities for applications, learning and recreation. In fact, HMI helps in the rapid acquisition of skills for users. A good HMI is able to provide realistic and natural interactions with external devices.

The advantages provided by incorporating HMIs include error reduction, increased system and user efficiency, improved reliability and maintainability, increased user acceptance and user comfort, reduction in training and skill requirements, reduction in physical or mental stress for users, reduction in task saturation, and increased economy of production and productivity, etc.

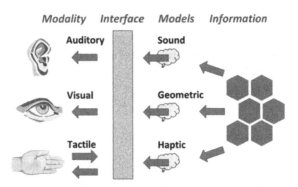

FIGURE 5.7   Human machine interface (*O'Hare, 2014*).

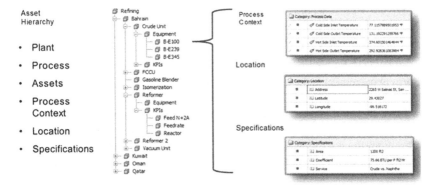

FIGURE 5.8 **Real-time data requires context.** (*Crompton*).

Touch screens and membrane switches can be considered as examples of HMIs. HMI technology is also widely used in virtual and flat displays, pattern recognition, Internet and personal computer access, data input for electronic devices, and information fusion. Professional bodies like GEIA and ISO provide standards and guidelines applicable for human-machine interface technology.

5.7.1   Asset performance management systems (APM)

The objective is to move from the traditional monitoring and surveillance processes where the operator wants to know what happened (descriptive analytics) and why it happened (diagnostic analytics) to a new state where the operator can predict what will happen under certain operating conditions (predictive analytics). It may even be possible in the future to move toward prescriptive analytics where the operator can "tune" the processing system to produce optimum results most of the time (how can we make only good things happen). This end state is sometimes called Asset Performance Management and is visualized in Fig. 5.8.

"Asset Performance Management (APM) is an approach to managing the optimal deployment of assets to maximize profitability and predictability in product supply, focusing on real margin contribution by asset by product code. Rather than looking at an asset on the basis of market value or depreciated value, companies can see how the asset is contributing to their profitability by looking at how individual assets are performing—whether inventory or Plant, Property, and Equipment (PP&E)—and developing a vision of how they want to allocate resources to assets in the future. APM is not necessarily purely financial or even operational, but it will cross functional lines. It combines best-of-breed enterprise asset management (EAM) software with real-time information from production and the power of cross-functional data analysis and advanced analytics. More broadly, it looks at the whole lifecycle of an asset, enabling organizations to make decisions that optimize not just their assets, but their operational and financial results as well,"

(Miklovic, 2015).

One of the basic steps of Asset Performance Management is process control and alarm management.

5.7.1.1 Process control and alarm management

Alarm management is the application of human factors (or "ergonomics") along with instrumentation engineering and systems thinking to manage the design of an alarm system to increase its usability. Most often the major usability problem is that there are too many alarms annunciated in a plant upset, commonly referred to as alarm flood (similar to an interrupt storm), since it is so similar to a flood caused by excessive rainfall input with a basically fixed drainage output capacity.

There can also be other problems with an alarm system, such as poorly designed alarms, improperly set alarm points, ineffective annunciation, unclear alarm messages, etc. Poor alarm management is one of the leading causes of unplanned downtime, contributing to over $20B in lost production every year, and of major industrial incidents, such as the one at the Texas City refinery in March of 2005. Developing good alarm management practices is not a discrete activity, but more of a continuous process (i.e., it is more of a journey than a destination) (Mehta and Reddy, 2015).

5.7.2 Key elements of data management for asset performance management

Process historians are the "system of record" for sensor readings, but there are other data management tools that are important as well. Historians store tag names for identification of where a sensor is located in the operations environment, but an operator needs additional contextual information to place the sensor in the operating or plant system. The contextual data, sometimes called master data, is often found in an Asset Registry.

5.7.2.1 What is an asset registry?

The Asset Registry is an editor subsystem, which gathers information about unloaded assets asynchronously as the editor loads. This information is stored in memory so the editor can create lists of assets without loading them. Asset registers are typically used to help business owners keep track of all their fixed assets and the details surrounding them. It assists in tracking the correct value of the assets, which can be useful for tax purposes, as well as for managing and controlling the assets.

There are different ways of building the data model for this kind of information; taxonomy and ontology.

5.7.2.2 What is the definition of data taxonomy?

Data taxonomy in this case means an organized system of describing data and information. The information exists. The challenge is to help executives, analysts, sales managers, and support staff, find and use the right information both efficiently and effectively. Many enterprises extract value from the business information they accumulate by organizing the data logically and consistently into categories and subcategories; therefore, creating a taxonomy. When information is structured and indexed in a taxonomy user can find what they need by working down to more specific categories, up to a more inclusive topic, or sideways to related topics (Walli, 2014).

5.7.2.3 What is the definition of data ontology?

In both computer science and information science, ontology is a data model that represents a set of concepts within a domain, and the relationships between those concepts. It is used to reason about the properties of that domain, and may be used to define the domain.

5.8 Data management for equipment and facilities

In addition to the sensor measurements and control system readings, there are many other important data records available; however, many of these data types are in the form of drawings, permits, documents, inspection records, certificates, P&ID drawings, process diagrams, etc. A data management system designed for documents is another type of data management technology that is needed (depicted in Fig. 5.9).

5.8.1 What is a document management system?

Document management systems are essentially electronic filing cabinets an organization can use as a foundation for organizing all digital and paper documents. Any hard copies of documents can simply be uploaded directly into the document management system with a scanner. Oftentimes, document management systems allow users to enter metadata and tags that can be used to organize all stored files.

Most document management software has a built-in search engine, which allows users to quickly navigate even the most expansive document libraries to access the appropriate file. Storing sensitive documents as well? Not to worry! Most document management systems have permission settings, ensuring only the appropriate personnel can access privileged information.

These are some of the most important document management features:

- Storage of various document types, including word processing files, emails, PDFs, and spreadsheets
- Keyword search
- Permissioned access to certain documents
- Monitoring tools to see which users are accessing which documents
- Versioning tools that track edits to documents and recover old versions
- Controls regulating when outdated documents can be deleted
- Mobile device support for accessing, editing and sharing documents (Uzialko, 2019).

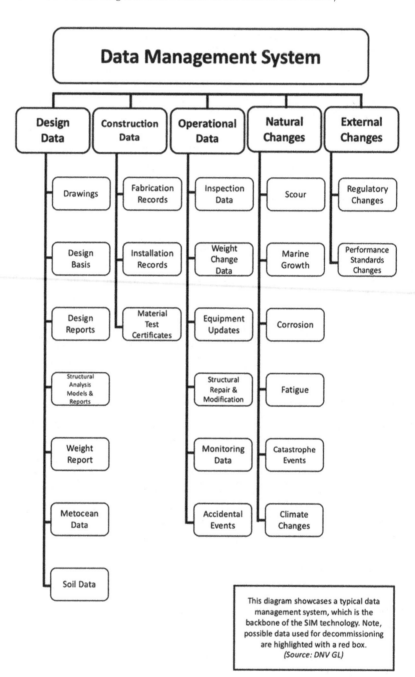

FIGURE 5.9   Data management system for equipment and facilities. (*Crompton*).

5.9   Simulators, process modeling, and operating training systems

Now that you have a basic understanding of the data involved from field instrumentation and process control systems, let's take a look at some of the applications that the sensor data is input to starting with process simulation, or training simulation systems.

Process simulation is a model-based representation of chemical, physical, biological, and other technical processes and unit operations in software. Basic prerequisites are a thorough knowledge of chemical and physical properties of pure components and mixtures, of reactions, and of mathematical models which, in combination, allow the calculation of a process in computers.

Process simulation software describes processes in flow diagrams where unit operations are positioned and connected by product streams, as illustrated in Fig. 5.10. The software has to solve the mass and energy balance to find a stable operating point. The goal of a process simulation is to find optimal conditions for an examined process. This is essentially an optimization problem, which has to be solved in an iterative process.

Process simulation always uses models which introduce approximations and assumptions; but allow the description of a property over a wide range of temperatures and pressures, which might not be covered by real data. Models

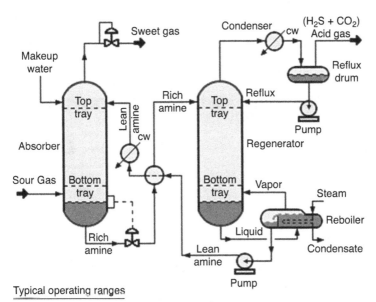

Typical operating ranges
Absorber: 35 to 50 °C and 5 to 205 atm of absolute pressure
Regenerator: 115 to 126 °C and 1.4 to 1.7 atm of absolute pressure
            at tower bottom

FIGURE 5.10   Process simulation. (*Mbeychok, 2012*).

also allow interpolation and extrapolation, within certain limits, and enable the search for conditions outside the range of known properties (Rhodes, 1996).

5.10 How to get data out of the field/plant and to your analytics platform

5.10.1 Data visualization

One area of rapid technology development in this area is the challenge of visualization of time-series data. Getting beyond the alarm stage and looking at the data historically to recognize patterns of equipment or process failure and building predictive models to help operators take action before catastrophic failure occur. You can see from this slide, included as Fig. 5.11, one of the many new data visualization technologies that are improving the HMI (Human Machine Interface).

5.10.1.1 From historians to a data infrastructure

While traditional process historians have served an important purpose for control systems for many decades, the limitations of historians are starting to become a barrier for providing data to advanced analytics. The limitations include:

- The data is "tag based" creating integration problems with enterprise data access
- One-off data feeds to specific solutions promote data silos
- This architecture is difficult to scale and requires expensive maintenance
- Solutions depend on analyst bringing data together to develop solutions
- There is limited use of automation to speed up data processing and data exchange

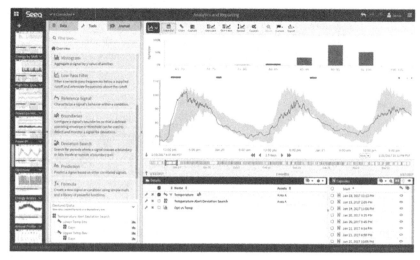

FIGURE 5.11 Data visualization of time series data. (*Seeq, 2018*).

One leading vendor is this space defines the data infrastructure as a "system of insight" versus the traditional "system of data." The new architecture includes the traditional process historian but adds real-time operational data framework capabilities such as an asset framework to provide asset-based hierarchy context and a library of appropriate data templates to help organize diverse inputs from sensor to metadata streams.

5.10.2 Data analytics

The end goal is to make the operations-oriented sensor, time-series data available for:

1. Monitoring and Surveillance
2. Addition of IIoT devices, edge, and cloud computing
3. Alarm Detection (process alarms more than just static set points)
4. Automation of routine tasks
5. Mobility solutions and advanced visualizations
6. Integrated Operations
7. Simulation Modeling
8. "Big Data" and Data-Driven predictive, "digital twin" models to drive manage-by-exception processes

These objectives have been coming in stages. The evolution of industrial analytics can be viewed in these three stages of development and maturity.

5.10.3 Three historical stages of industrial analytics

The **first stage** was defined by employees walking around with pencil and paper reading gauges on various assets in the plant; over time walking around was replaced with wired connections to sensors which used a pen in the control room to write sensor levels on a roll of paper, the "strip chart."

The **second stage** came with the digitization of sensors and the electronic collection of data, presented to operators on monitors and then stored in historians. Data was viewed with trending applications that started out as electronic versions of strip charts. A connector imported historian data into a spreadsheet and enabled engineers to do calculations, data cleansing, etc. The history of the spreadsheet and historian are closely intertwined; both were invented in the mid-1980s and used in conjunction ever since. Every historian includes at least two application features; a trending application, and an Excel connector.

The **third stage** is currently emerging and assumes changes to the software and architecture for data storage and analytics, tapping advances such as; big data and cognitive computing, cloud computing, wireless connectivity, and other innovations.

5.10.3.1 Where is data analytics headed?

Advanced analytics refers to the application of statistics and other mathematical tools to business data in order to assess and improve practices. In manufacturing, operations managers can use advanced analytics to take a deep dive into historical process data, identify patterns and relationships among discrete process steps and inputs, and then optimize the factors that prove to have the greatest effect on yield. Many global manufacturers in a range of industries and geographies now have an abundance of real time shop-floor data and the capability to conduct such sophisticated statistical assessments. They are taking previously isolated data sets, aggregating them, and analyzing them to reveal important insights. The process to move from new technology into optimization of use in the field is shown in Fig. 5.12.

Automation increases productivity and helps engineers spend more time on translating data into meaningful information and valuable decisions. Real-time analytics for equipment and operations is providing significant bottom line value. Multiple predictive analytics are underway to predict compressors and valves failure. The objective is to reduce cost and increase uptime—(i.e., compressors trips/failure is one of the top bad actors). Moving to condition-based maintenance will help reduce OPEX and deferment.

> **Valves example**: *Valves maintenance is mostly time-based; if you are too late, the operator ends up with unscheduled deferment and health, safety, and environmental risks. If your maintenance routine is too early, the operator ends up with scheduled deferment and unnecessary costs.*

> **Artificial Lift Optimization example**: *The data comes from the dyna-card measurement, or the pump cycle. The algorithms attempt to separate out the patterns of good performance (constant fluid getting moved up the wellbore) or bad performance. For the patterns of poor performance (vibration of the rod, no fluid*

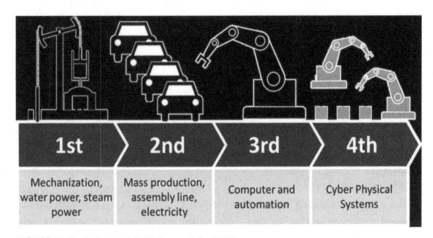

FIGURE 5.12 **Industry 4.0.** (*Balyer and Oz, 2018*).

being lifted, etc.) the operator wants to algorithms to suggest possible causes and mitigation or repair steps.

5.11   Conclusion: do you know if your data is correct?

Are there dark clouds on the horizon, holding your data hostage? As organizations look to adopt the new wave of coming technologies (like automation, artificial intelligence and the Internet of Things), their success in doing so and their ability to differentiate themselves in those spaces will be dependent upon their ability to get operational data management right. This will become increasingly important as connected devices and sensors proliferate, causing an exponential growth in data—and a commensurate growth in opportunity to exploit the data.

Those that position their organizations to manage data correctly and understand its inherent value will have the advantage. In fact, leaders could pull so far in front that it will make the market very difficult for slow adopters and new entrants.

References

Anderson, M. (2019). What is fieldbus? *RealPars*. Available from: https://realpars.com/fieldbus/.

Balyer, A., & Oz, O. (2018). Academicians' views on digital transformation in education. *International Online Journal of Education and Teaching (IOJET)*, 5(4), 809–830. http://iojet.org/index.php/IOJET/article/view/441/295.

Bentek Systems. (2019). Wireless SCADA. *SCADALink*. Available from: https://www.scadalink.com/support/knowledge-base/wireless-scada/.

Inductive Automation. (2014). Old SCADA Vs. The New SCADA Reinventing the SCADA User Experience. https://www.inductiveautomation.com/resources/webinar/old-scada-vs-new-scada.

Lin, Jessica, et al. (2003). A symbolic representation of time series, with implications for streaming algorithms. In *Proceedings of the 8th ACM SIGMOD Workshop on Research Issues in Data Mining and Knowledge Discovery*: (pp. 2–11). New York: ACM Press.

Maksim. (2006). Random-data-plus-trend-r2. *Figure 4.2. GNU Free Documentation License*. Available from: https://commons.wikimedia.org/wiki/File:Random-data-plus-trend-r2.png.

Marffy, S. (2019). Home. *Novotek*. Available from: https://www.novotek.com/en/solutions/kepware-communication-platform/opc-and-opc-ua-explained.

Mbeychok. (2012). AmineTreating. *Figure 4.10. GNU Free Documentation License*. Available from: https://en.wikipedia.org/wiki/Process_flow_diagram.

Mehta, B. R., & Reddy, Y. J. (2015). Alarm management—an overview. *Industrial process automation systems: design and implementation*. ScienceDirect Topics. https://www.sciencedirect.com/topics/engineering/alarm-management.

Miklovic, D. (2015). What is asset performance management? *Industrial Transformation Blog*. Available from: https://blog.lnsresearch.com/what-is-asset-performance-management.

Moreno, H. (2017). The importance of data quality—good, bad or ugly. *Forbes Magazine*. Available from: https://www.forbes.com/sites/forbesinsights/2017/06/05/the-importance-of-data-quality-good-bad-or-ugly/#6b0b9f9d10c4

O'Hare, B. (2014). Enactive_human_machine_interface. *Effective Interfaces Figure 1. GNU Free Documentation License*. Available from: https://en.wikipedia.org/wiki/Enactive_interfaces.

Pearlman, S. (2019). Sensor data—talend. *Talend Real-Time Open Source Data Integration Software.* Available from: https://www.talend.com/resources/l-sensor-data/.

Piovesan, C. M., & Jess, K. (2009). An intelligent plaform to manage offshore assets. *SPE Annual Technical Conference and Exhibition*doi: 10.2118/124514-ms.

Rhodes, C. L. (1996). The process simulation revolution: thermophysical property needs and concerns. *Journal of Chemical Engineering Data, 41,* 947–950.

Rouse, Margaret, et al. (2015). What is sensor data?—definition from whatis.com. *IoT Agenda.* https://internetofthingsagenda.techtarget.com/definition/sensor-data.

Seeq Corporation and Processing Staff. (2018). Seeq secures $23M in funding for IIoT strategy. Available from: https://www.processingmagazine.com/news-notes/article/15587489/acquisitions-and-expansions-seeq-secures-23m-in-funding-from-altira-group-others-jbt-acquires-ftnon.

Techopedia. (2019). What is human-machine interface (HMI)?—definition from techopedia. *Techopedia.com.* Available from: https://www.techopedia.com/definition/12829/human-machine-interface-hmi.

Tomczak, P. (2019). How VEE processes ensure data quality for IIoT systems and applications. Kx. Available from: https://kx.com/blog/kx-for-sensors-data-validation-estimation-and-editing-for-utilities-and-industrial-iot/.

Ullah, M. I. (2013). Time series analysis. *Basic Statistics and Data Analysis.*

Unified Architecture. (2019). *OPC Foundation.* Available from: https://opcfoundation.org/about/opc-technologies/opc-ua/.

Uzialko, A.C. (2019). Choosing a document management system: a guide. *Business News Daily.* Available from: https://www.businessnewsdaily.com/8026-choosing-a-document-management-system.html.

Walli, B. (2014). Taxonomy 101: the basics and getting started with taxonomies. *KMWorld.* Available from: http://www.kmworld.com/Articles/Editorial/What-Is/Taxonomy-101-The-Basics-and-Getting-Started-with-Taxonomies-98787.aspx.

Winslow, R. (2019). What is a data historian? *Data Historian Overview.* Available from: http://www.automatedresults.com/PI/data-historian-overview.aspx.

Chapter 6

Getting the Most Across the Value Chain

Robert Maglalang
Value Chain Optimization at Phillips 66, Houston, TX, United States

6.1 Thinking outside the box

In today's global and fast changing business environment, the urgency with multinational companies to find readily implementable digital solutions has increased significantly as the underlying science yielding operational and business improvements has matured over the past decade. Companies lacking innovative ways to extract incremental value from existing and new business ventures will be left behind.

Manufacturing plants, in general, manage billion dollars' worth of raw and finished products daily across the US—and globally. The quantities of data analyzed in a product life cycle, capturing cost-to-produce, logistics, working capital, and other ancillary costs are massive and difficult to both consolidate and integrate. Many production companies still rely heavily on manual processes to evaluate opportunities and economics.

Adopting new technologies has been slow in many industries, and much more so in the oil and gas sector. The traditional, siloed infrastructures to drive quantitatively based decision making are not up to the challenge as the processes in-place are simply inefficient, thereby creating business risks with regards to data consistency, accuracy, and completeness. In a production plant setting for example, replacing a paper-based system with intrinsically safe gadgets or storing data in a cloud is not an easy transition due to the sensitivity in the security level and risks of causing disruption in the operations. However, since many architecture foundational systems are near the end of useful life—and with renewed focus on improving productivity and yield while cutting down costs—more energy companies are opening their world to the digital technology era.

The production plant is the universe around which the organizational improvement focus revolves—a wise investment of resources, as they are the outright cash cows. Emphasis on the plant level optimization and improvements can help a company realize significant value creation through asset reliability

Machine Learning and Data Science in the Oil and Gas Industry.
http://dx.doi.org/10.1016/B978-0-12-820714-7.00006-6

111

and integrity and increase in energy efficiency and productivity. However, putting all strategic efforts alone in the production margin may not move the needle in terms of dollars, especially for the oil majors. Outside of this huge "plant box," the greater portion of the business operates in a very competitive market environment where significant opportunities lie. Decision makers require intelligent models with real-time information not only to be immediately available but also to provide valuable insights.

6.2   Costing a project

When implementing ML methods in the industry, we naturally ask: Is it worth it? Measuring the benefit of the technology and its outcomes is not easy. In fact, not even its cost is easy to measure. Fig. 6.1 presents a common situation in large corporations when a significant new initiative is launched. While it is comical to portray the misunderstandings in this fashion, the dire reality is that they often prevent the value from being generated and, in turn, give the technology a bad name.

The point is that before we can talk about measuring the added value of machine learning, we must be very clear and transparent in our communication what we expect it to deliver. Often, industry managers are unsure about what they expect and need help to fully understand even what they could expect. Many vendors choose to dazzle with vocabulary, acronyms, technologies and dashboards without ever getting down to properly defining the situation—the last image in Fig. 6.1. As John Dewey said, "a problem well stated is a problem half solved."

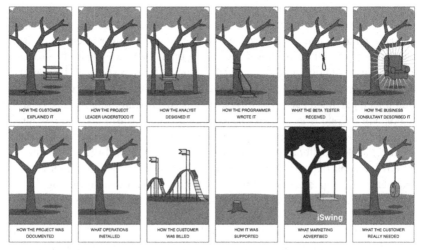

FIGURE 6.1   **Common problems in communication when implementing a complex project, such as machine learning initiative.**

In costing an ML application, consider

- Providing sufficient domain expertise to the machine learners so that they can model the problem in its full practical complexity.
- Involving the end users to define in what manner they need the answer delivered so that the answer becomes actionable and useful.
- Obtaining good quality, comprehensive, representative, statistically significant, and high-frequency data to enable the data-driven learning to take place.
- Employing expert machine learners to design the algorithms to calculate a precise and accurate answer.
- Spending enough time and effort not just in building the system but testing and debugging it while involving everyone concerned.
- Investing in change management to transition the organization's procedures into being able to utilize the new technology in a practical way.

Vendors focus on the technological costs of the project, the effort and cost of testing and change management are significant. In fact, most projects that fail, fail because of poorly executed change management. We will discuss this issue later in this chapter.

Most projects that run over the time and financial budget, do so because testing and debugging the system takes longer than expected. One way to mitigate this problem is to use the agile methodology of software development—discussed in Chapter 7 of this book—which incorporates feedback into the development cycle. No matter how the project is run however, sufficient attention must be paid to gathering practical feedback from the end users and adjusting to that feedback. The second most common reason for project failure, after insufficient change management, is a lack of real-world understanding flowing into the analysis. Whenever a vendor claims to be able to solve a problem without detailed and deep domain knowledge, it spells doom from the start.

6.3   Valuing a project

After knowing what it would cost to realize a project, we can think about what it would yield. The reason to think about the benefit after the cost is that the process of costing involved a detailed definition of what the project entails. From experience, projects that were valued prior to a detailed planning were overvalued because of the hype of machine learning that supposedly promises a panacea to all ills in no time at all.

Based on the project definition, what is the kind of benefit being derived? In many cases, there are several benefits. Common dimensions include (1) an increase of revenue, (2) a decrease in cost, (3) a higher production yield, (4) a higher production efficiency, (5) a reduction in staff, (6) a speed-up of some kind, (7) consumption of fewer material resources, (8) a reduction in waste or

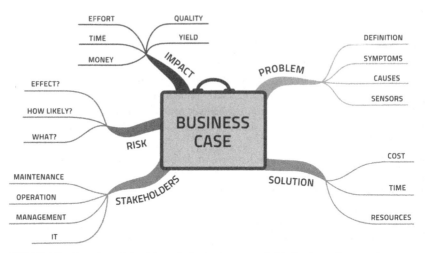

FIGURE 6.2 Overview of the drivers in the business case behind a data science project.

scrap, and (9) a novel business model. Most of these can be measured if we have realistic estimates available.

The last item, a novel business model, is interesting because it is often cited and recognized as very exciting. It is also quite uncertain. Nevertheless, artificial intelligence has enabled numerous new business models that are very profitable and are based essentially in the fast speeds and high complexities that ML models support. Cars without drivers are perhaps the most commonly envisioned transformation in this respect.

In considering the benefits of a project, consider all the stakeholders and their individual benefits as well. They may align with the corporate goals, or not. Also consider the risks if the project does not work out or works less well than imagined. See Fig. 6.2 for an overview of all the factors involved in valuing a project.

A case in point is the famous "pain point" of the customer. Frequently, the issue that generates the most frustration for certain individuals is at the top of their mind and quickly raises the idea that ML could make this issue go away. Enthusiasm is quickly built and a project is envisioned. Vendors are tripping over themselves to fulfill the need and make the customer's pain go away. Finally, a real customer has a real pain point! A sale might be near. Celebrations! But hold your horses. Two separate issues must be examined before too much enthusiasm can be spent on this idea by either the vendor or the people who feel the pain.

First, it must be determined how the benefit is going to be measured. Second, the benefit must then be measured and determined to be large enough. Many situations conspire together to make these two issues work out against doing the project. Let's address both points.

6.3.1   How to measure the benefit

The example for this is predictive maintenance. This is the most discussed topic in the oil and gas industry where applications of ML are concerned. By the number of articles published and webinars held—by top managers of oil and gas operators—on this topic, one would think that this is applied ubiquitously across the industry and world. Far from it. Only a very few companies have deployed it, and even then only in special cases and restricted geographical areas.

One common reason is that maintenance is almost always viewed as an annoying cost item and entirely separated from production. Assessing a maintenance solution by the amount of additional production it enables, is obvious. However, most companies do not do their accounting in this way. As a maintenance solution, the maintenance department will have to pay for it and so it must yield a financial benefit within the confines of the maintenance budget. This benefit may be large but the effect on production is almost certainly far greater. Can your organization think broadly enough to consider cross-departmental benefits?

6.3.2   Measuring the benefit

At first glance, measuring the benefit is just an estimate of what the ML project can deliver and putting a financial value to it. However, it is not quite so easy.

All ML methods will produce some false-negatives and false-positives. In the cases of a regression or forecasting task, the equivalent would be an outlier calculation. No matter what method you employ, it will sometimes make a mistake. Hopefully, this will be rare but it will occur and you must consider it.

Statistically speaking, the most dangerous events are those that are very rare and cause large damage. It is very difficult to accurately assess the cost of such events because we cannot reliably assess their likelihood or cost. The most famous example is Deepwater Horizon where a simple cause leads to a global disaster. Clearly this type of event is rare and very costly in multiple ways. Fear of such events often leads to an operator not implementing, or not rolling out, an ML initiative. It may represent the main obstacle to full adoption of ML by the industry. The perception of risks threatens to undermine the entire value consideration and so it is important that this be dealt with early on in the discussions, see Fig. 6.3.

6.4   The business case

The business case, then is a combination of five items

1. Definition of the situation, challenge and desired solution
2. Cost and time plan for producing or implementing the solution.
3. Benefit assessment of the new situation.

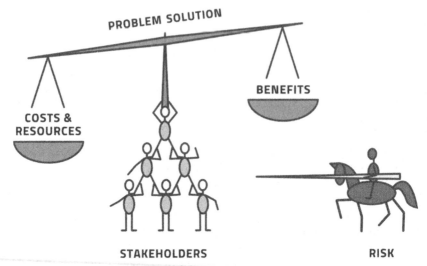

FIGURE 6.3   Risk, or the perception of them, may derail any careful consideration of costs and benefits.

4. Risk analysis of what might go wrong and lower the benefits or increase the costs.
5. Change management plan of how to implement the solution once it has been created.

If the benefits outweigh the costs—plus or minus the risks—it is worth doing. The assessment is very sensitive to one's point of view as illustrated above with the manner in which benefits are calculated.

Chapter 8 in this book analyses the position of pilot testing in this process in some detail. It is often desired by oil and gas operators to conduct a pilot for an ML project in which the ML method is tested out. There are three common dangers in pilot programs. First, pilots often assess only the technology and not the benefits. Second, some pilots try to fully gain and assess the benefits without spending the costs and then obviously fail. Third, most pilots are started using a concrete problem only as a stand-in because the real vision is a vague desire to learn what ML can offer and how it works. This last danger is particularly significant because it means that the organization never intends to put this particular project to productive use.

The business case of ML is similar to all business case arguments. It differs in only two essential ways.

First, the uncertainty of what the solution will look like, which makes the cost and benefit analysis more intransparent.

Second, the expectations and fears of ML in general. Often both are overly inflated by marketing from vendors (who inflate expectations) and cautionary tales by utopians (who are afraid of the robot apocalypse).

The first can be solved easily by some level-headed thinking. The second by talking with some machine learners who are not afraid to say what is realistic and what is squarely within the realm of Hollywood movies.

6.5   Growing markets, optimizing networks

A simple example of an application of ML in the oil and gas industry is a plant operation with multiple terminals and retail stations, see Fig. 6.4. Each node on the graph represents either a terminal (if arrows are leaving it) or a retail station (if arrows arrive at it).

The product distribution and placement are bounded by many challenges. There are logistics constraints, bottlenecks, new routes, and infrastructure restrictions among others that require predictive analytics for real-time decision support. With the sheer volume of information moving at lightning speed, lacking intelligent details and degree of granularity is not optimal for agile, yet vetted decisions.

In an ideal environment, plant production is well matched with secure placement demand. The volume allocation in a pipeline or terminal is at the optimum integration and any additional barrels will fill the line up-to capacity. However, even in this ideal logistics set-up, price fluctuations in the market can represent significant risks. In addition, operating plants normally have unplanned outages and unit upsets that can disrupt the supply and demand equilibrium, and the integrated systems would need to absorb the impact in a dynamic market.

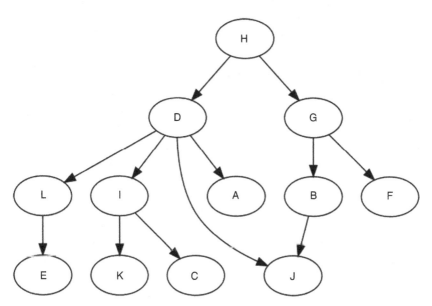

FIGURE 6.4   Integrated network of terminals and retail stations.

The availability of advanced tools such as Machine Learning (ML) provides a better way to analyze data more quickly and accurately than ever—and unlock potential for margin improvements by incorporating optimization and growth economics to maximize value chain profitability. Machine learning allows manipulation and analysis of "Big Data" to generate key insights such as: (1) what happened in the business, (2) what is happening now, and (3) what is likely to happen in the future. Likewise, optimization using ML offers more detailed analytics and new possibilities to answer important strategic business questions such as:

- What is the optimum integration target that maximizes overall margin?
- What is the impact to the net margin if new products or blend stocks are produced to meet the market needs?
- What is the overall net effect to re-supply a network of terminals to cover production shorts?
- Where is the optimum volume allocation if new route is added?
- Where is the next best alternative?
- What happens to the net margin if there are production change impacts or market shifts?
- How much volume should be sold at the spot market or exchange partners?
- Where are the growth opportunities based on industry views or market intelligence?
- Is a project investment justified to meet market demand?

The answers can be derived in multiple ways, and the level of accuracy and consistency can vary significantly depending on the methods utilized. While any forecasted metric is only an estimate, the decision criteria are tied mathematically and can be scientifically computed using real-time optimization to capture market opportunities.

6.6 Integrated strategy and alignment

In a multi-channel network, the applications of data driven analytics to make supply and demand decisions are integral to efficiently deliver consistent and transparent optimization processes and to increase margin capture. Across business units, the data mined to generate actionable insights do not provide a complete picture without the ability to view all the components together as a system. The diagram in Fig. 6.5 is a good example of a quantitatively dependent systems that can significantly benefit from business driven analytics:

In complex, multi-functional processes, using conventional ways of mining hundreds of thousands of rows of data, analyzing the information, and presenting the results is a huge challenge without a reliable ML model. It is enormously time consuming to clean-up, sort, or transfer data before it can be accessed and consumed by decision makers. The most cost-effective solution to solve this common problem in the industry is to build a supported IT architecture anchored by ML algorithms in order to align the value chain capture.

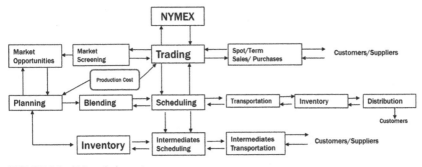

FIGURE 6.5   Value chain system.

The ML model integrates, correlates, and computes answers to strategic business questions as well as screen opportunities dynamically. The traditional method would take several hours or days to get to a degree of granularity needed for vetted decisions and could be error prone due to its inability to handle this size of information with limited functionalities. Extracting useful data and moving the information to where it is needed quickly would add significant value to the financial bottom line.

6.7   Case studies: capturing market opportunities

The downstream refiners deal with a number of challenges across the energy value chain, from crude oil purchases to product placement all the way to the retail business. Employing ML as a tool to make many detailed analyses in real-time and respond to external factors that are moving too quickly enables refiners to adapt their go-to-market strategy and operation to support profitable business decisions.

One of the strategic opportunities is on crude optimization, which can be in billion-dollar transactions for refiners with multiple sites. The main challenges are crude transport and delivery period which can be roughly 1–2 months transit time; and there is no real-time dashboard for crude selection and trading decisions available that are "plug-and-play." The process starts when a price forecast is published as the basis for a refining crude run. The site planning team determines how to re-optimize the plant on a weekly cycle according to the prior week's forecast. A facility will run its own Linear Programming (LP) modeling which is oftentimes done via a simple spreadsheet program, and the results are shared back to the trading and commercial teams to show the optimized crude rates based on the incremental margin. If there are multiple refineries in a region, the synergies are often ignored, as the LP analysis is limited to a single site. Table 6.1 summarizes the process from data collection to dashboard development utilizing ML and analytics for crude selection and trading decisions on most recent LP reports.

TABLE 6.1 Crude/feedstock optimization.

Data sources	Database	Crude supply/trading (user inputs)	Results
Crude assays	Integrates all relevant data to find correlations and empirically develop the model without the involving human expertise	Selects available crudes and corresponding volume	Dashboard
Refinery LP (unit yields, capacities, product blending, rates, etc.)	Mathematical representation to compute the "state" of the refinery based on the historical datasets, factoring in time and cause and effect	Adds price "real-time"	Compares base slate versus new/optimum slate in terms of volume swapped and margin versus reference
Price forecast ($/bbl cost/margin/crack, RCV, BE, etc.)	Goal is to maximize profit by comparing available crudes to purchase versus alternative	Enters transit time	Shows current tank levels (OIS connectivity), crude rates
OIS—plant monitoring system			Other relevant info for Traders/Supply Team to make decision real-time
Upcoming T/A plans—unit impacted	**Independent variables:**	**Refinery**	
	Crude Type	Update unit limitations, as needed	
Note: At least 5 years of Historical Data are available to train the model	Price of Crude		
	Transportation Cost		
	Transit Time		
	Product Prices		
	Market Crack		

TABLE 6.1 Crude/feedstock optimization. (*Cont.*)

Data sources	Database	Crude supply/ trading (user inputs)	Results
	Dependent variables:		
	Refinery Crude Value—margin or incremental relative to a base slate (generally determined via LP—change in variable margin per barrel of crude substituted). Delta objective function divided by barrels substituted		
	Breakeven Value— full product value without regards to its cost (a.k.a. cost at which you are indifferent)		
	Product Mix— expected production yields based on parameters		

In a nutshell, the software tool streamlines the following process:

1. Crude assays are integrated into the data sets and merged with the refinery model and trader's price inputs or forecasts, considering all the data sources as a single, complex system.

2. Models are created to correlate refinery LP results (i.e., crude base slate, available crudes for substitution, margin/or incremental value relative to a crude base slate, breakeven, products sales, etc.) to determine the mathematical relationships of the parameters.

3. Each crude will have a corresponding assay, crude/freight costs (changes over time) and relative crude value (value or margin of the candidate crude relative to a crude base slate) and break even (price at which the refinery is indifferent to the crude purchase) and other key measurement from the LP run.

4. Once the "state" of the refinery is set, crude buyers can then make crude selections based on current prices and product margins to capture opportunities.
5. The goal is to compute for the highest margin from the available crude in the market to maximize profit across the value chain.
6. The model is refreshed whenever the refinery updates the LP run (to incorporate process unit constraints, among others).

The process simplifies the data handoffs, improve agility and speed-to-deal, and consistent and transparent transactions when buyers term-up commitments and optimize crude purchases.

Another opportunity for refiners is on product placement economics. Rebalancing volume allocations to account for strategically advantaged markets and seasonal market liquidities are huge value creation—if all pieces of information are analyzed collectively. Combining hundreds of thousands transactional data points into a coherent scorecard, while examining the market conditions to figure out whether the benefits outweigh the risks, can be a game changer (Fig. 6.6).

Divisions within commercial or marketing organizations can rely on an optimization tool anchored by ML that determines the optimum integration level in specific regions, based on maximum return to the business unit. The model assesses logistics, identifies bottlenecks, incorporates seasonal market liquidity, and covers production shorts to re-balance volume allocation. Product placement economics is evaluated real-time to adjust for production changes and market shifts.

The first step is to understand the historical volumes produced at specific site, including seasonal market liquidities, see Fig. 6.7.

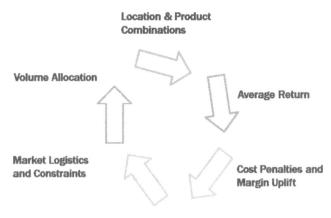

FIGURE 6.6   Marketing system.

Terminal	Product	Average (GPY)	2013: Summer (GPY)	2013: Winter (GPY)	2014: Summer (GPY)	2014: Winter (GPY)
A	Gasoline	63,000.00				
B	Diesel	13,615,698.00	6,605,088.00	6,581,148.00	7,117,304.00	6,273,673.00
B	Gasoline	2,507,653.67	104,004.00	474,338.00	384,381.00	1,187,932.00
C	Diesel	210,000.00				105,000.00
C	Gasoline	190,428.00		95,214.00		
D	Diesel	2,837,352.00		2,228,520.00	1,806,798.00	420,000.00
D	Gasoline	2,083,410.00	294,000.00		2,234,904.00	462,000.00
D	Conventional	968,072.00		505,260.00		568,848.00
D	Biodiesel	32,545,359.00		17,424,582.00		14,199,612.00

FIGURE 6.7   Understanding seasonality.

Then, the capacity constraints and capabilities at the terminals and pipelines to meet fuels requirements are incorporated into the model, including sourcing and blending, to establish baseline numbers, see Fig. 6.8.

Once the baseline is established, the tool can evaluate the net margins at the terminal and product level, see Fig. 6.9.

One of the fundamental drivers to maximize net margin is the flexibility to allow for pricing sensitivities and variations. These can be manually inputted into the software interface, see Fig. 6.10.

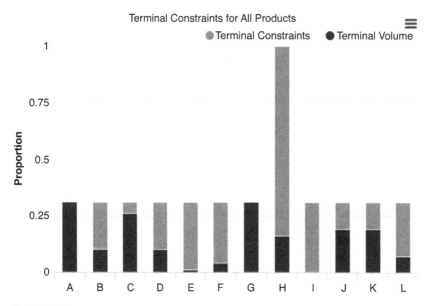

FIGURE 6.8   Visualizing constraints on the system.

Terminal	Product	Average (USD)	2013: Summer (USD)	2013: Winter (USD)	2014: Summer (USD)	2014: Winter (USD)
A	Gasoline	0.01				
B	Diesel	−0.03	−0.04	−0.04	−0.05	−0.03
B	Gasoline	0.03	0.06	0.01	−0.12	−0.03
C	Diesel	−0.17				−0.17
C	Gasoline	−0.12		−0.12		
D	Diesel	−0.03		−0.06	−0.04	−0.05
D	Gasoline	0.00	−0.05		−0.03	0.07
D	Conventional	−0.05		−0.08		−0.01
D	Biodiesel	−0.40		−0.99		−0.58

FIGURE 6.9 Analyzing margins.

	Terminal	Product	Re-Supply Adjustment	Market Penalty	Grade Adjustment	Market Uplift
1	A	Gasoline	0.00	0.00	0.00	0.00
2	B	Diesel	0.00	0.00	0.00	0.00
3	B	Gasoline	0.00	0.00	0.00	0.00
4	C	Diesel	0.00	0.00	0.00	0.00
5	C	Gasoline	0.00	0.00	0.00	0.00
6	D	Diesel	0.00	0.00	0.00	0.00
7	D	Gasoline	0.00	0.00	0.00	0.00
8	D	Conventional	0.00	0.00	0.00	0.00
9	D	Biodiesel	0.00	0.00	0.00	0.00

FIGURE 6.10 Specifying the drivers and decisions.

Once all the information is combined—and applying robust methodology to improve margins—the tool can determine the most advantaged market and incremental economics on a consistent basis. Key decisions, such as project investment, alternative supply availability, and integration targets, would require advanced predictive analytics to anticipate market shifts and production changes. The ML platform can view the entire market as a whole and allows for aggressive marketing penetration as local placement grows; or becomes more competitive.

As shown in Fig. 6.11, significant margins can be realized when the overall network is optimized.

Relevant dashboards and data visualizations can also be built on the front-end to support real-time decision making, with seamless integration with other systems, see Fig. 6.12.

The digital foundation can be a mash-up of multiple applications built over a single optimization ML tool. Leveraging existing systems for crude or products data analytics can help achieve the profitability goals more efficiently, as long as the digital engine is backed by the right technology platform.

Terminal	Product	Historical Volume (GPY)	Netback (USD/GPY)	Adjusted Netback (USD/GPY)	Optimal Volume (GPY)
A	Gasoline	31,500.00	0.01	0.01	693,000.00
B	Diesel	6,807,849.00	−0.03	−0.03	12,254,129.00
B	Gasoline	1,253,826.00	0.03	0.03	27,584,192.00
C	Diesel	105,000.00	−0.17	−0.17	189,000.00
C	Gasoline	95,214.00	−0.12	−0.12	171,385.20
D	Diesel	1,418,676.00	−0.03	−0.03	2,553,616.75
D	Gasoline	1,041,705.00	0.00	0.00	22,917,510.00
D	Conventional	484,036.00	−0.05	−0.05	871,264.81
D	Biodiesel	16,272,679.50	−0.40	−0.40	29,290,824.00
Total Network		10,091,932.12 USD			13,126,838.72 USD
Profit					3,218,770.84 USD

FIGURE 6.11   Improved margins after optimization.

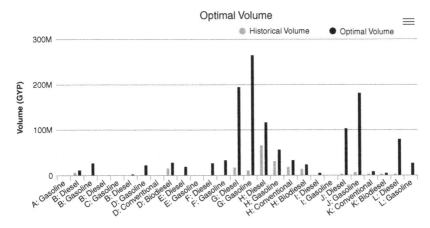

FIGURE 6.12   Visualize actions that must be taken.

6.8   Digital platform: partner, acquire, or build?

In most cases, the software solutions for industrial applications are not readily available. Outside of the plant environment, business driven analytics and models with accompanying reports and dashboards have not been implemented on a large scale to augment decision-making. If vendors have plug and play software that works, the estimated cost is typically a tiny fraction compared to the potential earnings that can be realized if the right solution results in a timely and efficient execution.

To continuously improve and deliver sustainable business performance, companies have put together dedicated resources and formed digital analytics teams within the organization. Relying on internal capabilities is critical to ensure there are clear accountabilities and long-term success. The main players are the data engineers and data scientists combined with subject matter experts and consultants to develop a pipeline of projects around system improvements and automation. With the support from industry consultants, clients evaluate the project trade-offs to set the framework for capital allocation. Also, by going through discussions around effort vs. impact, for instance, they can prioritize high value projects, quick wins, and strategic activities aligned with the overall company objectives and aspirations.

One of the main challenges in launching new applications is timely implementation. Staff needs to understand the complexity of the problems and the right solutions while also figuring out whether to build the software, partner with vendors that have existing or similar type solutions, or acquire the technology outright, if available. There will be trade-offs and conflicting perspectives with these options, and the required implementation timeframe is an important parameter in the decision criteria.

Another consideration is developing a network of excellence with data science and engineers in-house, which is not easy since the roles require programming and IT skills, math and statistics knowledge, and scientific expertise. For oil and gas companies, it would take several months or even years to build internal capabilities as these roles are neither their bread-and-butter nor reside within the organization. Plus, the importance of the full-time internal experts may diminish over time when the technology has been implemented company wide.

A hybrid approach (in-house experts with outside consultants) is more suitable for skill reinforcement and coordination; and to distribute the load evenly. While it is costly to use consultants to provide on-site expertise and support, utilizing them to varying degrees and fostering knowledge sharing could fast track the implementation of the key projects across the enterprise, allowing the company to improve cross functional collaboration and enhance competitive position in the market in a timelier fashion.

6.9   What success looks like

Applying machine learning in the decision process is business critical, most especially when the key factors are unknown, and the quantification of uncertainties and risks can be formulated using existing data. ML has been fully tested in various applications not only to measure opportunities but also to predict and prescribe solutions that enable rapid decisions and deliver profitable growth and enhance returns.

Adoption of ML in the workplace environment is not going to be easy, like any new tool to get the users' buy-in and trust. Change management is fundamental in this process to communicate the scope definition and expectations

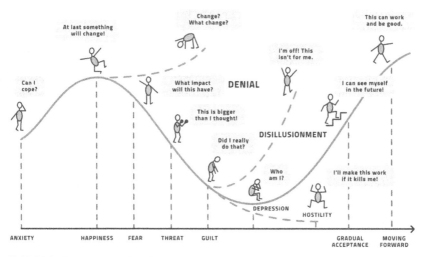

FIGURE 6.13   An adaptation of John Fisher's personal transition curve.

up-front and drive awareness and establish connection points with the impacted stakeholders. With dedicated change management resources, they can support the tool implementation by providing required training, gaining alignment, and creating an open channel for feedback.

Change management is nicely displayed in Fig. 6.13 where we see the evolution of an individual over the course of some change. It begins with anxiety and transitions into happiness that something will change, bringing with it the hope that a pain might go away. But quickly fear settles in as it becomes unclear what impact it will have and might present a bigger change than anticipated. This is followed by guilt at having initiated or participated. At this point, many people chose to stop or become hostile. We must encourage people to see the project as an opportunity here so that they can see it working and see themselves as part of the brave new world.

Guiding people through this process is what defines change management. Without a dedicated change management process, too many people involved will either give up or develop hostility and this leads to project failure. Only with dedicated effort can we bring most, or all, people to the new situation with some enthusiasm and only then will the ML project succeed.

In this effort, it must be recognized that the number of people affected by the new technology is usually quite a bit larger than the group that worked on the project itself. Particularly in the oil and gas industry, it is the end users who are expected to use the product on a daily basis that are not involved in the process of making or configuring the product. This alone leads to tensions and misunderstandings. If the end users are the maintenance engineers who are expected to use predictive maintenance, for example, we may be talking about several hundred individuals who have to transform from reactive to proactive lifestyles at the workplace.

Ultimately, the business solution needs to be simple and streamlined, provide a higher level of accuracy, and give decision makers the right information at their fingertips at the right time. The company significantly benefits financially when existing or new data is examined from a slightly different angle— and derive actionable insights.

Chapter 7

Project Management for a Machine Learning Project

Peter Dabrowski

Wintershall Dea, Hamburg, Germany

Now that we have a better understanding of machine learning in the context of oil and gas, we can explore how to best execute these types of projects. A valid question to ask at this point might be: Is the management of projects with a machine learning component so different that it warrants its own in-depth discussion? The simple answer is "yes," due to the complex and unique nature of machine learning projects, especially when applied to the oil and gas industry.

To further answer this question, it is helpful to view machine learning in the context of digitalization. Regardless of chosen technology, cloud computing, smart sensors, augmented reality, or big data analytics, the driver behind digitalization will in many cases be a change in the business model to allow for higher efficiencies and new value adding opportunities. This is not to say the exploration and production (E&P) business has not been innovative in the past. The introduction of horizontal drilling, multilateral wells, deep water drilling and other new technologies pushed the envelope of what is technically feasible in E&P.

Now, with the application of digitalization, we are leaving the arena of our core business. As we work to marry the traditional rough-neck technologies with digitalization, what is home turf to Amazon, Google, and Microsoft, we might as well learn from their experiences.

You will notice that this digitalization approach is somewhat different. Buzzwords, such as *Scrum* and *Agile*, will be demystified and translated into our business processes. In the end, you will find that these new processes are yet another set of tools in your toolbox.

7.1 Classical project management in oil & gas-a (short) primer

Projects in the oil and gas industry are some of the most complex and capital intensive in the world. Major projects, like installation of an offshore production rig, can cost upwards of US$ 500 million and make up a significant percentage of a company's expenditures and risk exposure. Therefore, focusing on how to best manage processes, dependencies and uncertainties is imperative.

Machine Learning and Data Science in the Oil and Gas Industry.
http://dx.doi.org/10.1016/B978-0-12-820714-7.00007-8

129

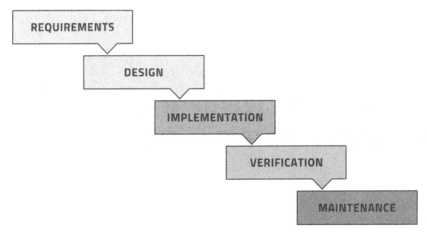

FIGURE 7.1 Example workflow of the Waterfall method.

Traditionally, large projects in engineering design are run sequentially. This sequential approach is phased with milestones and clear deliverables, which are required before the next stage begins. These stages can vary, depending on needs and specifications, but typically all variations contain at least these five basic blocks (Fig. 7.1):

With the workflow cascading toward completion, this approach is often referred to as the Waterfall method. This terminology was applied retroactively with the realization that projects could be run differently (e.g., Agile).

The Waterfall method is one of the oldest management approaches. It is used across many different industries and has clear advantages, including:

- Clear structure-With a simple framework, the focus is on a limited number of well-defined steps. The basic structure makes it easy to manage, allowing for regular reviews with specific deliverable checks at each phase.
- Focus-The team's attention is usually on one phase of the project at a time. It remains there until the phase is complete.
- Early well-defined end goal-Determining the desired result early in the process is a typical Waterfall feature. Even though the entire process is divided into smaller steps, the focus on the end goal remains the highest priority. With this focus comes the effort to eliminate all risk that deviates from this goal.

Although the Waterfall method is one of the most widely used and respected methodologies, it has come under some criticism as of late. Depending on the size, type, complexity, and amount of uncertainties in your project, it might not be the right fit. Disadvantages of the Waterfall method include:

- Difficult change management-The structure that gives it clarity and simplicity also leads to rigidity. As the scope is defined at the very beginning of the process under very rigorous assumptions, unexpected modifications will not be easy to implement and often come with expensive cost implications.

- Excludes client or end user feedback-Waterfall is a methodology that focuses on optimization of internal processes. If your project is in an industry that heavily relies on customer feedback, this methodology will not be a good fit, as it does not provision these kinds of feedback loops.
- Late testing-Verification and testing of the product comes toward the end of a project in the Waterfall framework. This can be risky for projects and have implications on requirements, design, or implementation. Toward a project's end, large modifications and revisions would often result in cost and schedule overruns.

Given its pros and cons and its overall rigid framework, the Waterfall method seems an undesirable management approach for machine learning projects. However, with so many management approaches from which to choose (e.g., Lean, Agile, Kanban, Scrum, Six Sigma, PRINCE2, etc.), how do we know which is best for machine learning projects?

One way to begin to answer this question is to categorize projects based on their complexity. Ralph Douglas Stacey developed the Stacey Matrix to visualize the factors that contribute to a project's complexity in order to select the most suitable approach based on project characteristics (Fig. 7.2).

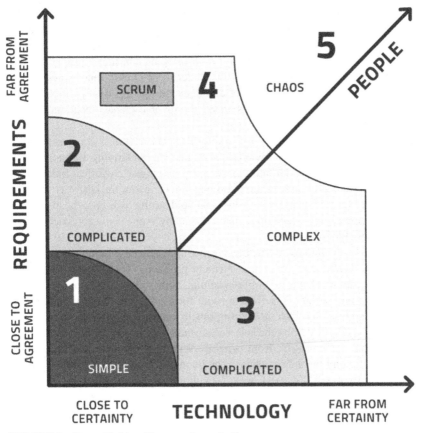

FIGURE 7.2   Stacey Matrix with zones of complexity.

On the y-axis, it measures how close or far members of your team are from agreement on the requirements and objectives of the project. Your team members might have different views on the goals of the project and the needed management style to get there. Your company's governance will influence the level of agreement as well.

Your project's level of certainty depends on the team's understanding of the cause and effect relationships of the underlying technology. A project is close-to-certain if you can draw on plenty of experience, and you have gone through the planned processes multiple times. Uncertain projects are typically challenged by delivering something that is new and innovative. Under these circumstances experience, will be of little help.

Based on these dimensions, we can identify five different areas:

1. Close to agreement/close to certainty-In this zone, we gain information from the past. Based on experience, it is easy to make predictions and outline detailed plans and schedules. Progress is measured and controlled using these detailed plans. Typically, we manage these types of projects using the Waterfall approach.

2. Far from agreement/close to certainty-These projects usually have high certainty around what type of objectives and requirements can be delivered, but less agreement about which objectives are of greatest value. In situations where various stakeholders have different views on added value, the project manager typically has difficulties developing a business case because of the underlying conflicts of interest. Under these circumstances, negotiation skills are particularly important and decision-making is often more political than technical. In these instances, the favored management approach is Waterfall or Agile.

3. Close to agreement/far from certainty-Projects with near consensus on the desired goals, but high uncertainty around the underlying technologies to achieve those goals fall into this category. The cause and effect linkages are unclear, and assumptions are often being made about the best way forward. The driver is a shared vision by stakeholders that everyone heads toward without specific, agreed upon plans. Typically, Agile is the approach chosen for these types of projects.

4. The zone of complexity-The low level of agreement and certainty make projects in this zone complex management problems. Traditional management approaches will have difficulties adapting, as they often trigger poor decision-making unless there is sufficient room for high levels of creativity, innovation, and freedom from past constraints to create new solutions. With adaptability and agility being the key, Scrum and Agile are useful approaches here.

5. Far from agreement/far from certainty-With little certainty and little agreement, we find the area of chaos. The boundary to the complex zone is often referred to as the "Edge of Chaos." Traditional methods of planning, visioning, and negotiating often do not work in this area and result in avoidance. Strategies applied to address these situations are called Kanban or Design Thinking.

Simplifying these different projects down to the degree of available knowledge and characteristics and the responsibilities of a leader highlights the inherent differences.

In Table 7.1, we see how important it is to choose the right process for each project. Although these categories are highly dependent on environment and the team's capabilities (a project that is complicated for an expert can be complex for a beginner), most oil and gas projects are typically categorized as complicated. They are characterized by best practices and focus on efficiency. Execution works fantastically well with top-down management and clean lines of authority for command and control. In these circumstances, the Waterfall method is the best management option.

What about machine learning projects? Application of machine learning and artificial intelligence to modern day problems is an innovative process. When compared with other industries, machine learning in the oil and gas industry has only recently found its application. Only governments lag farther behind oil and gas even further when comparing industry adoption of digitalization technologies [Source: World Economic Forum].

Machine learning is most effective when applied to complex problems. As outlined earlier, these are projects with many variables and emerging,

TABLE 7.1 Complexity in relation to management style according to the Cynefin framework.

Environment	Characteristics	Leader's job
Chaotic (little is known)	High turbulence No clear cause-effect True ambiguity	Action to re-establish order Prioritize and select work Work pragmatically rather than to perfection, act, sense, respond
Complex (more is unknown than known)	Emerging patterns Cause-effect clear in hindsight Many competing ideas	Create bounded environment for action Increase level of communication Servant leadership Generate ideas, probe, sense, respond
Complicated (more is known than unknown)	Experts domain Discoverable cause-effect Processes, standards, manuals	Utilize experts for insights Use metrics to gain control Sense, analyze, respond
Simple (everything is known)	Repeating patterns Clear cause-effect Processes, standards, manuals	Use best practices Establish patterns and optimize Command and control, sense, categorize, respond

interdependent interactions. With the interplay between these variables and dependencies being too complicated to predict, upfront planning is useless

As previously stated, Scrum is the most often used management approach to tackle complex projects. Soon, we will dive into the details of applying Scrum to projects, but before doing so, let us highlight the pitfall for managers, if this premise is not well understood.

The danger comes in the form of established processes and habits. As mentioned, the majority of E&P projects are being managed through the Waterfall method. However, a leader tasked with managing a complex machine learning project and using only familiar tools from simple or complicated projects is a recipe for conflict, failure, and misunderstanding.

From Table 7.2, we can see how the characteristics of a complex project, with uncertainty in the process and creative approaches, entails too many competing ideas that rely on the respective skills and competencies of the leader.

For a complex project, a good approach is to rely less on experienced professionals in the specific technical field, but rather, collect various theories and ideas and observe the effect of choices by using an Agile approach. (You can, of course, heavily rely on team members with experience with Agile projects). The project team must identify, understand, and mitigate risk as new results emerge. This often happens at a rapid pace, requiring a good leader to be an integral team player by enabling the rest of the team and driving cooperation and open communication. This type of leadership is referred to as "servant

TABLE 7.2 The importance of matching the right management style with the respective project type.

Environment	Characteristics	Leader's Job
Chaotic (little is known)	High turbulence No clear cause-effect True ambiguity	Action to re-establish order Prioritize and select work Work pragmatically rather than to perfection, act, sense, respond
Complex (more is unknown than known)	Emerging patterns Cause-effect clear in hindsight Many competing ideas	Create bounded environment for action Increase level of communication Servant leadership Generate ideas, probe, sense, respond
Complicated (more is known than unknown)	Experts domain Discoverable cause-effect Processes, standards, manuals	Utilize experts for insights Use metrics to gain control Sense, analyze, respond
Simple (everything is known)	Repeating patterns Clear cause-effect Processes, standards, manuals	Use best practices Establish patterns and optimize Command & control, sense, categorize, respond

Source: Adapted from Scrum.org/PSM.

leadership." In order to arrive at productive solutions for complex projects, teams must approach a problem holistically through probing, sensing and responding, as opposed to trying to control the situation by insisting on a plan of action (Fig. 7.1).

The potential mismatch between organizational requirements of a successfully managed, complex project and what a typical Waterfall environment provides as outlined in Fig. 7.2 is why we need a different project management approach to machine learning projects. In the next section, we explore the specifics of Agile and Scrum and learn how these approaches are best applied to the world of oil and gas.

7.2   Agile-the mindset

Agile, Sprint, Scrum, Product Owner, and Retrospective-welcome to the world of digitalization buzzwords. A world that from the perspective of traditional project management might come across as disorganized-a passing fad, not to be taken seriously. However, by the end of this chapter, you will be able to put meaning to the buzz and understand where and how this approach is best applied.

The term *Scrum* was introduced 1986 by two Japanese business scholars. They published the article "New New Product Development Game" (that is not a typo) in the Harvard Business Review, describing a new approach to commercial product development that would increase speed and flexibility. Scrum has its roots in manufacturing (e.g., Japanese automotive, photocopier, and printer industries) and made its way into the software development industry to establish a new process, as an alternative to the dominant Waterfall method.

In 1995, Jeff Sutherland and Ken Schwaber, two US American software developers, formalized the method in a paper they presented at the OOPSLA conference in Austin, Texas. Their collaboration resulted in the creation of the Scrum Alliance in 2002-a group of pioneers in Agile-thinking who came together to evaluate similarities in Agile methods. The result of the group's work was the *Agile Manifesto,* outlining the 12 principles of Agile software development.

In 2009, Ken Schwaber left the Alliance and became the founder of Scrum. org, which oversees the Professional Scrum Certification and publishes an open-source document called The Scrum Guide. The Scrum Guide is today's standard reference guide for Scrum project management.

So, what exactly is Agile and Scrum then? Agile is a general term that describes approaches to product development, focusing on incremental delivery and team collaboration with continual planning and learning. It is a set of principles and values with people, collaboration and interaction at its core and it can also be applied in other fields, such as organizations and training. While Agile provides the mindset, it is Scrum that outlines the concrete framework. It is one of various frameworks under the Agile umbrella, and the one that we will concentrate on for the purpose of managing machine learning projects.

7.3 Scrum-the framework

If you are a fan of the sport of rugby, you certainly have heard of the term Scrum before. This word choice is no coincidence. In rugby, a Scrum (short for scrummage) refers to restarting a play and involves players being densely packed with their heads down, attempting to gain possession of the ball and progress down the field.

The terminology is purposefully chosen by Takeuchi and Nonaka (1986), who already in their original paper pointed towards the analogy: "The traditional sequential or 'relay race' approach to product development [...] may conflict with the goals of maximum speed and flexibility. Instead, a holistic or 'rugby' approach - where a team tries to go the distance as a unit, passing the ball back and forth - may better serve today's competitive requirements" (Fig. 7.3).

Also, Schwaber recognized the similarities between the sport and the management process. The context is the playing field (project environment) and the primary cycle is to move the ball forward (sprint) according to agreed rugby rules (project controls). Further, the game does not end until the environment dictates it (business need, competition, functionality, or timetable). Interestingly, he also acknowledged that rugby evolved from breaking traditional soccer rules, alluding to how Agile is new relative to traditional Waterfall framework.

Fast-forwarding to the present and using *The Scrum Guide* as reference, we note the following definition: "Scum (n): A framework within which people can address complex adaptive problems, while productively and creatively delivering products of the highest possible value."

Scrum is a lightweight framework for enabling business agility. Rather than being a process or technique for building products, it provides the framework within which you can employ various processes and techniques. This framework is based on empirical process control theory, acknowledging that the problem

FIGURE 7.3   **A Scrum during a rugby match (© Luis Escobedo).**

cannot be fully defined or understood up-front. Instead, it focuses on the team's ability to quickly deliver value.

Scrum is best understood by investigating its elements. In the following section we will look at the 3-3-3-5-5 framework, which describes the following:

- 3 pillars of Scrum theory   what is Scrum based on?
- 3 roles   who is involved with which responsibilities?
- 3 Artifacts   what is being delivered?
- 5 Events   what happens when?
- 5 values of Scrum   which beliefs motivate actions?

Let us start with the foundation. Scrum is based on three pillars of process control in which certain assumptions are met:

1. *Transparency*-Important aspects of the process are visible to the entire team, and there is a common understanding of language, including the definition of "done." Everyone knows everything.
2. *Inspection*-Progress and deliverables (artifacts) are often inspected toward their goal to detect undesirable variances. Check your work as you do it.
3. *Adaptation*-If inspection shows that aspects of the process deviate outside acceptable limits resulting in an unacceptable product, the process must be adjusted as soon as possible to minimize further deviation. It is OK to change tactical direction (Fig. 7.4).

Inspection and adaptation are applied systematically throughout the process, as will be evident as we dissect all elements of the scrum framework.

7.3.1 Roles of scrum

There are three, and only three, roles within the Scrum team: the product owner, the development team, and the Scrum master. Ideally, they should be co-located to foster communication and transparency.

The *product owner* represents the product's stakeholder and voice of the customer. He is responsible to optimize towards adding most value to the

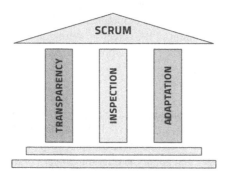

FIGURE 7.4   **Three pillars of Scrum.**

TABLE 7.3 Services of the scrum master.

Stakeholder	Services of scrum master
Product Owner	• Ensures goals, scope and product domain are understood by everyone • Helps arranging and prioritizing the product backlog
Development Team	• Removes impediments • Coaches team in self-organization • Facilitates scrum events
Organization	• Help employees and stakeholder understand and enact scrum

product and manages the product backlog. He does not have the responsibilities of a typical manager (i.e., he is not managing a team or a project).

The *development team* typically has three to nine members and focuses on delivering increments of added value to a product during a specific amount of time (a *sprint*). Although members of the team are often referred to as developers, they can often be from various cross-functional backgrounds, such as designer, architect, or analyst.

Most importantly though, especially relative to setups in E&P, is the fact that the team is self-organized. This means that while all work comes through the product owner, all responsibility to deliver its work falls solely on the team itself-no one tells the development team how to do their work. The product owner and the Scrum master are not part of the development team.

The *Scrum master* facilitates the Scrum methodology. He promotes Scrum by making sure everyone understands its practices, rules, and values. By removing possible impediments to the ability of the team to deliver the product goals, the Scrum master acts as the servant-leader of the Scrum team. Table 7.3 summarizes the Scrum master's most important services toward stakeholders.

The Scrum master's role differs from that of a project manager because the Scrum master does not have any people management responsibilities. Since the responsibility of managing the development team lies within the team itself, the Scrum master's involvement in the direction setting process is limited. In fact, Scrum does not recognize the role of a project manager.

7.3.2 Events

Scrum has a set of well-defined events to minimize the need for other unnecessary meetings. All events are time-boxed, that is, they have a specified maximum duration. As you will see, all these events (except the sprint itself) provide opportunities to inspect and adapt the ongoing process of development and add to transparency (Fig. 7.5).

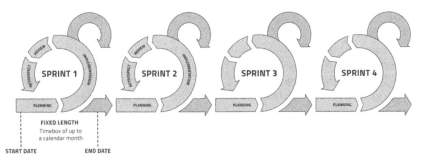

FIGURE 7.5   Development cycles are done in sprints.

In Scrum, everything revolves around the *sprint*. It is typically between two to four weeks long (never more) with the goal to produce an increment of useable, potentially shippable product by the end. At the beginning of each sprint (during the *sprint planning*) the team agrees on the scope and defines the goals of the sprint by moving items from the *product backlog* into the *sprint backlog*. During the sprint, no changes are made that endanger the sprint goal, although the scope may be clarified and re-negotiated between the product owner and the development team as new insights arise.

Every sprint ends with the *sprint review* and *sprint retrospective.* As soon as one sprint ends, the next begins immediately (Fig. 7.6).

The *sprint planning* meeting marks the beginning of a sprint. A sprint planning, capped at 8 hours, focuses on answering two main questions: (1) Which product increment can be delivered in this upcoming sprint (sprint goal)? and (2) How can this chosen work be done?

In practice, the first half of the meeting involves the entire Scrum team (Scrum master, product owner and development team) suggesting product backlog items practical for the upcoming sprint. Ultimately, the development team

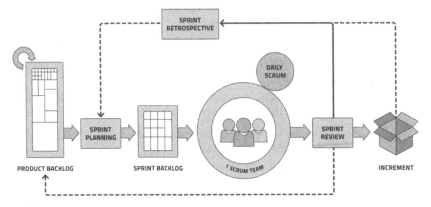

FIGURE 7.6   Scrum framework.

selects the specific items for the sprint, as they are best equipped to judge what is achievable.

In the second half of the meeting, the development team breaks down the chosen product backlog items into concrete work tasks, resulting in the sprint backlog. To refine the sprint backlog, the development team might negotiate the selected items with the product owner to agree on the highest priority tasks. In the end, the development team explains to the Scrum master and the product owner how it intends to complete the work.

The *daily Scrum* is a meeting that is held every day of the Sprint and is time-boxed to 15 minutes. Everyone is welcome, but it is the development team members that contribute. It should always happen at the same time and location and start promptly, even if some team members are not present. The goal of the meeting is for the whole team to answer three questions:

1. What did I do yesterday that contributed to the team reaching the sprint goal?
2. What will I do today that will contribute to the team reaching the sprint goal?
3. Do I see any impediment that could prevent me or the team from reaching the sprint goal?

It is the Scrum master's responsibility to keep the development team focused and to note potential any impediments. The daily Scrum is not to be used for detailed discussions, problem solving or general status updates. Once the meeting is over, individual members can meet to resolve any open issues in breakout sessions.

The *sprint review* and the sprint retrospective are held at the end of each sprint. They provide the opportunity to assess progress towards the goal and the team and collaboration, respectively. For the purpose of reviewing the sprint progress, the team does the following:

1. Reviews and presents the completed work (e.g., in the form of a demo),
2. Address which items were not "done" (completed),
3. Collaborates on what to work on next, and
4. Reviews of timeline, budget, and required capabilities.

Output from the meeting is a revised product backlog and dialog to inform the next sprint planning.

The *sprint retrospective* is an opportunity for the team to reflect and improve. Three questions facilitate this process, including:

1. What went well in the last sprint with regards to people, relationships, processes, and tools?
2. What did not go so well?
3. What can be improved to ensure better productivity and collaboration in the next sprint?

The entire team participates in the sprint review and sprint retrospective, which for four-week sprints, are time-boxed to two hours and 90 minutes, respectively (and scales proportionally for shorter sprints). For both meetings, the Scrum master is responsible for ensuring all participants to stick to the rules and time limits are honored.

Should it become apparent that a sprint goal is obsolete (due to company strategy, market conditions, technology changes, etc.) during the sprint itself, the product owner may cancel the sprint. However, due to the short duration of most sprints, cancellation is rare.

7.3.3   Artifacts

Another component contributing to transparency in Scrum is the *artifact*. Artifacts provide information on the product, forming a common understanding and allowing for progress on inspection and adaptation. Typical Scrum artifacts include the product backlog, sprint backlog and product increment.

A *product backlog* is an ordered list of all that is initially known about the product. In addition to foundational knowledge and understandings, it includes a breakdown of work to be done. Business requirements, market conditions or technology can cause modifications to the backlog, making it an ever-evolving artifact. The listed items can be features, bug fixes or other non-functional requirements. The entire team has access to the product backlog, but the product owner is solely responsible for it. Only the product owner can make changes and set priorities for individual items. Typically, the product owner will gather input and feedback and will be lobbied by various stakeholders; however, it is ultimately his decision on what will be built (and not that of a manager in the organization).

The *sprint backlog* is a set of product backlog items selected for the current sprint. It serves as a forecast of what functionality will be delivered by the end of the sprint and is being modified throughout the sprint to represent progress. Backlog items can be broken up into smaller tasks, which, rather than being assigned, team members will tackle based on priorities and individual skill sets. When the team finalizes the sprint, they analyze and reprioritize the remaining product backlog to have it prepared for the next sprint.

The team's definition of "done" results in a *product increment*, which is a list of all product backlog items that were completed during the sprint. When combined and integrated into all previous increments, the product increment embodies a potentially shippable product that is functional and usable. It is up to the product owner to release it or not (Fig. 7.7).

7.3.4   Values

From the earlier figure, one can tell how all the different elements of Scrum enable a feedback-driven, empirical approach. The three pillars of Scrum-transparency,

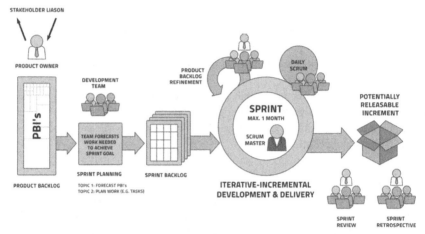

FIGURE 7.7 All elements of Scrum during a Sprint.

inspection, and adaptation-require trust and openness. To enable this trust and openness, Scrum relies on following five values:

- *Commitment*-Individual team members personally commit to achieving the goals of the scrum team.
- *Focus*-Team members focus exclusively on the work defined in the sprint backlog; no other tasks should be done.
- *Courage*-Team members have the courage to address conflict and problems openly to resolve challenges and do the right thing.
- *Respect*-Team members respect each other as individuals who are capable of working with good intent.
- *Openness*-Team members and stakeholders are open and transparent about their work and the challenges they need to overcome obstacles.

These values are essential to successfully using the Scrum method.

7.3.5 How it works

By now you have a rough understanding of Scrum. But if we break it down, what does it mean to execute a project in an Agile way, using the Scrum framework? Let us translate the buzzwords into practice.

Short iteration cycles provided by time-boxed sprints afford high transparency and visibility. With this visibility, and the mandate to self-manage, it is quite easy to maintain adaptability to business and customer needs as well as steer your product in the right direction. Early product releases at the end of these short iterations help generate value relatively early in time, reducing risk, and unwanted cost overruns.

Returning to our comparison of the Scrum framework to traditional project management techniques, Fig. 7.8 illustrates how the differences transpire

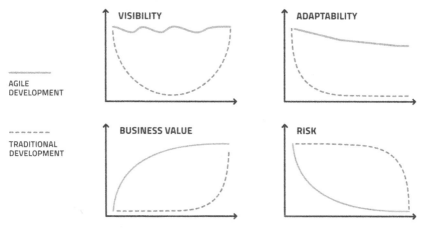

FIGURE 7.8   Value proposition of Agile development compared to traditional development.

between the two approaches. Predicting all features of a project with large uncertainty is much more challenging at the beginning of the project, often resulting in surprises at project end when the customers are presented with results when deadlines loom. Agile development allows for better adaptability, lower risk early on and earlier creation of business value.

The process of creating a car is one commonly used example to illustrate how Agile is applied in the real world and how it differs from big bang deliveries (i.e., build it until it's 100% done, and then deliver it).

Let us assume we have a customer who has ordered a car. Delivery of the first iteration of the product is often misunderstood as delivering an unfinished product. In the case of our example, it would mean the delivery of a tire at iteration (1) (Fig. 7.9). Naturally, the customer will not be happy-after all he has ordered a car, not a tire!

The customer's reaction is not going to be much different at iterations (2) and (3), when the product is still a partial car at best. Although the product is getting closer to its final state, it is not until the final iteration (4) that we have a satisfied customer. And in this example, he is happy because it is what he ordered. A lot of time passes before the customer gets to see any of the product, resulting in the product being based on a lot of assumptions and design flaws.

Now, let us contrast this with the Agile approach. Although the customer ordered a car, we start by focusing on the underlying need he wants to see fulfilled. In this case, conversation with your customer might show that "getting from A to B quickly" is the underlying need, and a car is one of many ways to do that. (Keep in mind that the car is just a metaphor for any kind of customized product development.)

After the first sprint, the team would deliver the most minimalistic thing to get feedback from the customer, for example, a skateboard. We could call this an **MVP** (**Minimum Viable Product**). Naturally, the customer will likely be

FIGURE 7.9 Building a car-not using proper Agile development.

unhappy about the result, as it is nowhere near the car, but the goal at this stage is not to make the customer happy, but to get early feedback, learn and make adaptations for product development down the road.

The next iteration could have a different color (or other changes based on early feedback) and a way of steering, to make it safer to drive around with. The third iteration could be a bicycle, which the customer might start using, resulting in even more valuable. Based on this learning, the 4th sprint might be a motorbike as a deliverable. And at this stage, it may be that the customer is much happier with this product than he thought he would be with the car he originally ordered (Fig. 7.10).

Or, in the final iteration, he sticks to his original idea and gets the car delivered at last, but it's better tailored to the feedback from previous iterations, perhaps resulting in a convertible rather than a coupe. In the end the customer is overjoyed-he got a car and a better one than he originally envisioned!

With the underlying question being "what is the cheapest and fastest way we can start learning?" it is possible to reduce risk early and provide the customer with business value that meets or exceeds his exact needs.

7.4 Project execution-from pilot to product

With our newfound understanding of Scrum and Agile, we can now identify opportunities to apply this approach within the context of an E&P company. Bringing this approach to a business world defined by Waterfall management will be a challenge that is best tackled with the introduction of pilot projects.

A pilot project, by nature of its test-setup, gets the mandate of experimentation. It is usually run in parallel to day-to-day operational business and focuses on the validation of a technology or process. In our case, this is the application

FIGURE 7.10 Building a car-using proper Agile development.

of our machine learning algorithm. Let us use an example to examine a possible pilot setup and see how the different roles and stakeholders would interact. We will end the chapter outlining how you can introduce Scrum as a possible management framework within your company.

7.4.1   Pilot setup

ABC Oil Ltd., a small oil and gas company, is interested in exploring machine learning in the context of a virtual flow meter. Under normal circumstances, ABC Oil conducts well tests to inspect flow rates periodically, at best. In the interim, production engineers rely on extrapolating from previous test results, which leads to uncertainty and inaccuracies in data estimates.

By incorporating machine learning, advanced analytics, and all available production data (e.g., pressures, temperatures, fiscal flow meters, well-test data) ABC Oil sees an opportunity in using virtual flow meters to calculate real-time rates for all wells. This innovation would allow for informed decision-making based on live data.

ABC Oil has identified SmartML Ltd. as a potential partner for developing and testing this virtual flow meter technology. SmartML has experience working with companies in the petroleum industry, but its main expertise lies in the application of artificial intelligence and machine learning. ABC Oil and SmartML agree to cooperate and perform a pilot investigating the effectiveness of applying machine learning to all available production data to create a virtual flow meter.

Both companies are aware of the uncertainties, accept them, and agree to run this project in using the Scrum framework, including filling roles of product owner, Scrum master and development team.

7.4.2   Product owner

A technical representative from ABC Oil Ltd serves as the product owner. As an ABC employee, she possesses the best understanding of the requirements of the product (a virtual flow meter) and thereby, she is responsible for setting priorities for the backlog items and making sure the team understands the product goal.

However, given that this pilot project is occurring concurrently with existing ABC Oil operations, both companies see benefit in identifying a proxy product owner who serves as the go-between for the product owner and the development team. The proxy product owner is part of SmartML and thus, is more available to dedicate time to the project than the product owner, who is also managing ABC Oil's day-to-day operations. The product owner and the proxy product owner require a relationship based on trust and open communication.

7.4.3   Development team

In the pilot, SmartMl's developers and designers make up the core of the development team. Generally, the composition of the team depends on the size and capabilities of the E&P company. The larger the oil-company, the more capabilities internal capacity it can provide, while smaller companies may only possess the capacity to supply the product owner.

In the case of ABC Oil, the more inter-dependencies that exist between software solutions and the existing infrastructure, of ABC Oil the more important it is to involve ABC IT and OT engineers and data architects on the development team. Connecting relevant data sources and maintaining cyber security, while simultaneously allowing an external partner access to data, can be challenging and requires prioritization and teamwork.

Depending on the skill set available at ABC Oil, it may or may not involve its own staff during the development phase, but certainly will involve staff in the testing and validation process when the team finishes the first iterations of the solution. To this end, there is a notable benefit to having the team co-located. The larger and more complex the project, the more important co-location becomes in terms of facilitating interactions and improving transparency and clarity-especially at the beginning of a project when trust is being established. Remember, that the team is self-organized, so, neither the product owner, her proxy, nor the Scrum master, act as managers or superiors in this setup.

7.4.4 Scrum master

The Scrum master focuses on facilitation, meaning this role can be filled by ABC Oil, SmartML or an external organization. With ABC Oil providing core business expertise and SmartML providing machine learning respectively expertise, an external Scrum master with Scrum and Agile framework expertise would be a valuable asset to the pilot project.

How development teams conduct meetings is dependent on the degree of co-location. With an external development team, most meetings will be virtual, including the 15-minute daily Scrum meeting. In case of virtual meetings, artifacts such as product backlog, sprint backlog and product increment can be adapted by using virtual collaboration tools.

The length of the sprint review, sprint retrospective and sprint planning correlate with the length of the sprint. With back-to-back scheduling, an in-person meeting, where feasible, is ideal. The kind of communication, interaction, and feedback intended for these meetings are invaluable and will work significantly better in a traditional workshop environment. Typically, these meetings take place every two to four weeks, making in-person meetings achievable, even if team members are not co-located.

7.4.5 Stakeholders

Now, we will shift gears from our ABC Oil Ltd. example to a more general examination of Scrum applied to oil and gas operators. Outside the core team, you will have many stakeholders to collaborate with and the way you manage these stakeholders will be particularly important if your project competes with daily operations.

IT/OT and data management might be your most important collaborators. Machine learning projects are heavily reliant on data and your team needs access to this data. Enabling data availability with consideration of cyber security, compliance, access rights, data exchange protocols and consumption will often be your most time-consuming task. Onboarding and integrating data security staff onto your team will save you future headaches and delays.

Further, getting management buy-in is necessary to execute these projects; however, it makes sense to differentiate between management levels. Generally, top management understands the added value of digitalization projects and provides support, whereas middle management often calls for more massaging. A large part of these machine learning projects comes down to change management, especially when deploying a Scrum approach (where management does not have a designated role). Thus, it is imperative that middle management is aware, involved and has ownership that has been communicated top down.

Further, it is important that your customers are fully aware of their role as customers. In a setting where a project is initiated from corporate headquarters into other operational organizations, the product owner might be someone who is not part of the actual operation's team. In fact, more often than not, this is the case.

As much as your project will be competing your customers' daily tasks and duties, their involvement will be required. Customer feedback will be critical to ensure user-friendliness and ultimately, user-acceptance. Your product could be the best from the perspective of the development team, but if your end customer is not using it, the product has failed. Your customer needs to be aware of his role, know how important his feedback is, and recognize that the development team and product owner are ultimately working to satisfy his needs.

7.5 Management of change and culture

New processes require change and change management, which can bring with it resistance. Consequently, your attention to the various stakeholders and your company's change culture will be crucial.

As previously mentioned, it is common for pilot projects to run in the form of rogue projects, that is, in parallel to your core business activities. This approach has several advantages, including:

- Setting context to test an assumption The context of a pilot project gives you more liberties to experiment and test the underlying technology you want to implement to solve a certain problem. It is usually easier to get approval for smaller projects.

- Implementing quickly, managing risk, and failing fast-A quick test implementation will allow you to better manage risk and minimize cost. In case of failure, you will have failed fast and gracefully, rather than blowing up. Experimenting on a smaller scale educates you about what works and what does not without the level of risk of a larger project.
- Discovering dependencies-You will learn more about how your project interacts with various stakeholders as well as other devices, technologies, or data sources. Some of these stakeholders might not have been in your original scope. You will also learn more about the accuracy of your planned resource allocation.
- Fine-tuning the business case-The learnings from the pilot will allow you to better estimate your business case should you want to scale and implement your project on a larger scale after successfully completing the pilot.
- Getting feedback from the users, your customers-Next to understanding whether the technology itself works, your most important insights will be whether your idea will be accepted by the users.
- Refining your solution and spur new ideas-Based on the user feedback, you will be able to fine-tune your product, so it matches your customer's needs. This might result in new ideas and concepts. Further, smaller endeavors allow you to change direction quickly, if needed.
- Preparing for rollout-Experiential feedback and learning will be the best possible preparation for a later rollout and scaling of your product. A small-scale rollout is the perfect testing ground and makes a later, larger release less risky and stressful.

With pilots like these, you are actively engaging your colleagues as testers. For many, this kind of involvement will be a new experience, resulting in various levels of enthusiasm. Some will embrace this level of engagement and participation in the development of something novel, while others will not. In the context of Scrum, its requirements of inspection and adaptation, such an approach might be viewed as disorganized and chaotic. If you are used to Waterfall project management, shifting to Scrum-like methods may not be so intuitive. You will need to invest significant time into setting the stage by explaining the approach and coaching your colleagues on the importance of their roles.

Since you are not directly contributing to the core business, the pilot could also be perceived as a useless distraction. With potential headwinds like these, it is important to get the right mandate from the start. The more novel the approach, the more important endorsement from higher management levels, such as CTO or CEO, will be. Management can be supportive by not only stressing the relevance to the organization for technical reasons, but also in the context of new ways of working and collaborating, including motivating the team to be brave and giving them the room and permission to fail. A basic understanding among top management of how Scrum and Agile works will give you yet another advantage.

Whatever the outcome of your pilot, you should not be afraid to share the results. With success or failure, there will be stories to tell. Be transparent, open, and honest. Use success stories that highlight your colleagues and customers and have them reflect, not only on the technical aspect, but on the way the team collaborated. Successful pilot projects can generate momentum that can be used to drive and multiply a solution into the organization.

7.6 Scaling-from pilot to product

With the completion of your pilot, hopefully you have a successful MVP. Assuming the product concept has been validated, you now have a better understanding of the technology and the risks and cost involved. Based on the positive feedback from your colleagues who serve as customers, you know you have a great product. With the product being operationally ready, it is now time to scale.

A great idea and productive pilot are only half of a successful product. Even more important is making sure you can scale your solution. Up until this point, the pilot has been a local, validation experiment-it is only after you scaled the solution to more assets that you can multiply and reap the full potential and benefits of the business case for the entire organization. Let us have a look at some important aspects of this process.

7.6.1 Take advantage of a platform

No matter which business units or assets you will scale your product to next, you will always need data and usually lots of it. Company wide, standardized data architecture will give your scaling efforts significant leverage. If your product only needs to learn to connect to one type of data source (e.g., a standardized data layer) your connectivity issues become close to plug-and-play, reducing scaling complexity significantly.

In the beginning of your scaling process you might find that the effort of getting relevant data connected to your data-layer is a task greater than connecting your application. Once your product is up and running, it becomes a lot easier to replicate, whereas hooking up different data-sources to your data layer can be tedious.

Considering data provision as part of your scaling requirements and as part of a global digitalization effort will give you additional leverage. Data availability will not only be relevant for your product, but hopefully, to future endeavors. These products will serve as business cases that will justify the data layer itself, since data provision alone does not create additional business value. Typically, data is moved from local data historians to global cloud solutions, facilitating access from anywhere in the world.

7.6.2 Establish a team and involve the assets

In the pilot phase, your team members might have been an ad-hoc ensemble of colleagues that were sourced to execute the project. As you move from pilot to product, your engagement will consequently be more long-term, including the creation of a self-sufficient, committed core team that can grow as needed.

Remember to Involve the asset from the start. Involving the user at the beginning of each roll-out assures you meet requirements and that you understand expectations. Your products will fail if you do not create co-ownership.

7.6.3 Keep developing

Be prepared to keep developing your product. Your users will find bugs and request new features. The product will evolve with implementation and use, and unforeseen issues will need to be addressed. Staying open-minded with an eye toward improvement, will help you meet these challenges.

As your product grows, you will need to ensure you scale its support. In the pilot phase, most issues will be handled by the development team. As the number of users increases, your capacity to support your solution will need to grow. Depending on the setup, this growth may be managed within your IT department or in conjunction with an external partner.

7.6.4 Involve UX expertise

While the pilot focuses on the technical feasibility of your product, you will have more time to fine-tune the user experience (UX) and design when you conceptualize the final product. In the end, your solution must not only work properly, but the user experience must be positive and sustainable. Again, the user's input will be critical to create an intuitive interface while maintaining high functionality. Finally, the design of the tool can also be aligned with the corporate design language of your company.

References

Takeuchi, H., & Nonaka, I. (1986). *The New New Product Development Game*. Harvard Business Review.

Further reading

This was but a glimpse into machine learning projects employing an Agile framework. If you want to dive deeper into any of the topics discussed earlier, we recommend the following reading list.
Schwaber, K., & Sutherland, J. (2017). *The Scrum Guide*. https://www.scrumguides.org/.
Verheyen, G. (2013). *Scrum-A Pocket Guide*. Van Haren Publishing.
Schwaber, K. (2004). *Agile Project Management with Scrum*. Microsoft Press.
Cohn, M. (2004). *User Stories Applied: For Agile Software Development*. Addison-Wesley Professional.

McGreal, D., & Jocham, R. (2018). *The Professional Product Owner*. Addison-Wesley Professional.

Lencioni, P. (2002). *The Five Dysfunctions of a Team: A Leadership Fable*. Jossey-Bass.

Adkins, L. (2010). Coaching Agile Teams: A Companion for ScrumMasters, Agile Coaches. *and Project Managers in Transition*. Addison-Wesley Professional.

Stacey, R. D. (2011). Strategic Management and Organizational Dynamics: The Challenge of Complexity Sixth Edition. In D. S. Ralph (Ed.), *Strategic Management and Organizational Dynamics,*England: Pearson Education Limited, Harlow, Essex.

Fierro, D., Putino, S., & Tirone, L. (2018). The Cynefin Framework and the Technical Leadership: How to Handle the Complexity. *CIISE 2017: INCOSE Italia conference on systems engineering, 2010*.

Lynch, W. (2019). *The Brief History of Scrum*. https://medium.com/@warren2lynch/the-brief-of-history-of-scrum-15efb73b4701.

Kniberg, H. (2016). *Making sense of MVP (Minimum Viable Product)-and why I prefer Earliest Testable/Usable/Lovable*. https://blog.crisp.se/2016/01/25/henrikkniberg/making-sense-of-mvp.

Jordan, J. (2018). *Organizing machine learning projects: project management guidelines*. https://www.jeremyjordan.me/ml-projects-guide/.

Zujus, A. (2019). *AI Project Development-How Project Managers Should Prepare*. https://www.toptal.com/project-managers/technical/ai-in-project-management.

Singh, D. (2019). *Things to Consider While Managing Machine Learning Projects*. https://cloudxlab.com/blog/things-to-consider-while-managing-machine-learning-projects/.

Foxall, D. (2019). *Scaling Your Product-thriving in the market*. https://www.boldare.com/services/scaling-your-product-thriving-in-the-market/.

Newlands, M. (2017). The Perfect Product is a Myth. *Here's How to Scale the Almost-Perfect Product*. https://www.entrepreneur.com/article/293010.

https://en.wikipedia.org/wiki/Scrum_(software_development).

http://Scrum.org/PSM.

http://agilemanifesto.org/.

Chapter 8

The Business of AI Adoption

Geoffrey Cann
MadCann Alberta Inc., British Columbia, Canada

Artificial intelligence and related technologies such as machine learning, edge computers, deep learning systems, and neural nets create new opportunities for value creation, as well as pose upsetting changes to established ways of working. Those interested in adopting these new technologies will need to consider the importance of creating trust in these new tools, and the nuances of successfully assisting organizations with adoption.

8.1 Defining artificial intelligence

A discussion of artificial intelligence must begin with some definitions and terminology. AI is a computerized capability to execute work requiring cognitive skills that are normally associated with humans. Examples include natural language processing, translation of languages, visual perception, auditory interpretation, and tool creation. The supreme ability of AI is its capacity to ingest massive amounts of data in whatever form or source that data originates, interpret the data, and take some action. All human senses, including touch and taste, and intangible things like emotional state, are also being worked on by the AI industry. The common ingredient is the availability and utility of data in extraordinary quantities.

AI will be a work in progress for some time as the range of data and quantity of data are continuously expanding. Trend lines suggest that technology will continue to advance its ability to replicate a full range of human cognitive skills. Many highly compensated roles and complex ways of working in oil and gas are contingent on these skills, and over time AI tools will progressively impact these roles.

8.2 AI impacts on oil and gas

AI is poised to have three direct and dramatic impacts on the oil and gas industry. First, AI is being applied to subsurface data and, through enhanced interpretation, petroleum engineers are confidently expanding the amount of oil and

Machine Learning and Data Science in the Oil and Gas Industry.
http://dx.doi.org/10.1016/B978-0-12-820714-7.00008-X

gas that is recoverable. Second, AI is helping reduce demand for hydrocarbons by transforming transportation markets. Finally, AI is reducing costs and improving productivity throughout the entire value chain, encompassing upstream production, midstream processing, and downstream refining and distribution.[a]

Combining AI with other digital technologies has the potential to unlock new business models in the industry.

8.2.1 Upstream impacts

By far the greatest opportunity for AI anywhere in oil and gas is in the upstream. Through better and faster interpretation and analysis of available subsurface data, and superior modeling of data, petroleum engineers can improve the recovery rates from resources. The principal beneficiary of this innovation is the tight oil and gas resources, such as shale deposits. Production wells in shale formations enter into decline more quickly in the life of the well (perhaps 3 years after initial production), than with wells in conventional reservoirs.

Conventional wells generally have a much shallower decline curve, and the amount of product recovered is much greater. By shifting the decline curves for shale wells to more closely match those of conventional wells (basically, heighten, lengthen, and flatten the curve), petroleum engineers will improve the productivity of shale wells, and add 5% to global reserves (or 500 billion barrels of oil equivalent). This is primarily a math problem, to understand and model out permeability and porosity, and to use that insight to complete shale oil and gas wells differently, a job for which artificial intelligence is well suited.

The prize is substantial. Gas wells typically produce up to 90% of the recoverable resource in a conventional reservoir. For oil, the recoverability rate is about 40% in a typical conventional reservoir. For shale, recoveries are typically 20% or so. A 5% increase in global reserves by improving recoverability is in the order of USD20 trillion in value.

Shale resources are primarily a North American phenomenon so this wealth will accrue principally to the US producers, who have access to the resource, a hotly competitive and well-developed services industry, the leading cloud computing companies, and the data scientists. Many countries have exceptional and similar shale resources but lag the United States in these additional attributes.

AI needs a lot of data to work optimally, and upstream companies have accumulated enormous data sets over the years from conventional oil and gas production. That data could also be fed into AI engines to enhance existing production or to extend the life of existing wells. Older wells, with low levels of productivity, could be brought back to life via AI. There are now data service companies that offer AI-based well remediation as a service.

a. International Energy Agency. "Digitalization and Energy 2017," Iea.org (November 5, 2017): http://www.iea.org/digital.

8.2.2   Downstream impacts

The second major impact from AI is looming, but not yet manifest. As transportation technologies evolve, the demand for fuel may be dramatically impacted. There is still considerable uncertainty surrounding the timing, uptake rate and consumption levels of more connected, autonomous, shared, and electric cars and trucks (CASE for short). Car and truck manufacturers around the world are all rapidly converting their manufacturing supply lines to produce these new transportation technologies. CASE vehicles, particularly the autonomous varieties, are dependent on AI engines to interpret the data generated from their onboard cameras and sensors.

Demand for transportation fuels could triple, according to one modeling effort, if urban populations abandon public transit services in favor of ride-sharing services. On the other hand, for businesses that are dependent on logistics and freight transportation (manufacturing, wholesale, retail, travel, and leisure), carbon is a big concern. Companies in these sectors are already highly motivated to remove carbon from their businesses (driven by tax policies now in force in many jurisdictions, by sweeping policy changes such as the EU Green Deal, and by outright bans that are coming to big cities in Europe), in which case demand for transportation fuels could be cut in half, according to another model. So far, companies who have implemented AI-enabled trucks and haulers report fuel savings as the trucks run optimally all the time.

The impacts of AI on the downstream illustrate the dual nature of this technology on the industry. AI is both expanding the supply of hydrocarbons and destroying the demand for petroleum at the same time.

The looming challenge for oil and gas is how to best manage the infrastructure that is presently dedicated to distributing petroleum products in a market facing possible and uncertain decline. Some oil companies now acknowledge that their retail businesses are effectively obsolete as a business model.

8.2.3   Production and midstream impacts

In the rest of the oil and gas value chain, AI is being deployed primarily to augment human decision making, rather than to displace humans, and in ways that help optimize asset execution, and predict asset performance.

An Australian producer[b] has loaded the entire engineering history of their offshore platform into a query-ready AI system. In response to an English language oral query, the system can instantly search all the filed content (engineering diagrams, consultant studies, status reports, maintenance history, asset register, logs), rank order the content candidates that match the query, and present the findings. Engineers that use the system estimate that it saves them some 40% of their work time that, prior to the AI solution, was devoted simply to searching for documents in their archives.

---

b. Woodside Energy presentation

A Canadian pipeline company[c] uses an AI system to model complex network assets such as its pipelines and tank farms in the cloud (creating a digital twin of the network asset), to help the owner optimize the network. The pipeline can be optimized to maximize throughput, improve customer ratability, lower carbon impacts, or maximize tolling margins. These calculations cannot be easily conducted by human operators but are simple for the AI solution.

A producer applies AI to interpret land contacts to extract the contractual terms that are the basis of royalty calculations. These contracts are many pages thick, take minutes to read through for a human, and number in the thousands. Using AI coupled with other digital technologies has shortened the processing time per contract from 800 seconds (achievable only by the best contracting professionals, and unsustainable over time), to 45 seconds. This producer reduces accounts payable costs, eliminates disputes between royalty partners, and reduces the month end work stress of some 50 professionals in contract administration.

A technology company[d] has used machine learning to create an algorithm that optimizes beam pump performance. A beam pump, also called a rocking horse, is one of the most common forms of artificial lift systems in use on onshore oil wells. The performance of the beam pump is influenced by the quantity of hydrocarbon in the well to lift, the presence of water, the depth of the well, the diameter of the well, the presence of accumulated fines in the well, and many other factors. The algorithm, housed in a small edge computer or controller attached to the beam pump arm, runs continuously, and optimizes every stroke. Eventually, such an AI controller could also consider market-pricing data or power costs to further optimize the pump.

A supermajor producer[e] has turned over some of its oil and gas field operations in the United States to AI. Marginal producing assets may not generate enough value to warrant full time, dedicated and expensive human oversight. These assets can gain extended life through the use of AI tools that can tightly optimize production and operations.

8.2.4 New business models

Reliable, 100-year-old industry structures are being overturned by new digitally enabled business models. The key technology is cloud computing, which is having roughly the same impact on industry that fossil fuels had on the development of flight. In the decades before the famous Wright brothers' efforts, engineers had broadly figured out the intricacies of flight. The missing technology was a fuel of the right density to drive a sufficiently powerful motor to pull the

---

c. Kinder Morgan pipeline and Stream Systems
d. Ambyint
e. BP deploys Kelvin Inc in the Great Divide Basin

aircraft through the air for liftoff. Once that fuel was mastered, airplane technology began rapidly advancing.

Cloud computing similarly provides enough storage for large data sets and the compute density to allow artificial intelligence, sensors, and automation, to "take off."

For the oil and gas industry, AI enables new business models, which are very disruptive. The analytic horsepower from cloud computing now rivals the best in-house compute data centers in the biggest oil and gas companies and is available to anyone on a variety of economic models. Cloud computing is rented by the cycle and AI algorithms are available on a per use basis, whereas in-house data centers are largely fixed cost assets. Smaller firms are able to create the massive kinds of data sets that benefit from AI by aggregating data from multiple companies using cloud computing. The operations of entire oil producing fields are now run by artificial intelligence engines. Sophisticated artificial intelligence interpretation capabilities, which would be otherwise accessible only to the largest firms, are now in reach for the whole of the industry.

The impacts of these innovations are widespread throughout the industry, deeply impactful on work processes, and targeted as much on professional roles as on operational roles. Jobs and work that have been stable for the better part of two generations are now in the cross hairs of a wave of change. Understanding and anticipating how workers will react to these tools and preparing them for the change will accelerate adoption.

8.3   The adoption challenge

Walter was a math wizz and correctly tallied up long columns of numbers in his head. One day the boss issued new battery powered calculators to his workers. To Walter's son, the calculator looked very cool, with its space-age red symbol read out, tiny keys and endearing form factor. But Walter did not trust it. First, he added up the numbers using the calculator, as the boss wanted, but then he redid the entire math by hand. He was suspicious that the calculator, the algorithm of his day, might not be right.

The oil and gas industry has worked diligently for decades to create a stable world. Engineers rely on the control technology model designed 40 years ago, SCADA systems, to help supervise assets. Analog sensors are built to known industrial standards, embeded directly in the equipment, and hardwired to a control panel, or to a front-mounted gauge. There are no uncertainties in the industrial standards, or in the tests of compliance that the devices and gauges must pass. Copper wire is highly reliable and resistant to compromise, and data delivered over copper is widely accepted as uncorrupted. The data is extracted from the historian directly into an Excel spreadsheet onto an employee's computer for analysis and interpretation. The professionals involved have been recruited for their expertise and are considered as trusted hands. This overall system design has been part of the education curriculum and engineering disciplines for decades.

Fast forward to Winnie, an oil and gas process engineer who has discovered that Engineering Services has strapped a novel digital sensor to some operating equipment under her control. She is then asked to rely on the results of an unseen cloud-based AI algorithm that is interpreting the data from the sensor.

8.3.1 The uncertainties of new technology

Winnie relies on the technical reliability of the process environment. This new sensor technology introduces fresh uncertainties into her stable world. For example:

- The sensor itself, its technical features, and its compliance with industrial standards
- The sensor mounting and how reliable it is to capture data correctly
- The power supply to the sensor and its reliability
- The behavior of the sensor to variability in power quality, temperature, vibrations, and other factors
- The data the sensor generates and the potential for compromise
- The integrity of the wireless network that moves the data from the sensor to the cloud analytics engine
- The technical competence of the algorithm's author
- The integrity of algorithm itself
- The results that the algorithm generates

Overcoming these mechanical uncertainties will be required to improve the trustworthiness of the AI solution. Even social norms governing human behavior can block progress in surmounting the challenges. For example, algorithm authors sometimes cannot, or are unwilling to, clearly explain how their AI algorithm technically works, out of concern that their IP will be compromised, or because they may not fully understand it either.

8.3.2 AI in the field

Assume that the technical and process engineers have satisfactorily resolved the problems of sensor provenance, network reliability, power supply, and connectivity for Winnie's sensor. The final remaining uncertainty is the inner workings of the algorithm.

Consider four scenarios:

One: Correct predictable analysis

The algorithm correctly interprets the data and the interpretation matches what the engineer thinks should be the result. The AI machine is improved, but she is irritated that the company spent money on something that she could already do.

Two: Correct unpredictable analysis

The algorithm correctly interprets the data, but the interpretation differs from what the engineer thinks should be the result. Now Winnie is in a dilemma: is this a false positive? Does she take the recommended action for an uncertain outcome, or ignore the recommended action and rely on her training and experience, or does she, like Walter and his new calculator, embark on a sidebar manual exercise to try to replicate the algorithm?

Performance metrics and targets compel Winnie to weigh the business and personal consequences to determine the safe path. What if the machine is correct and proves she has been wrong all this time? Will there be repercussions?

Winnie responds to the performance measures and ignores the results of the algorithm. She loses a learning opportunity for both the human and the machine, and she will spend needless time and money trying to replicate the algorithm. She can plausibly claim to have avoided a possible catastrophe, which she would have avoided anyway, and she will not be embarrassed by a machine.

Note that had Winnie followed the machine's recommendations, she is better off in both the short term (because of a smarter decision and fewer wasted resources) and in the long run (because the machine is made smarter).

Three: Incorrect predictable analysis

The algorithm incorrectly interprets the data and Winnie agrees that the interpretation is incorrect, forcing the engineer to rely on previous knowhow and analysis. She is likely irritated that the company spent money on something that does not work and she is no better off. Her choices then inform the algorithm, which improves the algorithm and might pay off later, but at the risk of doing herself out of a job. She is not comforted by the possibility that her bosses will question why she did (or did not) use the algorithm's analysis and recommendations.

Four: Incorrect unpredictable analysis

The algorithm incorrectly interprets the data, but the engineer cannot tell that the interpretation is incorrect and has no better analysis to leverage. Again, the engineer is in a dilemma: what if the machine is wrong? If she follows its recommendations and it does not work, is she to blame? Does her boss have her back?

Winnie is in a bind, and has no option but to execute, and there is a failure. She may be sanctioned at the next performance review, but at least she can point out the failings of the algorithm. In the end the algorithm is a little smarter.

8.4   The problem of trust[f]

Winnie and her organization have long had positive work relationships with the kinds of automation developed since the 1950s. The arrival of artificial intelligence and machine learning algorithms embedded in a new class of smart machines is now forcing a rethink of the trust equation with technology.

---

f. The contribution of Kevin Frankowski to this section is gratefully acknowledged.

8.4.1 Work is evolving

The building blocks of the latest wave of automation, such as mobility, the Internet of Things, and cloud computing, are very democratic, expensive, and available to every business. They find opportunity in heavy and light industries alike, varying only by the pace of adoption. Automation is now plainly visible in agriculture where farmers rely on AI solutions to inform planting, irrigation, pest control, and harvesting. The shipping industry is trialing unmanned cargo vessels. Automated vehicles are navigating many city streets in many different cities so as to prove the technology under a variety of social settings.

The energy sector, in particular, is undergoing a second shift as the world looks to embrace more energy from electrons and less from molecules, particularly in transportation. This shift has huge implications for work. The molecules that yield energy when combusted require a healthy level of human supervision to keep the mechanical equipment that contains those molecules from gumming up, overheating, and wearing out. Electrical gear has many fewer moving parts overall, and a dramatic reduction in the number of parts that need to be lubricated, cleaned, cooled, and replaced.

Industry will require far fewer people to maintain this equipment. Anticipate a leaner workforce on site to supervise the assets, use tools to measure how equipment is behaving, lubricate parts, and detect aberrant sounds or odors. Simpler electrical equipment lends itself to machine learning, robotics, remote control, and remote monitoring, more so than molecule-based machines.

8.4.2 Driverless transportation

Humans do not extend much trust in machines, particularly where the machines are responsible for making very important decisions. For example, the aerospace industry has developed all the technology required for an aircraft to fly without on-board pilots. Drone technology is well advanced and in wide military usage. Similar autonomy technology exists for submersibles, helicopters, trains, and many other applications. Despite the economic benefits from removing both pilot and co-pilot from the cockpit, no airline yet offers flights without at least one human pilot at the controls at all times.

Pilotless aircraft demonstrate that any successful technology must meet the MASA principle—most advanced but socially acceptable.

Many mission critical assets, especially in heavy industry, still feature a sizeable, but expensive and technically unnecessary, on-site human management and operations team. The social pressure that might block the adoption of modern intelligent solutions is not present in some obscure plant far from human settlement, and yet the workforce remains. It begs the question why management teams, particularly in the energy sector, facing critical competitiveness challenges, are resistant to embrace greater levels of automation in anticipation that automation is coming. Is it a matter of trust?

8.4.3   Trust and the machine

Why don't humans trust new smart machines to perform mission-critical tasks (even for tasks that machines are clearly better at)?

There may be several reasons.

Building trust within an interactive and mutually dependent relationship with other humans is something innately human. People rarely have that same emotional relationship with a machine. Machines lack the human-type needs that drive their decisions, and this dispassionate, cold, and calculating approach creates considerable emotional distance with humans.

For the most part, humans care about being reliable to those that depend upon them. Machines do not care about how they are perceived by humans— they just follow their instructions. The developers of those machines are dis-incentivized from being transparent about how their complex and proprietary algorithms behave.

Humans are endlessly adaptable and can problem-solve in the moment when things go wrong (as with pilots at the helm of an aircraft or boat). Machines are getting much better at this, thanks to fleet learning, but humans still distrust these quickly evolving abilities.

Humans can exercise judgment, such as when to break a rule or create something novel. Machines are bound to their instructions, and in fact, the notion that a machine is empowered to break its governing rules is viewed with considerable suspicion.

Humans can be held accountable for their actions and react well to rewards and punishments. Mankind has lengthy if imperfect experience levying consequences against other humans in social structures that stretch back through the whole history of humankind, and it is central to the experience of being human.

Humans are more accepting of having a human "at the switch" (or at least in the loop, keeping watch over the machines and having supervisory control). While machines can be programmed, using gamification theory, to respond to rewards and to avoid punishments, their decision framework is still math based, and not sensitive to emotions. Shunning a human for a transgression might cause a change in their behavior, but shunning a robot has no effect.

8.4.4   The human-smart machine trust gap

To attain the highest performance from relationships with smart machines, humans require trust in that relationship. To gain the trust of users in an industrial setting, smart machines must overcome a number of reasonable human concerns.

Lack of performance: Is the machine even going to work? Will it meet the requirements?

Lack of long-term track record: Even though the machine has performed acceptably in the demo or pilot setting, what level of confidence is there that the machine will work over the long term, under every conceivable condition (e.g., cold and hot weather extremes; loss of power or communication; unexpected operating conditions; emergencies and exceptions)? Can the machine be relied on when thoroughly unpredicted conditions prevail, and creative problem solving and adaptability are needed?

Lack of integrity: Since machines lack the human sense of morals, what is the confidence level that the smart machine will be truthful? In an industrial setting where safety really matters, how will the machine handle conflicting decision-making criteria, such as the famous Trolley Problem[g] that self-driving cars must deal with?

Lack of clarity: Why does the machine make the decisions it does? Under what conditions will those decisions change? How can those decisions be influenced in real-time as conditions or drivers change (e.g., oil price fluctuation, currency volatility, input price variability)?

If the machine makes a poor decision, how is it held accountable? What kinds of consequences will even matter to a smart machine? How can the contribution of the machine's decisions relative to decisions that others make be distinguished, a particular concern in complex industrial settings where many small decisions taken by many participants add up to larger outcomes?

8.4.5 Trusting a smart machine

To achieve high performance in a working relationship with smart machines, managers and users need to be able to trust them, which begins with the five components of trust:

- **Competence**: All the participants in the relationship must have confidence that the parties have sufficient capabilities to fulfill their respective responsibilities.
- **Reliability**: All participants in the relationship must have confidence that the parties will fulfill those responsibilities reliably, on a consistent basis.
- **Transparency**: All participants in the relationship must hold a shared goal that guides the decisions that are to be taken. Each must believe that decisions taken by the others will be predictable, mutually compatible, and acceptable. Each must hold a level of influence over the others to impact decisions. Each must publish their decisions and the results achieved.
- **Integrity**: All participants in the relationship must be fully truthful with each other and adhere to any agreed-upon principles or norms of behavior. Each participant must have a compatible set of priorities.
- **Accountability**: All participants in the relationship must seek the same outcomes through an aligned set of interests. Each must be subject to a consistent set of meaningful motivators (e.g., bonuses) or deterrents (e.g., consequences) to help maintain and nurture this alignment.

In human terms, there must exist shared values, morals, and ethics with smart machines.

---

g. An ethical thought experiment for autonomous vehicle design, which may require programming to choose whom or what to strike when a collision appears to be unavoidable

It is important to note that these five components of trust (competence, reliability, transparency, integrity, accountability), are subject to both hard quantifiable facts as well as perception. If the smart machine is competent (or reliable, etc.), but the human manager or user harbors some doubts, then there is no trust. Similarly, if humans mistakenly believe that any of these components of trust exist when they do not, eventually the truth will come out, and the mismatch results in broken trust. Trust is hard to build, and erodes quickly, in the presence of this mismatch between fact and perception.

Not only must these five components be present to create the trusted relationship, but all the participants in the relationship must be aware that the components are present and must be confident that all participants are authentically aware.

It is also important to realize that all five of these components of trust need to be present at all times, or trust is compromised. Many kinds of interactions and relationships take a "grow in trust" approach, where trust is built over time, as evidence mounts that these components are in place. When even one of these components goes missing, even temporarily, then things take a step backward and trust needs to be rebuilt.

Trust needs to be both earned and maintained, in both human relationships and in human relationships with smart machines.

8.4.6 Trusting the smart machine developer

Behind every smart machine is a team of people (designers, programmers, operators, etc.), and building and maintaining trust begins with this team. Users of smart machines must make it very clear to the technology designers and vendors of the expectation that the five components of trust are to be addressed in a very concrete manner. Call it the social contract for digital innovation.

To earn and maintain trust in their smart machines, the designers and vendors of digital technology need to address very explicitly the five components of trust:

- **Competence**: This is the easiest aspect to prove. Technology authors must demonstrate that smart machines, algorithms, and computations achieve the promised performance under a range of test conditions.
- **Reliability**: Performance assurances from the authors must address "normal" operating conditions, and all the various extremes that might be encountered (the "edge cases"). Consistency of reliability can be very situation specific—social standards for consistency might be very different for a personal navigation and traffic app (80% reliability might be good enough) versus the navigation and flight control software on an aircraft.

  One solution is to allow humans to intervene in the decision-making process only if needed, so that humans, who excel at creative problem solving, can address the edge cases. Humans need not occupy every single pilot seat. Retailers have recognized that self-check-out lanes—one human teller oversees 6–8 tills—allow for exception only intervention by a human operator and capture an 80% improvement in efficiency.

- **Transparency**: AI developers must share how their software works, in ways that are understandable to the users.
- **Integrity**: Settling on aligned principles of behavior can be challenging, especially for some of the edge cases. However, if developers do not significantly open up the conversation to customers on how they approach integrity, then strong trust will continue to elude these relationships.
- **Accountability**: Giving a meaningful consequence to a machine is difficult. However, once there is agreement on the outcomes that the smart machine is working towards, plus alignment on decision-making principles and how the outcome will be achieved, then accountability can flow through those agreements to the smart machine's human supervisors.

8.4.7 Making it real

The social contract will play out in the real world in a variety of ways.

If a self-driving car gets a speeding ticket because the owner stipulated conditions that caused it to make poor driving decisions, the owner should pay the ticket. However, if the software was written in a way that allowed it to speed of its own volition, then the software vendor should be held accountable.

If a drilling management program enabled by artificial intelligence achieves better than expected financial returns due to new upgrades from the vendor, does the contract discuss how these additional benefits will be shared? If the drilling software fails to anticipate an unexpected high-pressure zone, and drilling activity causes a well blowout, is the algorithm designer jointly accountable for any environmental incidents?

Incorporating a human touch in smart machine design and operation provides a symmetry of deterrence and incentives that not only encourages reliability and spurs innovation, but also fits with how humans are wired and motivated. Technology authors should be clear and honest about the underlying principles on which the machine bases its decisions. For example, what will the algorithm optimize for—operating efficiency and speed? return on financial investment? health and safety? Developers should be transparent about how their business model works. An intention to mine collected customer datasets for insights to monitor market performance or sell more solutions should be explicit. Designers should incorporate the capacity for human intervention in the operations of a smart machine for those instances when the machine owner feels the need to intervene.

8.4.8 Getting to trust

Michael, founder of an AI technology solution, sought some advice on how to grow his business in oil and gas. His clever technology consists of a small box of sensors that attaches to an oil or gas well, gathers data about the performance of the well, leverages a machine-learning algorithm, and adjusts the equipment in real time to run more efficiently. One of its key features is that it eliminates the need for a worker to drive to the well and carry out manual inspections, and manual adjustments of the well equipment. This new solution helps move an oil well from a highly manual operation to one that is substantially more automated.

Michael is keen to discover the best practices for convincing the field to take on new digital tools. The oil and gas managers in the head office state that it is his job to go to the field and convince the supervisors, field engineers, service contractors, and local admin to embrace this new technology.

Compare this normal practice in oil and gas with the situation in the retail industry, led by Amazon. At the time of writing, Amazon is rolling out new automated pick and pack robots for its fulfillment warehouses. These gigantic facilities are probably one of the few remaining labor intense aspects of the warehouse and logistics businesses. Pickers gather odd shaped packages from the shelves, and packers arrange them into a cardboard shipping box with the ubiquitous Amazon logo emblazoned front and back, stuff the box with packing materials for the long journey to the customer, tape, label, and dispatch them to shipping.

Packing has long been the exclusive job of humans who are solely able to figure out the optimal way to fit odd shapes into a box. But the job is monotonous, turnover is high, and the constant employee churn is a constraint on growth. Once this job is figured out, the day of an almost lights-out warehouse operation is in sight since Amazon also has figured out how to use drones to count inventory. The robot that unloads containers is within sight. Only the picking task remains with human workers.

Other complex jobs, such as driving vehicles on public roads, or running automated trains on complicated rail lines, are targeted in the on-going drive to digitalize.

8.5   Digital leaders lead

How do the digital leaders drive change in their organizations? It is widely accepted that Amazon is a leader in driving innovation throughout its business. The company releases 50 million software updates each year (a pace of one software change every second, around the clock). Its market capitalization is among the highest in the world. The company is a leader in retailing, book sales, cloud computing, home automation, and many other fields that leverage their core competencies in big data, analytics, artificial intelligence, language processing, and robotics.

Amazon drives its experiments in work automation from the top down. There will most certainly be job impacts from automated packing: fewer human packers, for example, although the growth in on-line shopping suggests there's probably not enough people to do the job. The role of the warehouse supervisor changes to be less about people management, shift coordination, load balancing, and training, and more about robot optimization and operation. Facilities set aside for human workers, such as lockers, lunchrooms, and parking, are freed up. Even the design of the warehouse shifts to accommodate 24/7 operations that can function at very high speed and allows for the interaction of different drones doing different jobs.

Amazon's efforts to drive deeper automation in the warehouse are not the work of the warehouse teams who do the picking and packing. It is doubtful that the workers hired to pack shipping boxes are also automation experts. These workers are also unlikely to start up a trial that is almost certain to do away with their existing manual jobs. Such trials and experiments, if left to the warehouse teams to decide, would be swiftly rejected.

Michael's small digital start up, with limited resources, no expense budget, and no authority, cannot successfully take on the job of visiting oil and gas field facilities and engaging with the local offices to convince them of the benefits of a technology that could result in work and headcount impacts.

The field is deeply aware of all the reasons why new technology might not work. They are quick to point out the incompatibility with other onsite technology. They highlight the lack of time for training. They stress the high production and productivity targets and the impossibility that they can both hit all the existing targets while introducing something new and unproven. They must tend to the operating equipment rather than have a meeting. They need to build up trust in the technology, and that takes work. They do not even have the time to invest in building a basic understanding of these technologies.

In other words, the field perceives all the threats and senses none of the benefits.

Successful oil and gas businesses generate their own kind of change aversion. Successful managers can easily convince themselves that they have mastery of the levers that drive economic performance. As a successful business, there is little benefit from the risk of trialing new ideas that might not work out.

But the past is no longer a certain guide to the future. Consider the impacts on the telecoms business (phone calls and text messages are now free); media (most news services are also now free); taxis (ride-sharing services offer superior customer experience); retail (on-line shopping has bankrupted hundreds of bricks and mortar shops); books (thousands of small retail shops have disappeared); music (streaming services have taken over compact disc technology); banking (the latest digital banks sign up millions of customers in a few weeks).

8.5.1   Finding the digital leader

How does Michael identify those companies in the oil and gas industry who are truly ready to work collaboratively to deliver a solution?

Digital innovators should learn what they can about how a potential oil and gas customer approaches digital change. Press releases and analyst presentations are useful sources of validation. Those companies that lack clear public statements about their digital strategies should be questioned about their commitment.

The digital innovator should highlight the important role of management in driving the change as a key part of validating the likely benefits to be gained from the initiative. Managers should be very clear about how much time and energy they can commit to the success of the work.

With the right conditions in place (a compelling business need, a manager willing to trial or pilot a new capability, a solution that appears to deliver meaningful results), the innovative digital solution may be invited to run a trial, proof of concept or pilot with a customer. Building a base for future success into the design of the trial or pilot is the next step of the change journey.

8.5.2   Moving beyond trials and pilots

Michael is successful in his efforts to find an oil and gas business unit manager willing to run a trial of his technology. The business unit is not performing to plan, prior actions to improve performance have run their course, and the manager is still under pressure to bring results back in line with company expectations. Months later, Michael finds himself in a familiar situation—trapped in a never-ending cycle of pilot projects and trials with an oil and gas company that struggles to progress beyond the pilot stage to full deployment.

8.5.3   The role of trials and pilots

Pilots and field trials are exceptionally valuable for both the digital technology startups and the customer. Developers use precious on-the-ground experience to fine-tune their solutions under real world conditions, identify key additional features that customers really want, and sharpen the fit of the solutions to the industry. Pilots also reveal shortcomings in scalability, support, deployment, and value that startups really need to solve if they are to succeed. Having a number of pilot sites is a badge of credibility with potential new customers, investors, and funding agencies.

Pilots and trials are also useful for oil and gas companies. Customers gain real experience with the latest technologies, and then make credible claims to investors and boards that their companies are innovating with the times. Pilots often surface deeper insights about customers, reservoirs, processing equipment and business processes, which are valuable in themselves. Pilots can reveal just how backward and regressive some business practices in oil and gas really are.

The main governance construct of a pilot project is contingency. If the technology or solution achieves specified milestones or benefits, then in theory the company will allocate funds to take the technology from pilot stage to a bigger pilot or to enterprise roll out. This model works well enough with many kinds of lab bench chemistry and processing technologies and explains why an oil and gas buyer is often quick to agree to a pilot or trial. The customer knows that there is an exit ramp if milestones are not met or benefits fail to materialize.

The role of technology incubators and technology accelerators that are now commonplace in many business centers is to provide the contacts, support, and tools that a startup needs to land a trial or pilot with a customer. Pilots are a necessary first step into the commercial world of creating and nurturing a viable product and business.

But this model hides certain problems.

8.5.4 The economics of pilot projects

The typical local buyer in a large oil and gas company is an unlikely enterprise champion. They have limited authority to make enterprise purchases. They are even constrained in their authority to broaden the pilot to other assets or adjacent opportunity areas, because of how oil and gas companies are organizationally structured. Individual spending authority is low. Champions of change need to secure hard-to-obtain approvals from corporate heads of engineering, lines of business, operations, and maintenance. Leaders in the company may suspect that the results of the trial are incorrect, optimistic, and unrepresentative of the rest of the organization. They must be convinced that the pilot has enterprise merit, in the face of performance measures in oil and gas that reinforce the imperative for safe operations, reliable execution, and minimal disruption.

Michael, typical of many digital innovators, does not leverage the pilot properly to create the conditions for "land and expand." In the eagerness to get a pilot going, he may not clearly set out the journey map that delivers increasing benefits in exchange for a larger presence. He rarely explains the merits of an enterprise solution in terms of better pricing or better performance to the initial small customer. The initial customer may suspect that they will be quickly abandoned in favor of a more powerful enterprise-level executive. Few innovators attempt to measure with precision the baseline into which their solution is to be deployed and lack the expertise or the methods to link the benefits captured back to their solution.

Pilot projects are prone to bookkeeping games, particularly those that claim to create a cost reduction. As soon as a cost benefit materializes in a specific line of business, line managers can quickly reallocate those savings towards other budget line items. A reduction in the cost of a services contract transforms into an increase in a consulting budget.

Perpetual pilots serve neither party well. Technology startups lack the resources to properly support and sustain dozens of pilot sites, which typically require more intensive care and attention. Some start ups are likely burning cash in their drive to scale up. The more pilots underway, the more likely that the technology start up is spreading itself ever more thinly over its growing base. Pilots often fail to convince technology investors, who are seeking the annual recurring revenue (ARR) that comes from committed enterprise customers.

Customers are not well served by the pilot that never ends. By not moving to an enterprise deployment, with its steady cash flow to the start up, the customer creates a risk that the technology company will struggle to grow. The customer fails to capture all the benefits they are entitled to, since many technologies are really platforms whose benefits grow with scale. The message that perpetual pilots communicate internally, that the technology is not ready for wider deployment, is toxic to the business and the technology company

8.5.5   Moving to enterprise deployment

To expect the technology supplier to solve this problem on their own is unlikely to succeed.

Both sides of this relationship, the technology company supplier and the oil and gas customer, need to change their approach to achieve a better outcome. Technology companies need to recognize the limitations of their local customers to drive enterprise purchases, and customers need to acknowledge the risks they face by letting pilot projects run without a defined end game.

Customer tactics

Capital allocated to digital innovations is now on the Board agenda. Boards are becoming increasingly aware of the transformative power of technology. Shareholders are asking Boards to account for the disposition of shareholder capital that is allocated to pilots. Certainly, some pilot initiatives are sensitive and should not be in the public domain, but many digital trials do not fall into this category.

Boards should take an interest in what innovations are happening in the business, and work with management to properly fund the pilots. It makes little sense to allocate shareholder capital to pilot projects and then let them run as pilots forever. Boards should press management for regular updates on pilot projects and important field trials.

Executives should ask for a regular accounting of the various pilot projects and trials that are underway, regardless of scale. That accounting should include the nature of the benefits at stake, how long the pilot has been running, the end game, and lessons learned. Executives should dial back their tolerance of pilots that do not progress and should actively fund the work necessary for line managers to build out the enterprise plan.

Companies should consider taking a stake in key technologies or technology companies to drive heightened attention on the pilot. Nothing creates interest more than an ownership stake.

Finally, managers should make time i to share the state of pilots across the company to signal that change projects are encouraged and rewarded.

Technology supplier tactics

Technology suppliers should build more clarity in the design of the trial and the pilot about the mechanisms for transitioning out of pilot mode.

The innovation team should invest in measuring the baseline of the benefits at stake at the outset of the pilot. The customer or an independent consultant should do this task so that the bias inherent in measurement is reduced. The baseline should consider indirect as well as direct productivity gains, cost reductions, carbon impacts, asset utilization and throughput, safety gains, pricing effects, and volume growth.

The pilot design should clearly specify the performance targets that the pilot is intended to achieve, based on the baseline measurement. That could be based on percentage or absolute gains, both quantitative and qualitative.

The pilot should set out the conditions under which the pilot will progress from stand alone to either a larger pilot, a second pilot site or enterprise roll out. The innovation team should identify the actions it will take to assist the organization through the transition from pilot to production and beyond.

To motivate the move to enterprise deployment, the innovation should create value for the initial customer that is only available if the pilot grows. This value could be additional features that are available through fleet learning, or new pricing, or both.

The pilot design should set out an end date to the pilot. It should be sufficiently distant that the company has enough time to learn to use the solution and embed it into their ways of working.

The innovator should support the customer in creating the growth pathway. For example, oil and gas companies typically run a steady series of internal knowledge-sharing and update sessions, which help to drive change throughout the organization. The innovation team should help the customer get on the agenda, craft the messages, and coach them to success.

8.6   Overcoming barriers to scaling up

Once past the pilot stage, digital innovations face a new set of hurdles. Scaling up is challenging as evidenced by the number of small technology suppliers and the relative absence of fast-growing digital companies serving the oil and gas industry.

8.6.1   The scale mismatch

The difference in scale between very large oil companies and very small digital companies or innovation startups hamstrings mutual success. With their multiple layers of management, tight procurement rules, waterfall design philosophies, and intergenerational masses of infrastructure, big oil companies inherently move slowly towards the future. There will continue to be steady demand for hydrocarbons for the foreseeable future, the industry is very mature, and the product is mostly undifferentiated. Capital markets thoroughly understand the business model and can easily rank one firm against another. Urgency is low.

Digital startups, on the other hand, are lean and agile. The founder is often still coding. Decision-making is nearly instantaneous. Work methods are based on new approaches, such as design thinking and agile. Digital companies start out in the cloud, collaborate closely with many third-party technology providers, expect costs to fall steadily over time, and race to achieve network mass as quickly as possible.

Digital solutions innovate constantly to stay competitive, while staying well clear of the competency zones of the digital giants.

This mismatch of scale is a permanent structural problem of the industry and requires structural solutions.

8.6.2   Supplier consolidation

One solution to the structural problem is through market consolidation, where promising digital solutions are acquired by the large technology incumbents.

Large incumbents solve the scale problem as they are already at scale. Their large installed base creates a natural pathway to drive market penetration. Their extensive product catalogues deliver valuable and more complete solutions for their customers. They already have the implementation teams to drive deployment, avoiding the need to recruit a large fixed cost service team. They likely have the integration know how to deliver interconnections. They overcome the trust challenges in the market that the solution will work, in all situations, under pressure.

For the digital start up, joining forces with an established technology player to the industry may appear to be an attractive answer to the problem of scaling up. At the time of writing, there are no unicorns (startups with a valuation of a billion or more), serving the digital oil and gas market, possibly because of this exit strategy.

In the long run, the oil and gas industry may not be well served if the bulk of the promising digital companies disappear and are absorbed into the ranks of the incumbents. The pace of innovation in large mature technology firms may match that of the oil and gas industry. The methods that have accelerated the growth of a digital company may give way to the traditional ways of working in older technology companies. Traditional technology companies have themselves struggled to embrace concepts like cloud computing, open systems, and agile methods.

8.6.3   The corporate accelerator

Another solution to scaling up is the customer accelerator. This concept has long existed in industry. For many years, mature industries have no longer experienced big jumps in productivity through some clever innovation. Gains come from harvesting the hundreds of tiny improvements made at the various plants and figuring out how to deploy those changes to all the other plants. A fractional reduction in energy use at a single plant translates into big energy gains across an entire the business if that reduction could be replicated.

Customer accelerators and internal engineering teams supervise a portfolio of pilots and trials across the fleet of plants. Successful innovations are carefully studied in a central lab to figure out how to productize the innovations and enable them to scale out across the fleet.

Industry leaders who are serious about scaling up digital often put in place their own internal digital accelerators. Dedicated corporate teams work with a portfolio of digital solutions to sort out scaling issues, promote change management, figure out deployment pathways, and measure value.

An example might be an oil field unit who trials a machine learning solution that optimizes artificial lift performance. The solution is piloted at the unit and achieves some success. The corporate accelerator pulls the solution into the lab and replicates it to other pump problems, generalizing the solution, and redeploying it back into the original unit for additional testing and proving. The original business unit is the test ground because it was their innovation in the first place, and they are generally committed to its success.

Once the solution is proven out, the corporate accelerator then figures out the enterprise deployment strategy, to take the innovation across the full company. The enterprise strategy considers the overall solution architecture, the pace and timing of deployment, the communications strategy, the migration approach, the implications for turn arounds, and the longer-term sustainment strategy.

8.6.4 The oil company investor

There is an increasing interest among some oil and gas companies to run their own portfolio of investments in small promising digital companies, usually by setting up an independent venture fund. These funds take an active interest in driving scale among innovation startups.

There is a compelling financial logic to this idea. The oil company investor can help shape solutions to better fit their needs. The oil company as investor lends strong credibility to the solution and helps reduce resistance to internal deployment. And as an investor, the oil company stands to gain should the innovation become a unicorn (which will absolutely not happen if the innovation is purchased early by a technology incumbent), and can likely exit by eventually selling their interest to a technology incumbent. Studies suggest that the typical digital innovation startup has only a 1% chance of survival, but when a customer takes a very strong interest in helping the startup, the chance of survival grows to 30%.

8.7 Confronting front line change

During and beyond the pilot stage, many digital champions struggle to overcome the resistance to change in work processes and practices in oil and gas, and indeed in many heavy industries. The hard part of digital adoption is not technology, but the need to change cultural norms, which might not be the competence of the digital innovator.

There may be many offered reasons why a change driven by technology will not work but at heart, the one good reason for not embracing a change is that people like things the way they are. Underneath this reason are the three very human drivers of greed, fear, and pride that must be considered in driving front line change.

Greed

Employees quickly figure out how to make an existing system work to their advantage and maximize their benefit. They know the levers to pull, how they interrelate, where to hide their mistakes, how to inflate the results. "Under promise and over deliver" is the theme on many an inspirational work poster. When mishaps happen, employees know who to blame. Lastly, if they are successful at what they do, they have no incentive to change.

Fear

Many employees are close to their next promotion or, more typically in oil and gas, retirement, and they are unlikely to risk losing their bonus on something they do not understand. If something goes wrong, they might not be able to fix it, or they might not have time to fix it, and they may be unsure who to blame. And if it does work, it exposes shortcomings in the well-run machine that they have successfully run for years. Innovation might even result in job loss.

Pride

Employees in oil and gas are justly proud of their well-deserved reputations as top operators. But appearing vulnerable in heavy industry does not inspire confidence, particularly in oil and gas. Employees do not wish to appear foolish, or out of touch. There is little chance employees will willingly display a lack of understanding about digital innovation at work.

8.7.1 The corporate parallels

These deeply seated human motivators have three parallel motivators in the workplace setting. These are growth, risk, and brand.

The corporate equivalent to greed is growth. Oil and gas companies in the western capitalist systems are based on growth—in revenues, reserves, production, assets, capital budgets, dividends. Managers in this work setting will have put in place performance measures to achieve growth goals. A system of bonuses and sanctions reinforces the attention to the performance measures.

The organizational equivalent to fear is risk. Oil and gas companies are particularly tuned to the risks of the industry since hydrocarbons are inherently dangerous to people and the environment. Safety measures and metrics are always paramount, uncompromising, and well ahead of financial returns. Production is hedged to offset the risks of price erosion.

The equivalent to pride is brand or reputation. Oil and gas companies guard their brands very carefully, especially in a world that views carbon emissions as a climate hazard. Most players in the industry invest to advocate their credentials to investors, employees, and regulators. Many are aggressively balancing their investments to include more renewables and alternative fuel sources in their portfolio.

Successful efforts to implement change in oil and gas include attention to both the corporate and the human drivers. An alternative way to express this balance is that change must appeal to the head, or the cold industrial logic of growth, risk, and brand, as well as to the heart, or greed, fear and pride. Many unsuccessful digital deployments can be traced back to an inappropriate balance in addressing these change drivers, and in particular, an over investment in highlighting the corporate benefits, and an under investment in addressing the personal drivers.

8.7.2 Early warning signs

Change champions tasked with rolling out a digital solution should be attentive to early signs that the change initiative may flounder.

The digital narrative

The first step for the change leader is to control the narrative about the digital innovation, its vision, purpose and expected outcomes. Presentations or training about digital will need to take place on site, since many oil and gas facilities are continuous operations running multiple shifts on a 24/7 basis. The change team will work to influence the narrative, manage the conversation, promote fresh lines of thinking, and deal with objections about the proposed change and digital innovation.

Unfortunately, many key oil and gas facilities are well away from civilization. Small digital innovators should not be expected to provide briefings for free, at awkward or inconvenient times, which will minimize the number and duration of such sessions, as well as the number and kinds of experts that attend. Reasonable questions related to raising trust in the solution will go unanswered without the right expertise available.

Corporate sponsors of change will note the value in traveling to places like Silicon Valley as an important way to teach the organization about the opportunities presented by digital. Capital and cost constraints often block requests for travel to conferences or to courses.

Should funding become available to send people to trainings, conferences, or benchmarking tours, it becomes important to carefully pick the attendees. Those employees who are most influential team members should be given leave to attend. They are better positioned to promote the digital innovations on their return, as they are well regarded by their peers, and will be more able to convince their colleagues that digital is something to embrace.

Manage the pace

Change champions should anticipate that the pace of change in oil and gas seems slow compared to the pace of change in the digital industry. This is to be entirely expected in an industry that weighs risk so highly. Maintaining pace is the second ingredient for a successful effort. Unfortunately, any early enthusiasm that the change champion can generate may quickly erode if the change is delayed.

Field organizations have many compelling reasons why timing is inappropriate for implementing digital innovations. Perhaps some new equipment is being delivered, or some other corporate initiative that must take precedence is occupying people. Managers and users in oil and gas often require compelling evidence that the proposed digital solution is largely risk free. It is very challenging, particularly for novel digital solutions such as smart machines, to prove that the solution is entirely without risk.

Other hard-to-satisfy requirements can contribute to slower progress. Most oil and gas facilities are brownfield in nature—that is, they are in production and have been for some time. It may be necessary for a new digital solution to integrate or interconnect with existing SCADA systems that predate the digital era. It is not surprising to discover that some site technology dates back 4 decades, to the very dawn of the personal computer age, and the skills to update the technology simply no longer exist.

It is particularly difficult to address perceived threats whose risks are portrayed as high impact and high probability, such as cyber threats. Sites may have little option but to put all digital changes, regardless of what they are, through the full management of change (MOC) process, which may delay progress for months.

The final and frequent obstacle to digital innovations particularly in oil and gas is the lack of network connectivity in the field. Many field sites, particularly offshore assets, lack sufficient network coverage, are served by costly satellite uplinks, or suffer from unreliable, or capacity constrained technology.

Execution challenges

The third and final challenge to a digital initiative is execution difficulties. Access to domain knowledge is one of the keys to developing an artificial intelligence solution. Key resources, such as a highly experienced engineer in high demand, will often be in short supply. Negotiating their involvement is challenging because their performance metrics for the year are negotiated in advance. Freeing them up means agreed commitments may not be met. Change champions will need to have in place the organizational support to undo these commitments, and any dependent commitments, to secure the availability of critical domain experts for the digital project.

Oil and gas is about doing what works, whereas digital is about trying things to see what works and what doesn't. This mismatch can quickly derail those solutions that have multiple experimental runs to work through, or that are in a minimally viable state. Workers in oil and gas generally have clear annual, quarterly, monthly and hourly objectives to meet, and they know how to meet those objectives with the tools they have at hand. Securing their support to experiment with or to teach a new machine learning system that might not work is not going to be greeted with enthusiasm.

A final execution challenge is the tendency to halt digital projects too quickly. To illustrate, an oil company started a trial to adopt robotic process automation tools, or bots, to carry out battery balancing. The initial bot failed dozens of times before its developers figured out all the nuances involved in this complex but routine task. Having critical help on hand at month end to bring bots back to life after failures was hugely disruptive to a business that does not run on agile methods. Many oil companies simply halt development after just a handful of failed attempts.

8.8 Doing digital change

The success of AI in oil and gas is not going to be based on how good the algorithms are but on how good management is at promoting change in the workplace.

8.8.1 A typical change champion

Brian is a 23-year-old engineer working for an oil and gas company. His job is to travel to the gas fields and record facility asset details about the as-built gas wells. The paper documents at home office about the wells are incorrect and inaccurate. The paper documents include equipment that is mislabeled or missing entirely. Parts numbers and tags are incorrect. Field assets include equipment that is not reflected in the paper documents.

Brian's instructions are to take the inaccurate paper engineering documents to the field and make corrective notes on the documents in red ink. This is normally a job for a highly experienced engineer.

Brian purchases a mini tablet and from a cloud service downloads a free app for building a home inventory such as might be used to complete an insurance application. He loads the corporate asset catalogue into the app and uses the app to build field data records. He scans the engineering document into the tablet for on screen annotations, takes high definition photos of the assets at site, and uses the GPS data to auto fill locations. Brian and a small team of new engineering recruits build highly accurate, perfectly correct, electronic, fully compliant, detailed site data supported with HD photos, dispatched to home office, at 10 times the normal pace of such work, generating 4 times the information, with near zero error rate, for the additional cost of a small tablet.

After 2 weeks he is recalled from the field. The home office is expecting incomplete, handwritten, paper files to trickle in once per week to be reviewed and corrected by risk reviewers, checkers, inspectors, and senior engineers, and then typed into the system. They are not expecting highly accurate, perfectly correct, electronic, fully compliant, and detailed site data, complete with photos and annotated electronic diagrams.

Brian is instructed to return to paper.

His plight underscores the most profound issue that many digital innovations face in oil and gas. It is not about the technology. Brian delivered a free technology success (a focus on data and the use of cameras, built quickly for free, and fun for his team), but a workplace failure (no management support, poor change management, and home office apathy, despite the dramatic productivity boost and a substantial overall cost reduction).

8.8.2 Organizational reaction to change

Brian's situation highlights natural organizational responses to change that both help avoid workplace failures, but at the same time kill off technology success. Here are seven tactics that are regularly employed as essential to drive change fast and with results.

Honor the past, define the future

Boards and executive management teams need to own the strategy and direction for digital innovation. Serious lasting change that upsets jobs and roles, disrupts long established business practices and relationships, and threaten short term results must have strength of support from the leadership team. At the same time, leaders need to acknowledge the accomplishments, successes, and results of the past. Troops need to feel appreciated for their past efforts and to hear that it is time to move to a new future. Without that clarity of vision, the change champions in the organization will bump continually into those locked into the past.

CEO as change leader

Digital innovation is not something that is best delegated deeply into the organization. The industrial revolution that digital represents requires the CEO to champion the journey. Strong messages stick longer when delivered directly by the most senior voice in the organization. At least 30% of the employees in every level, every unit and every location must understand the need for change to reach the tipping point where change efforts can accelerate.

Communicate

Digitally driven business changes achieve greatness when company leadership takes a personal and active interest in driving change at work. Business leadership, including the CEO, VP Operations, VP HR, and VP Corporate Services, invest their personal time in communicating the direction. Everything associated with the digital change—successes, trials, pilots, awards, results, promotions, failures—are shared. The team on the front line needs to know that management is serious about achieving success.

Committed leadership demonstrate their unwavering ownership of the change through their acts:

- going to the site or field in person, and not just once
- holding field meetings at the sites
- explaining the vision for the change, and why the company sees it as important
- answering directly and honestly the questions from the field
- adjusting the expectations so that the field has the time to invest in training
- providing leadership advice to supervisors on coping with resistance to change
- responding personally and directly to follow ups from the field
- allocating enough resources to help deal with the reasonable problems that will appear
- recording videos to circulate, championing the change and those most impacted
- celebrating successes personally by handing out badges/rewards/prizes in person
- adapting manager and supervisor performance metrics to include support for digital change

Be purpose driven

There are likely more opportunities for digital innovations than there is capital available. Management will need to make choices about where to invest and why. Managers should be able to answer questions such as:

- What is the vision for the change?
- What compelling picture of the transformed business lets others embrace the change?
- What is in it for the impacted?
- What positive benefits can be tied to the success of any digital change?
- How will productivity gains turn into personal growth and not job loss?
- How will the digital change journey be different?
- How will "fail fast" work in practice and who will stand up for the change when deadlines are missed in favor of a digital trial?

Those choices should be tightly tied to specific business results. Oil and gas managers have little time for digital efforts that feel more like science projects. Performance metrics should reward the behavior required: supportive supervisors, agility, and willingness to trial new concepts, and a smarter AI machine. If front line workers suspect that management will not support them as they work to evolve learning systems, or if the metrics reinforce alternative behaviors, they are unlike to invest the time required to master the new technology.

Think big, start small, be agile

Digitally driven change will be as big a disruption to the industry as deploying the major enterprise resource planning systems of the past, or adopting the internet, or introducing SCADA technology. The pace of change, driven by the constantly falling costs of technology and consistently improving capability, suggest that the solution design for today may be dramatically different in just a couple years later. Leaders should set forth an embracing vision of the future that will endure for some time, undertake small projects aligned to that vision, and execute them with agility.

Build cyber security in

Provision for cyber security should be built into any digital innovation since digital innovations create a far greater attack surface for cyber activity. Artificial intelligence and machine learning tools are already part of the arsenal of the cyber criminal. A message for helping drive change forward is that cyber awareness and protection is merely a variant of the safety culture.

Stay the course

Artificial intelligence and machine learning solutions are, by definition, learning systems. Much like other digital technologies, they are unlikely to work perfectly, immediately on installation, and under all possible operating conditions. Early versions might even be called "minimally viable." Change leaders must anticipate that the solution will need to progress through multiple iterations to reach a state of acceptable utility in the work environment. Oil and gas companies are well versed in seeking out that which does not work and shutting it down.

Chapter 9

Global Practice of AI and Big Data in Oil and Gas Industry

Wu Qing

Department of Science & Technology Development, China National Offshore Oil Corporation, Beijing, China

9.1 Introduction

Oil and gas exploration and production data have outstanding big data characteristics. The industry is in the petabyte-exabyte stage in terms of the size of its big data (Boman, 2015). This is likely a result of the industry requiring large amounts of data to determine new drill sites and ensure sustainable production, among other tasks. Moreover, reports indicate that a modern unconventional drilling operation produces up to 1 MB of data per foot drilled, and it is no secret that such information can be exploited to optimize drill bit location, improve subterranean mapping, enhance efficiencies related to production and transportation, and predict the location of the next promising formation, if analyzed appropriately via big data analytics (Martinotti, Nolten, & Steinsb, 2014).

The recent technological improvements have resulted in daily generation of massive datasets in oil and gas exploration and production facilities. It has been reported that managing these datasets is a major concern among oil and gas companies. A report by MR Brulé Group IBMS (2015) stated that petroleum engineers and geoscientists spend over half of their time searching and assembling data.

The integration of various data sources combined with fundamental studies can be used to enormously enhance the understanding of subsurface operations, thereby increasing well productivity. According to domestic Chinese journal articles of 2019, artificial intelligence research in the field of petroleum engineering applications are decades old, from application management to all aspects of the exploration and development of the construction site.

The followings are some typical global practices of AI and Big Data in Oil and Gas Industry.

Machine Learning and Data Science in the Oil and Gas Industry.
http://dx.doi.org/10.1016/B978-0-12-820714-7.00009-1

9.2 Integrate digital rock physics with AI to optimize oil recovery

9.2.1 The upstream business

Oil and gas field exploration, development, and production data have prominent big data characteristics. The preservation and utilization of big data in oil and gas fields usually requires completeness, regularity, and intensiveness, so that scale and cluster effects can be used. Big data reveals the true distribution of oil and gas and enables oil field modeling, production prediction, and oil field production management.

With the continuous deepening of the integration process of oil and gas exploration and development, the focus of China's oil and gas work has gradually shifted to complex structures and lithologic oil and gas reservoirs, unconventional oil and gas, deep offshore oil and gas reservoirs, and tight oil and gas reservoirs. How to use artificial intelligence to combine digital rock technology with big data to effectively reduce the difficulty and risk of exploration and development, improve oil and gas recovery, and control development costs is a topic that the industry urgently needs to tackle. Combining digital rock analysis with oil and gas field exploration and artificial intelligence, we can combine the characteristics of the digital rock database with logging and well data and implement various state-of-the-art classification techniques to determine the distribution and remaining oil mining potential.

For specific geological research objectives, there is usually a combination of multiple observation and analysis methods. Because of the traditional geological methods and quantitative requirements of reservoir characteristics, there will be some deviations. Oilfield mining research usually requires the use of multiple analysis methods to organically integrate multi-source information, to falsify various analysis results, further improve analysis accuracy, and reduce multiple solutions of reservoir models. Therefore, in general, the exploration, development and production of big data in oil and gas fields require deep data mining to extract more effective value.

In recent years, the development of digital core (also called digital rock physics, DRP) technology has made it possible to obtain more reliable physical information on pore-scale multiphase flow, thereby determining the cause of low recovery factor and improving reservoir performance for different injection schemes. In other words, with the new integrated multi-scale model, multi-scale physical properties can be integrated to find the most effective displacement scheme to optimize the recovery factor. The development of multi-scale integrated digital rock technology provides a new method and opportunity for deep mining of oilfield big data in order to extract more effective value. Through the establishment of a new multi-scale integrated model, the core and logging data can be integrated, the large-scale reservoir seepage model can be used to predict the best injection and displacement method, and the multi-phase flow simulation can be used to predict the maximum production using the most effective production plan under different scenarios.

We use deep machine learning to integrate the recently developed PAM (Pore Architecture Model) with core, logging, seismic, and production data to establish a comprehensive scheme for quickly and effectively predicting the distribution of multi-scale heterogeneous remaining oil and deciding how to improve recovery. Specifically, the cutting-edge technology used will involve continuous modeling of multi-scale pore structures to quantify the spatial connectivity of the matrix and fractures, and AI technology will be used to analyze the remaining oil distribution found in the big data from logging and seismic exploration. The distribution from the matrix, rock strata to the lithofacies system. These integrated reservoir systems are the actual characterization of complete multiscale flow physics and will be able to model multiscale and multiphase flow in heterogeneous and complex reservoirs for the first time.

We implemented a set of advanced tools. First, a multi-scale pore structure model is required. Second, a pore network multiphase flow simulator tool is developed. Third, big data analysis and AI deep learning tools are deployed. Using digital core technology can further quantify the spatial wettability distribution of the reservoir mix (between different pore types and positions within the oil column), which will allow us to quantify the matrix and fractures on the multi-scale movement of oil and gas.

We are working with partners to enrich the field exploration test and production data of complex reservoirs, give full play to the advantages of vendors in multi-scale digital core and multi-scale AI integration, deep machine learning algorithms, and jointly developing big data based on CNOOC oil field applications. We will obtain the accurate input parameters of complex reservoir modeling and flow simulation, predict the oil and gas reserves of complex reservoirs and the distribution of remaining oil, guide the formulation of complex oil and gas field exploration and development plans, and develop and produce complex oil and gas fields. This work has important academic significance and practical application value. More importantly, the cross-scale digital core AI technology can be applied to all complex oil fields, including carbonate reservoirs, unconventional oil fields, and shale gas fields.

Due to the heterogeneity of the reservoir, the biggest challenge facing energy companies is low recovery. There is a large gap between DRP technology and on-site production management applications. The main obstacle is that the detailed information obtained by DRP cannot be directly applied to existing reservoir simulation tools. If you can use DRP additional information, DRP can play an important role in the exploration and production of petroleum engineering. We use an integrated method that combines spatial changes in wettability and multi-scale pore information architecture, combined with digital petrophysics, using oilfield data for deep learning. Due to the high complexity of underground oil and gas flow, the relationship between macro/micro scale properties is nonlinear. Using advanced multi-scale imaging and calculation capabilities, the complete geometry (pore size and shape) and topology (pore connectivity) of the rock structure of the reservoir can be obtained and used to construct a representative multi-scale model to calculate effective multi-scale relative permeability at large scales. Specifically, it uses the most advanced tools to integrate data from structured and unstructured wells to identify and classify the remaining oil potential that may not have been previously identified or produced less than expected and seeks solutions to improve reservoir recovery.

Using advanced 2D and 3D high-resolution imaging and calculation capabilities, the complete geometry (pore size and shape) and topology (pore connectivity) of the pore/fracture structure can be obtained from reservoir rock samples, called pore structure models or PAM. We will use the new integrated multiscale PAM to fully integrate the multiscale matrix and complex reservoir systems. Develop multi-scale DRP methods and expand the new DRP model from small scale to field scale, integrate logging, well testing and even seismic data as much as possible to capture the large-scale characteristics of reservoir heterogeneity. Increasing the recovery factor can extend the development life of the oil field and reduce costs, as follows:

1. For the first time, a new and effective method is developed, which uses a set of coupled calculation methods (PAMS) to integrate all the multi-scale combinations of all pore information (from large pores, fractures to micropores). These methods integrate different scale networks into a parallel network system.
2. Quantify the variability of the interaction of the rock surface wetting state and further important control of multiphase flow, especially the control of the displacement efficiency.
3. Use a set of feature-based classification techniques to filter well data and images to identify the remaining oil potential after injection.
4. Design a more effective oilfield development strategy.

9.2.2 Digital core technology

Digital core technology dates back to before 2000. With advances in technology and improved imaging scans, this technology adds to the rapid development of lithology. Due to the evolution from two-dimensional to three-dimensional technology, the Nian Journal has 22 articles on digital cores in 2019.

A recent report from the Chinese Research Institute of Petroleum Exploration and Development, Qinghai Oilfield Company says: "Multiscale X-CT (commonly known as gamma knife) scanning imaging technology to establish different levels of 3D digital cores, using the maximum sphere algorithm and pore throat size correction method, extract and establish a nanoscale digital core pore network model of carbonate reservoirs, and realize the use of 3D pore network model to simulate a reservoir. The physical property parameters, mercury entry curve, and pore and throat distribution curves are compared in parallel with the results of conventional high-pressure mercury intrusion method. The research results show that the porosity and permeability parameters simulated by digital core technology are slightly different from the actual measurement results. The simulated mercury entry curve, pore throat distribution curve and the curve obtained by the indoor mercury intrusion experiment have high consistency, and the simulation results obtained by the digital core have high reliability. The digital core technology is microscopic in carbonate reservoirs, intuitive in its quantitative characterization of the pore structure described aspect and has obvious advantages."

The present state of research in digital core technology has already moved towards multi-scale and multi-level ranks. Due to the high cost of high-precision scanning instruments, the number of multi-scale high-resolution X-CT scanning imagers is limited in China, so the popularity of digital core technology in China is weak, and there are few practical digital core databases. Judging from the domestic research literature, there is no report of combining big data with digital core.

The development of digital core technology makes it possible to combine artificial intelligence with petroleum engineering big data, because digital core technology can obtain accurate reservoir attribute characteristics from the big data of petroleum exploration, so that the big data of petroleum exploration can be further utilized and upgraded. Judging from the domestic research literature, there is currently no research report on artificial intelligence and digital core.

This project aims to improve methods by which prediction of recovery for heterogeneous reservoir can be integrated from core-scale to larger scales via AI. The biggest challenge for energy companies is the low oil recovery due to the heterogeneity of the reservoir. The recent advances in digital rock physics (DRP) technology has enabled us to get more reliable and representative information of pore-scale multi-phase flow physics to identify the causes of low recovery, providing new approaches to improve the reservoir performances via different injection scenario. In other words, using the new integrated multiscale model we can integrate multi-scale physics to find the most effective displacement scheme to optimize oil recovery. There is a substantial gap between the DRP technology and its application to field production management. The main obstacle is that the detailed information obtained by DRP cannot be used directly in the existing reservoir simulation tools. The DRP can add significant value in the exploration and production of petroleum engineering if the extra information of DRP can be used. We propose an integrated method that combines spatial variance of wettability and architecture of multiscale pore information, in association with deep learning DRP with oilfield data. As the multiphase flow in the subsurface is highly complex, the relationships between macro and micro-scale properties are expected to be non-linear. Using advanced multiscale imaging and computing power, the complete geometry (pore size and shape) and topology (pore connectivity) of reservoir rock structures can be obtained and used to construct a representative multiscale model to calculate effective multiscale relative permeabilities at large scales. Specifically, state of the art tools will be integrating structured and unstructured well data to identify and classify the potentials of remaining oil, which may indicate previously unrecognized or under-performing petroleum productions and find solutions for the improved recovery of reservoirs. This will have significant impact where the oil fields are extremely heterogeneous and need new technology to improve reservoir recovery to boost production and reduce cost.

Although wettability is a notable characteristic of many of the most productive reservoirs known, its contribution to both hydrocarbon storage and flow is commonly ignored in multi-phase flow physics. The interacting effect of the wetting state of the reservoir rock surface, known to be highly variable and a further important control on multi-phase flow in general and displacement efficiency, will also be quantified. We will use a suite of feature-based classification techniques filtering well data and images, to identify the potential remaining oil after injection.

9.2.3 Modeling wettability at the pore-scale

The ultimate control on fluid flow occurs at the pore scale, since no matter how large scale an oil reservoir is, the injected fluid (water or gas) will displace the oil pore-by-pore. The nature of the reservoir pore structure is well known for its heterogeneous multiscale pore structures. The existence of such multiscale and highly complex 3D networks of connected pores makes hydrocarbon flow behavior difficult to predict. Therefore, comprehensive multiscale pore information is essential and the multiscale 3D pore network flow model with spatially various wettability may indeed make an important contribution to the understanding of both oil migration and residual oil saturation.

Current methods of hydrocarbon and aquifer reservoir flow performance use uniform wettability and constant permeability assumptions which are dependent upon quantification of the multi-phase flow characteristics for each grid block in the simulation. Typical grid sizes are of the order of 10s or 100s of meters and the properties assigned to such grids are porosity (ϕ), permeability (k), relative permeabilities (kro, krw, krg) and capillary pressures (Pcow, Pcgo). All information used to construct and operate such simulations is drawn from macro-scale understanding, usually from measurements at the core-scale. Expensive and time-consuming laboratory measurements are required to derive estimates of the relative-permeability functions, but these do not capture all the heterogeneity and multi-scale flow properties that often exist within the reservoir. Considerable uncertainty may be introduced into the flow predictions as a consequence of the upscaling approach, especially where multi-scale effects may contribute to flow in a complex manner.

The multiscale integrated digital core technology is a new technology that enables a more reliable macro-scale representation from pore-scale physics, leading to multiscale and multi-phase flow predictions that cannot be done experimentally. The workflow is shown in Fig. 9.1. More accurate multiscale 3D models of the entire flow system from micro-matrix to core-scales (and beyond) are required to better understand hydrocarbon flow in reservoir rocks. Indeed, one possible approach to this task is to use X-ray CT. A more efficient method is to derive 3D statistics of the spatial distribution of components of the rock using multiscale SEM images that will always, in principle, be of much higher resolution than 3D images. 2D statistics can then be used to develop a 3D model by the Markov Chain model (Wu et al., 2006). This method has been rigorously

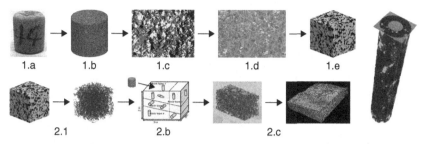

FIGURE 9.1   Workflow for 3D modeling of formation alterations and upscaling of the well inflow model. A description of the items follows:
1. Core scale
   a. Core Sample
   b. Micro CT Scan from Core
   c. Identify minerology, clay, and cement via SEM EDS images
   d. Pore classification: Green = Pore space
   e. PAM: 3D Pore Architecture model
2. Upscale to Wellbore
   a. Extract Pore Network from 3D pore architecture model
   b. Multiscale models to represent large scale flow property
3. Upscale to Reservoir scale model
   a. Near-wellbore inflow model

validated (Wu et al., 2007) and we propose to extend it to the multiscale pore system, see Fig. 9.1.

The wetting properties of reservoir rocks are also fundamental to an understanding of hydrocarbon flow in reservoir rocks. It has been suggested that most reservoirs are neither water-wet nor oil-wet, but mixed-wet. Also, the wettability can vary within a single rock sample which may, in part, be a function of pore size (Marzouk, 1999). The mixed wettability results bypass water flow since the water tends to flow in the water-wet flow path and leave the oil-wet region untouched. These mixed wettability or oil-wet reservoirs have yielded very low recovery using water flooding. Recent research on wettability observation use ESEM observations and NMR studies in conjunction with other detailed characterization of the pore system, e.g. multi-scale thin section data for both the macropore and micropore systems, mineralogical characterization, and mercury capillary pressure data. These ancillary measurements can be performed and will be analyzed by using them in constructing 3D pore models using the PAMs technique.

The recent advances in multiscale digital core technology have enabled multiscale information about the heterogeneity in the reservoir, providing integrated approaches to analyze different displacement mechanisms via different injection scenarios. In other words, using the advanced computing method we can integrate multi-scale microphysics into macro representation multiphase predictions to find the most efficient displacement scheme to optimize oil recovery. While there is a high uncertainty of subsurface structure due to the heterogeneity, a lot of exploration data such as well data and logging data are not fully utilized since there is poor understanding between the interpretation of logging data and potential recovery of the reservoir rocks. This cutting-edge technology offers great potential for utilizing the big data of reservoir information to identify the potential remaining oil. Combining digital rock technology with AI, the crucial reservoir field information via self-supervised learning of additional information can be linked with digital rock technology, and such a process must be underpinned by a set of smart and efficient tools for data search and retrieval, data-analysis, and deep data learning framework as well as multiphase flow simulation. A basis for all these is the ability to classify a large collection of logging data and well borehole images linking to the microstructure of rock type via advanced machine learning to identify remaining hydrocarbon potential and hereafter to predict how multi-scale heterogeneity impacts recovery efficiency using multiphase flow simulation (Figs. 9.2–9.4).

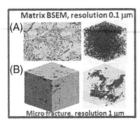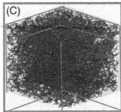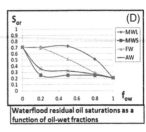

FIGURE 9.2 Illustration of integrating matrix micro porosity with mini fracture and predicting oil recovery. A description of the items follows:
(A) Micro-pore imaged in matrix rock from SEM and pore network constructed using PAM.
(B) Micro-fracture image from micro-CT and fracture network constructed using IM PAM.
(C) Integrated matrix and fracture network constructed using PAM.
(D) Inhibition of oil predicted using multiscale pore network flow models under different wettabilities.

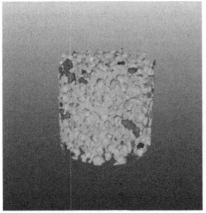

FIGURE 9.3 Air in a packed column of hydrophilic glass spheres before and after secondary water inhibition visualized by our X-ray micro-CT scanner.

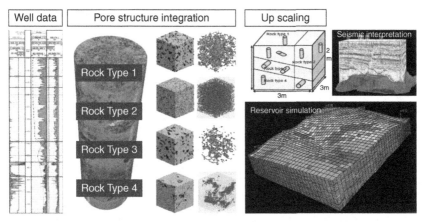

FIGURE 9.4 Illustration of integrating multiscale digital rock models from core-scale to reservoir-scale.

9.3 The molecular level advance planning system for refining

Through the refinery production planning and scheduling and collaborative solutions (optimization system), cost-effective calculation of potential processable crude oil varieties under different production schemes can be achieved, which is conducive to more accurate selection of crude oil and early understanding of different crude oils. The feed properties of the atmospheric and vacuum devices and subsequent secondary devices under the reconciliation plan are changed, which is conducive to ensuring the stability of the feed properties of the device and thus ensuring the safe and stable operation of the device. The product quality changes under different device operating conditions require the timely adjustment of the product-blending plan according to the economic benefits or the demand situation. It can also realize the reverse operation of the secondary device operating conditions required according to different product quality requirements, and even the mixing of crude oil.

9.3.1 Prediction of crude oil mixing and molecular properties

By analyzing the characteristic factors of unused crude oil and the PIONA value of naphtha, and performing model calculations on molecular data, it is possible to study and obtain important data to guide production. For example, a subsidiary company of China Petroleum processed mixed crude oil (manufactured by mixing three crude oils) for analysis and found molecular information for further studies; see Fig. 9.5.

In-depth analysis of the molecular information of crude oil C actually contains more benzene and toluene compounds in this process. Aromatic hydrocarbon molecules with a low carbon number are strong solvents, resulting in increased intermolecular forces, which will produce a series of chemical effects,

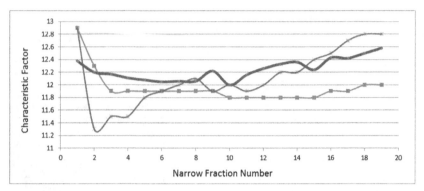

FIGURE 9.5   Three kinds of crude oil characterization factor of the distribution case.

including increased fraction overlap, colloidal molecules wrapped around crude asphalt macromolecule micelles and naphtha. Aromatic hydrocarbon molecules form a new micelle phase equilibrium, etc., which will lead to a decrease in the yield of naphtha and an increase in the yield of the diesel fraction.

The crude oil mixing model can also consider using multiple molecular properties such as solubility parameters and eccentricity factors to describe the influence of crude oil intermolecular forces on crude oil mixing, and to quantify the change in naphtha yield under the influence of such intermolecular forces. Fig. 9.6 displays the three mixed oils and the mixed result in terms of the eccentricity factor and oil solubility.

Table 9.1 is the prediction data of the yield and properties of a mixed crude oil based on the above three crude oils. Table 9.1 can be seen, in can be seen, in consideration of the intermolecular forces between oil, naphtha fraction

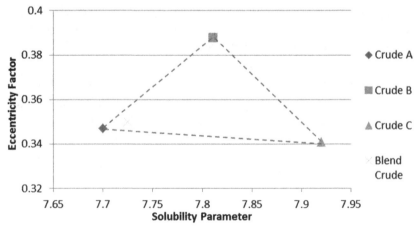

FIGURE 9.6   The solubility parameters and corresponding eccentricity factors of several crude oils (A, B, C) and their blend.

TABLE 9.1 Examples of prediction research on yield and properties of mixed crude oil.

The upper temp., °C	The lower temp., °C	Yield, wt%	Sum of yield, wt%	Error*, wt%	Density (20C), kg/m³	Sulfur, wt%	Nitrogen, wt%	Carbon residue, wt%	Acid number, mgKOH/g	Aniline point, °C	Characteristic factor
HK	15	0.78518	0.78518	−0.09238	546.33328	0.00164	0.00000	0.00010	0.02044	39.79136	13.24319
15	60	2.12597	2.91115	−4.38651	632.75513	0.00203	0.00000	0.00011	0.02079	40.51383	12.92433
60	80	2.13798	5.04913		699.29251	0.00334	0.00000	0.00010	0.02692	39.33381	12.10596
80	100	2.70435	7.75348		724.28515	0.00470	0.00000	0.00011	0.04032	40.27548	11.91664
100	120	2.71454	10.46802		740.21360	0.00647	0.00001	0.00010	0.04937	42.45500	11.87358
120	140	2.81088	13.27889		749.95032	0.00942	0.00001	0.00010	0.06025	46.14350	11.92179
140	160	3.15269	16.43158		762.30233	0.01349	0.00003	0.00011	0.08260	49.76346	11.92157
160	180	2.92622	19.35781		774.00647	0.01857	0.00005	0.00011	0.11045	53.59143	11.92550
180	200	2.70010	22.05791		786.23178	0.02590	0.00010	0.00011	0.16282	57.55987	11.91602
200	220	3.25891	25.31682	1.39375	801.77682	0.03559	0.00019	0.00012	0.22637	61.00075	11.85323
220	250	5.16249	30.47931		814.78822	0.05652	0.00042	0.00017	0.28043	63.84342	11.86355
250	275	4.63937	35.11868		830.06427	0.10754	0.00105	0.00027	0.33000	66.63230	11.85368
275	300	5.30893	40.42761		838.55745	0.19155	0.00244	0.00045	0.34729	71.25273	11.91452
300	320	3.41735	43.84496		847.56743	0.26681	0.00492	0.00073	0.35141	74.42572	11.94464
320	350	6.73245	50.57740		857.74840	0.36047	0.01021	0.00143	0.38877	76.58067	11.97038
350	395	8.28569	58.86309	4.24334	882.03159	0.47522	0.02602	0.00464	0.49472	80.07645	11.87808
395	425	7.14688	66.00998		885.88547	0.54258	0.04611	0.01515	0.52501	85.66021	12.05147
425	450	4.03830	70.04828		897.78467	0.61752	0.06380	0.04287	0.55502	89.38504	12.05033
450	500	8.71518	78.76345		912.46569	0.72546	0.09697	0.23113	0.60642	93.49545	12.06306
500	560	6.30044	85.06389		927.19236	0.91815	0.15595	2.08774	0.69256	98.52882	12.15694
560	KK	14.93611	100	−0.81176	991.03400	1.56466	0.57658	28.17862	0.83378	104.51967	12.07659
Crude oil					839.44619	0.42103	0.08603	2.45239	0.40832	71.65463	11.97996

segment (15–200°C) yields the predicted value of 21.273%, then the predicted value of 26.659% increases 5.386% without considering intermolecular forces.

Through the establishment of benchmark oils (including specifying the nature, certain fraction or fractions of crude oil, and the origin of crude oil), using crude oil rapid evaluation, molecular reconstruction, and the crude oil molecular information database (e.g., MAPS), we can conclude the best processing schemes, product schemes, and economic benefit calculations of new oil refining and mixing compared with the benchmark crude oil.

In addition, through the molecular information of the base crude oil, we can understand the molecular nature of the various oil blocks.

Combining this with the needs of each refinery equipment and the demand for peripheral products in the market forecast, allows us to maximize the value of rational transporting of crude oil. It also supports the procurement on the basis of molecular level group-level crude oil, crude oil resource allocation, optimization plan release, crude oil blending, multi-plant transmission and distribution and production management, which can also be used as a major aspect of the offline application of the crude oil molecular information database.

Fig. 9.7 is a comparison of the economic results of a company's screening of crude oil and the use of several different processing schemes for users to determine the choice of crude oil and the formulation and implementation of processing schemes according to the situation.

Therefore, the use of MAPS can achieve agile optimization of crude oil selection and can easily achieve optimization schemes such as oil benefit ranking and optimal crude oil combination. The specific implementation method varies, for example, according to the actual refinery equipment configuration, the maximum crude oil processing capacity, one or more base oil types (main crude oil), the number of processing units, and more. For each oil type to be

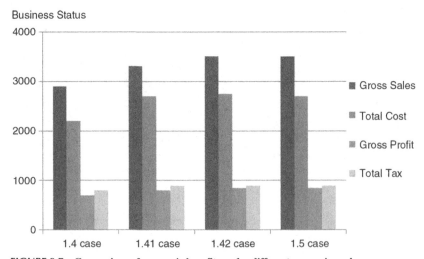

FIGURE 9.7   **Comparison of economic benefits under different processing schemes.**

tested and the ratio (or the proportion for the maximum processing capacity), we must compile the corresponding plan and sort the oil benefits. On the basis of the existing model, if the restriction on the number of crude oil varieties to be purchased is considered, and mixed integer programming is used to maximize profits, the optimal crude oil combination that meets the constraints is selected to achieve the optimal crude oil combination.

It should be pointed out that only MAPS can provide the possibility of agile optimization for enterprise crude oil selection optimization. Although many companies often use PIMS and RPMS to optimize the selection of crude oil, there may be a limited amount of crude oil selected, or it may be due to habits and other reasons. Often, the range of choices does not necessarily meet the goal of maximizing benefits. For example, some companies are used to choosing heavier crude oil combinations, and some companies are used to choosing high-sulfur oil combinations. But in fact, due to changes in crude oil prices, processing costs, product prices, crude oil physical properties, device technical changes, product quality standards, etc., the habitual choice is incorrect. The optimal selection of crude oil procurement at this time should be based on the matching processing capacity, and the rationality of the equipment matching needs to be calibrated according to the conversion ability of the main physical properties. MAPS provide very convenient technical means and possibilities for matching verification of crude oil and processing capacity, short-term evaluation of process flow capability, etc., in order to truly realize the optimization of crude oil selection.

9.3.2   Scheduling optimization at the molecular level

With the help of the crude oil molecular information database, the optimization of production scheduling at the molecular level can be achieved. From the plan level, the relevant plan content including crude oil molecular information is obtained. At the scheduling level, through the core modules of crude oil blending principles and precise molecular reconciliation, output including crude oil scheduling information (mobile quantity information) and crude oil composition information (mobile molecules information) is generated. The operation plan is obtained through the simulation of the oil at the molecular level, and then the composition analysis of raw materials, intermediate products and product molecules, combined with the refining reaction rule library to automatically generate a reaction network to form the quantity and quality of other devices (molecules). The scheduling plan includes information, together with the production information system, that can realize the optimization of the production scheduling at the molecular level.

The construction of MAPS also provides raw material information for the subsequent secondary processing device model, so that the molecular information can be transferred from the crude oil end to the secondary processing device, and further to the product, to master the "journey" of each molecule in the refining production line. After the molecular information is transferred to the product end, product-level blending such as product oil blending, lubricating oil blending, and asphalt blending can be performed based on the product molecular information. In the process of transferring molecular information, this molecular information is also the basis for establishing a molecular dynamics model, and with the molecular dynamics model, the prediction of product distribution, that is, "quantity" and product properties, that is, "quality," can be achieved. This transfer, and its essence is the "quantity" and "quality" of the visualization process, is the current popular "refining wisdom."

Therefore, MAPS can achieve collaborative optimization of refinery production planning and scheduling, can provide a more scientific and convenient tool for the preparation of various refinery production plans, and can improve the accuracy and operability of production planning. At the same time, it can also make the optimization and adjustment of production in an emergency more effective and more comprehensive and provide sufficient information support for the production management of the whole refinery.

9.3.3 Collaborative optimization of the entire industry chain

The full utilization of products and the maximization of product output benefits can be achieved by using the quick assessment of the base oil molecules, crude oil molecular reconstruction techniques, production systems such as advanced process control (APC) and real-time optimization (RTO) to manage the various production processes at the molecular level.

In the past, it was usually impossible to go from plan optimization to full-process simulation. The main reason is mainly that software systems such as PIMS and RPMS that are currently used by refinery companies to optimize production scheduling cannot do this. In essence, they are simplified models based on two-dimensional representations, while the whole process simulation is three-dimensional or higher. In addition, the database structure in existing production planning optimization-scheduling software (PIMS, RPMS, etc.) is mostly based on the original design data. Due to technical progress, changes in crude oil resources, technical modification of the device and catalyst, these database data are far from each other, and some even cause reverse benefits. In addition, these data are very rough and cannot meet the needs of agile and more accurate model calculations. That is to say, the past plan optimization system, whether in terms of process, algorithm or other, does not match the whole process simulation.

With MAPS, it can be directly matched with the whole process simulation. If you still want to use the original plan to optimize the production system, you can get the corresponding delta value from the MAPS physical properties and use the database model of the original optimization system through iterations. The convergence of this procedure results in an optimized production operation plan that can be actually executed.

The collaborative optimization of the entire industry chain also includes all aspects of production operation, management and control. For example, intelligent operations coordinate and integrate logistics, capital flow, and information flow resources of the entire industrial chain from a global perspective to achieve centralized and unified management of people, finance, and materials to effectively reduce production costs, promote product competitiveness, and achieve full control, real-time feedback, dynamic coordination, and best benefits. Establishing corresponding models and evaluation indicators at the enterprise operation execution layer, production operation layer, operation management layer, and decision-making layer, requires integrated technology to open up the corresponding business processes, provide accurate data for the models and indicators, and simulate and evaluate enterprise production operations in real-time performance, provide guidance and measures for enterprise improvement.

The use of desktop refineries and other technologies to formulate business development plans enables the refinery to adapt to the processing of crude oil with large changes in composition and properties, and can greatly change the product distribution to meet market demand or price changes in a cost-effective manner. Real-time tracking and analysis of the progress of production ensures that the production plan is completed on time and in volume. Rapid scheduling of production plans to respond to emergencies ensures production safety, achieves daily plant-wide material balance, reduces material loss, and increases the overall commodity rate. Analyzing water, electricity, gas, monitoring, analysis and optimization of wind and other public works reduces processing energy consumption. Accurately assessing and analyzing corporate bottlenecks in business operations and maximizing asset utilization ensures the acquisition or construction of assets with higher returns.

With the goal of reducing the operation and maintenance costs of the enterprise, through information forecasting methods, the maintenance plan and related resources are reasonably arranged to reduce maintenance costs and increase the availability of assets. Use key performance indicators (KPIs) as the basis for measuring the company's operating status, carry out quantitative management of enterprise performance, and track and supervise financial and business-related indicators.

9.4 The application of big data in the oil refining process

9.4.1 Principle and methodology

In order to solve economic benefit and production safety in depth, policymakers need to predict the future based on real data to make right production and management decisions, production layer needs to carry out the prediction and early warning of production safety in real-time. Refining enterprises need to improve enterprise core competitiveness by means of informatization in the existing good refinement management level.

Refineries are characterized by large scale, multiple processes, high industry concentration degree, complex management system, and so on. These characteristics result in huge production and operation data of structure, semi-structure and non-structure. How to find the internal law and optimize the business process? Refineries ever integrated various independent application information systems and built intelligent factories to break information island in the past, On the basis, how could refineries screen the value data, dig out the potential demand, and show the business trend globally now?

The solutions are data extraction, transformation, analysis, and modeling by using big data technology, from which the key data supporting production decisions are extracted to realize the mining and prediction of relations.

The solutions of big data application include data management, analysis and prediction, and decision control. Data management is the systematic discovery of useful relationships in a large amount of data, that is, the achievement of the repeatability of empirical rules, that is, the acquisition of effective data from the raw data. Analysis and prediction is to establish and fit different models to study different relationships until useful information is found, which is used to analyze the causes and solve problems. For decision control, it is to find potential value, foresee some "bad future" that may happen and give suggestions, that is, to predict and provide solutions. In other words, it is to make predictions and suggestions on process optimization, structural adjustment and exception handling for decision-makers and executors through abnormal warning and trend prediction.

Some algorithms are used in the application of big data, such as clustering, taxonomy, association and prediction, and so on. Among them, the clustering method is to divide the database into different groups, and the differences between groups are obvious, while the data between the same group should be as similar as possible. Unlike classification, it is not clear how the data will be divided into groups or how they will be divided before aggregation. By analyzing the data in the sample database, classification is to make an accurate description for each category or establish an analysis model or dig out the classification rules, and then use the classification rules to classify the records in other databases. Correlation is the search for correlations between different items that occur in the same event, such as the correlation between different items bought in a single purchase. The essence is to find strong association rules in the database; Prediction is based on the time series data, from the historical and current data to predict the future data. Based on the preliminary neural network prediction model, retraining method was added to improve the accuracy of the model continuously.

9.4.2 A case study of CCR process unit

It showed how to explore the application of big data analysis technology by taking the continuous reforming unit CCR of a petrochemical enterprise as an example.

1. Correlation analysis

The refined management of petrochemical enterprises needs more and more collaborative management, and collaborative management will certainly bring a lot of demand for correlation analysis. This demand can be between different majors within the enterprise or between different enterprises. Correlation analysis is an important branch of big data analysis, which can find the correlation between variables in the chaotic data. Therefore, the correlation analysis algorithm can be used to explore the potential factors beyond the traditional experience, and finally realize the potential synergies.

The specific research methods can be divided into data acquisition, data setting and standardization, and correlation analysis. Data acquisition is actually the process of importing all historical data such as operational data, quality data, corrosion data, cost data, material balance data and energy data into the cloud platform (such as Aliyun). Therefore, it is important to develop the interface between the relevant system and the cloud to realize real-time data import. Setting and standardization of data is to align the raw data in chronological order, such as operation data, data quality, corrosion data, equipment operation data, cost data, material balance and energy consumption data, according to certain setting alignment algorithm. And then the data are filtered, set, rejected abnormal values, and standardized to obtain qualified or valid data. Pearson correlation coefficient algorithm can be used to calculate the correlation coefficient matrix of each indicator, and to extract the variables strongly correlated with the key indicators to complete the correlation analysis, which include the positive correlation variables and the negative correlation variables.

The Fig. 9.8 is a schematic diagram of the correlation analysis of pre-hydrogenation unit and renormalization unit including the positive correlation variables (red) and negative correlation variables (blue) contained in. Similar methods can be used to obtain the effects of operating conditions and raw material properties on product yield and equipment operation, as well as other effects such as operating conditions, raw material properties and distillation outlet quality on equipment corrosion, production costs and environmental emissions.

2. Single indicator abnormal detection

Seven key indicators (such as octane number, energy consumption and aromatic difference index, pure hydrogen yield and thermal efficiency, flue gas SOX emission, sewage COD and unit cost) related to superior assessment, environmental protection and efficiency of continuous reforming unit are selected as the objects of anomaly detection. It involves several steps such as data setting and standardization, correlation analysis and characteristic selection, construction of prediction model and abnormal judgment of single index. Among them, characteristic selection refers to the extraction of variables with strong correlation from seven key indicators after obtaining the correlation coefficient matrix.

For the establishment of prediction model, the operational variables that are strongly related to the prediction index are selected as the input of the SVM prediction model to establish the SVM prediction model and to realize the real-time calculation of key indexes. For the abnormal judgment of a single indicator, boxplot algorithm can be used to calculate the range of each indicator and calculate the abnormal limit of each indicator. If the value exceeds the limit, the indicator is judged to be abnormal. For example, for seven indicators of a certain enterprise's reforming unit (unit cost, octane number, effluent wastewater COD value, the SO_2 emissions, energy consumption and aromatic difference

(A)

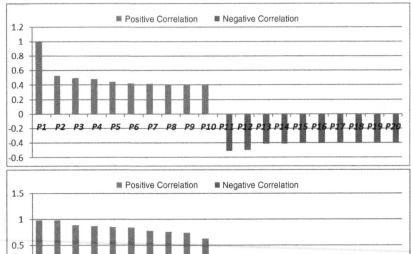

(B)

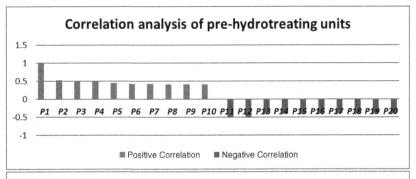

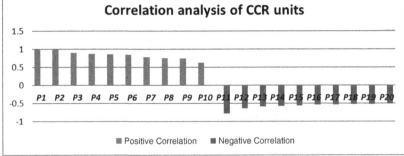

FIGURE 9.8 (A) Results of pre-hydrotreating unit by using correlation analysis (B) Results of CCR unit by using correlation analysis.

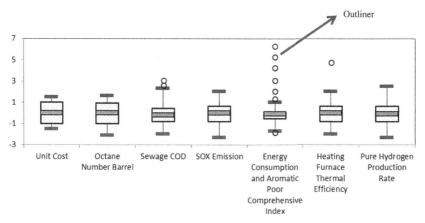

FIGURE 9.9 Box plots for determination of outliers of seven indicators.

index, thermal efficiency of heating furnace and pure hydrogen production rate), the input and output of real-time data based on the SVM prediction model are shown with boxplot in the Fig. 9.9.

3. Abnormal detection of multidimensional data

In the production process, it is possible that the whole deviates from the normal range while all single indicators are normal Just like all indicators are normal in a certain human body examination, but the medical person in the state is sub-health. Therefore, it is necessary to carry out abnormal detection of multi-dimension data.

Abnormal detection of multi-dimension data includes data setting and standardization, extraction of characteristic variables and dimensionality reduction processing, clustering analysis and abnormal prediction and warning. Data setting and standardization are to sort out the data set of seven indicates, to align them in chronological order, and then to standardize them to eliminate the influence of dimension and order of magnitude. In other words, the boxplot algorithm is used to judge the abnormal points of indexes for the index data. Characteristic variables and dimensionality reduction process can be extracted by principal component algorithm, so as to explain most variables with fewer variables and to achieve the goal of dimensionality reduction. During cluster analysis, the principal component is extracted as the data source of clustering, and k-mens algorithm is used for clustering to search for outliers. The method of abnormal prediction and warning is to calculate the Euclidean distance between each observed sample and its cluster center. Whether the sample is anomaly or not can be judged by the distance value, whose threshold is selected statistically based on historical data.

4. Single target parameter optimization analysis

In the operation sample library, the optimal value of the target and its corresponding strongly correlated operation variables under the condition of a certain kind of raw material are searched, which can mine the best operating experience in history, such as mining the experience of experienced operators and solidifying it, and can be used complementary with optimized software such as RSIM. The parameters optimization can be achieved by using raw material clustering analysis, raw material classification model, operating samples database. In the process of raw materials clustering analysis, the historical data of the nature of the reconstituted raw materials can be sorted out and reorganized. First, it goes through pretreatment and standardization, then principal component is used to reduce dimension, and finally K-means clustering is adopted to output the clustering result. Based on the clustering results of raw materials, the SVM model of raw materials classification was established, and the classification effect of the model was evaluated, which can automatically classify a new batch of raw material property data. For the optimization target strongly correlated variables, the category of raw material and its corresponding strongly correlated operation parameters are imported into the operation sample library, which is used as the sample of parameter optimization. In the operation sample database, the optimal values of the target parameters under the conditions of different types of raw materials are searched, as well as the values of the corresponding strongly correlated operation variables. Furthermore, the operation parameters can be recommended based on the raw material quality and optimization objectives.

In the operation sample database, the query statement is used to search for the value of strongly correlated operation conditions when the target parameter is optimal under the condition of a certain kind of raw materials. For example, under the condition of e type raw materials, the optimal yield of pure hydrogen is 3.392%, and the corresponding value of strongly correlated operating parameters is taken.

5. Multi-objective parameter optimization analysis

According to the selected multiple optimization objectives and their optimization directions, the optimal value of each objective under the condition of a certain kind of raw materials is determined, and the optimal value and the historical actual value are used as the coordinates of the theoretical optimal point and the actual point in the multi-dimensional space respectively. The actual point closest to the theoretical optimum is selected as the optimization result. For this purpose, it is necessary to establish an operation sample library first, that is, a set of optimization variables, raw material categories and their corresponding operation variables to form a multi-objective optimization operation sample library. The clustering analysis of raw materials should be completed to determine the corresponding raw material category every day. The raw material category, all optimization variables and their strongly correlated operation parameters are written into the operation sample library in units of days. The next step is how to determine the optimal theoretical advantage, that is, to select multiple optimization objectives and directions and establish the optimal value of each optimization objective under the conditions of each type of raw materials. That is to say, the optimal values of each optimization variable under the condition of a certain kind of raw material are searched in the operation sample library, and these values are taken as the coordinates of the theoretical optimum in the multi-dimensional space. Sorting out the sample points in different raw material categories, which are composed of the values of variables to be optimized as coordinates, and calculating the Euclidean distance between the optimized sample points in the multi-dimensional space under the condition of a certain type of raw material and the optimal theoretical advantages. In order to recommend operation parameters according to the size of Euclidean distance, it is to search for the minimum value of Euclidean distance under the conditions of different types of raw materials and the value of the corresponding strongly correlated operation variables in the operation sample database, so as to achieve the recommended operation parameters based on the properties of raw materials and optimization objectives. The following tables (Tables 9.2–9.4) are the optimization sample, the optimal value of the target and the corresponding operating parameters under the condition of A class of recommended raw materials.

6. Unstructured data analysis

Based on the text mining and analysis of the scheduling shift log, and the correlation of structural data such as the recovery rate of reformed gasoline, hydrogen production and aromatics content of reformed gasoline, it is possible to excavate the influence of crude oil types on the recovery rate of reformed gasoline and other technical and economic indicators which could guide the crude oil procurement.

The research methods include text feature analysis, transformation of unstructured data into structured data, correlation of structured data and final calculation and result display. For example, in the text feature analysis stage, the historical scheduling communication class log is exported to analyze the text features of the crude oil species and the processing amount of the atmospheric and decompression device, and determine the rules for extracting key information. Then, according to the unit of days, according to the text characteristics determined previously, the crude oil types and corresponding processing amount are extracted and stored in the database to complete the transformation of unstructured data. Continue to extract data such as gasoline yield, hydrogen production and aromatics content of reformed gasoline from MES and LIMS systems on a daily basis, and store them in the database after correlation with crude oil types to obtain relevant structured data. The weighted value of each crude oil corresponding to the gasoline yield and other indicators are calculated and listed from large to small, which could guide the purchase of crude oil.

TABLE 9.2 The Euclidean distance is sorted from small to large, and the minimum distance sample is determined as the optimization sample.

Sample point	Feedstock type	Hydrogen yield, wt%	LPG yield, wt%	Fuel gas consumption, kg/t	Gasoline yield, wt%	Euclidean distance, m
2	a	4.0362139	0.408051453	0.050288315	90.23035255	0.479166783
3	a	3.9697842	0.340080972	0.051629265	30.18622053	0.578984526
4	a	3.9121334	0.356683345	0.051493443	90.15026447	0.635410454
1	a	3.830809	0.311651179	0.050302572	90.46376459	0.647438127
6	a	4.1264003	0.503603403	0.054528679	89.81878764	0.70938834
7	a	4.0945915	0.57745754	0.052688831	89.76294647	0.769286569
5	a	3.9398124	0.494639028	0.048057041	89.8464045	0.783815043
8	a	4.1276415	0.50393138	0.05783485	89.70752919	0.810905065
9	a	4.4172044	0.530840676	0.056266014	88.80308645	1.657665121

TABLE 9.3 The optimal value of the target parameter.

Hydrogen yield, wt%	4.0362139
LPG yield, wt%	0.408051453
Fuel gas consumption, kg/t	0.050288315
Gasoline yield, wt%	90.23035255

TABLE 9.4 Recommended operating parameter.

Naphtha flow regulation of heater exchanger E701, m³/min	175.0105814
Reforming reaction temperature, °C	525.1799787
Input pressure of the fourth reactor R704, MPa	0.36949156
Top temperature of stabilize T701, °C	58.26197232
Bottom pressure of stabilize T701, MPa	0.80265751
Output temperature of furnace F701, °C	527.4304937
Output temperature of furnace F704, °C	535.8905265

7. Prediction analysis based on the properties of raw materials

Based on the historical data, the prediction models of the properties of raw materials and the yield of gasoline, hydrogen production, dry point of gasoline, conversion rate of alkane and conversion rate of cycloalkane were established. After raw material property data being input, the value of the above five indicators can be accurately predicted to guide production.

The research methods include data acquisition, model training, and model prediction. The data acquisition process is to import all historical data such as operation data, quality data, corrosion data, cost data, material balance data, and energy data into aliyun platform. The interface between the relevant system and aliyun is built to realize real-time data import. In the model training stage, the raw material quality data, and the forecast index data such as gasoline yield were derived from the raw data as the input and output of SVM model training. The model should be retrained every day to ensure the prediction accuracy. In the model prediction stage, the model should predict technical and economic indicators such as reforming gasoline yield, hydrogen production and conversion rate after 31 major laboratory analysis data of reforming raw materials are imported.

Before the input of raw materials, the laboratory analysis data of raw materials are input into the prediction model, and the model automatically calculates five technical and economic indexes of gasoline yield, hydrogen production, gasoline dry point, alkane conversion rate and cycloalkane conversion after the input of raw materials. Based on the forecast results, technicians can adjust the relevant operating parameters according to the production plan, such as the mixing amount of heavy naphtha. This can save valuable resources of heavy naphtha.

The above studies have been verified on industrial devices and obtained satisfactory results. In addition, there are new findings, such as the adjustable parameters ignored by enterprises after big data analysis. For this purpose, a one-month test was carried out with a special cycle of every two days, and the results were obtained.

There are usually three methods to verify the correlation between the gasoline yield and the operating parameters: (1) instantaneous value verification: the correlation between the instantaneous value of the gasoline yield and the instantaneous value of the verification variables can be seen. This method takes into account the time delay factor, because the gasoline yield cannot change immediately after the adjustment of operating parameters. Therefore, transient value validation can be used as a reference for validation results. (2) MES data validation: MES data validation is to verify the correlation between the gasoline yield calculated by dividing the gasoline output quantity within 24 hours by the reformed feed quantity in the MES system and the mean value of the validation variables during this period. This method eliminates the effect of time delay and is relatively accurate in calculating the gasoline yield. (3) Process simulation verification: since it is impossible to independently adjust a certain variable in an industrial device and keep the properties of raw materials and other operating parameters unchanged, the process simulation method can be adopted to keep the properties of raw materials and other parameters unchanged, and only change the verification parameters to observe the change in the gasoline yield. According to the practice results, MES data validation can be selected to verify the adjustment results, and instantaneous value validation and process simulation validation can be used as auxiliary references. In addition, the key parameters affecting the gasoline yield should be ensured to be relatively stable as far as possible during the validation process, so that the validation results can truly reflect the gasoline yield affected by the validation variables.

In practice, the operating parameters will not be adjusted greatly, and the potential content of aromatic hydrocarbons (aromatic potential) and heavy naphtha will vary greatly. Therefore, we should choose the samples for comparison in which aromatic potential and heavy naphtha refining capacity are similar.

Good practice in continuous reforming units has also been extended to other refinery units and chemical units, such as catalytic cracking and ethylene cracking. The knowledge from other fields, such as mechanism model, pattern recognition, system identification, are added to the analysis of data from multiple business areas, such as the influence of process fluctuations on the state of equipment and so on, so as to make the model gray box.

9.5 Equipment management based on AI

Integrated application of Internet of things, deep learning, knowledge map, such as technology can be used to build distributed equipment health monitoring and early warning system, real-time online for early detection of equipment hidden danger, early warning, early treatment of providing effective means, which can guarantee equipment run healthy to make it stable for a long period of time, and to reduce unplanned downtime losses.

Predictive maintenance strategies are based on the combination of traditional condition monitoring enhanced with analytics algorithms that enable the prediction of machine failures before they occur. IoT and advances in analytics are widely adopted in market, which give the users a 25%–30% benefit.

9.5.1 Equipment hazard monitoring and warning

Extensive collection of field equipment operation data, the use of deep neural network autonomous learning equipment operating parameters, the formation of a deep learning-based equipment fault fuzzy prediction model, to realize the functions of shaft displacement, shaft fracture, shell cracking, power overload and other equipment failure of effective monitoring and early warning.

The introduction of early warning mechanism to equipment management is one of the effective means to avoid major accidents caused, so it can ensure safe production of the enterprises to a greater extent. The real-time evaluation of the running status of the equipment, can greatly reduce equipment operation cost of failure and unplanned downtime. Continuous dynamic tracking of device status lay the groundwork for predictive maintenance. The process requires efficient artificial intelligence algorithm and data fusion modeling technology.

The Fig. 9.10 is a schematic diagram of an industrial installation. Sensor data from different parts of the equipment is an important foundation and basis for judging the status of the equipment. However, the sensor data has the characteristics of dynamic, mass, high concurrency, nonlinearity, strong noise, and heterogeneous source. Fig. 9.11 is an example of data from a sensor.

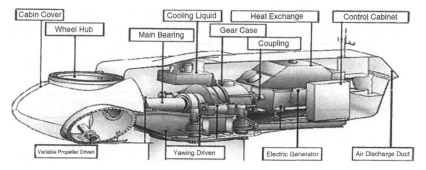

FIGURE 9.10   **A schematic diagram of equipment composition.**

```
09/01/16 00:00:00,61.692001,56.923996,58.404999,56.130997,56.416998,55.559999,45.843,46.452999,47.48,46.778004,47.610001,39.269005,64.224995,57.494997,36.43,37.255997
09/01/16 00:01:00,61.743,56.923996,58.404999,56.130997,56.416998,55.559999,45.843,46.452999,47.48,46.778004,47.610001,39.269005,64.224995,57.494997,34.678001,37.102001
09/01/16 00:02:00,61.743,56.975998,58.404999,56.130997,56.520998,55.559999,45.843,46.452999,47.48,46.778004,47.610001,39.269005,64.224995,57.494997,36.386997,41.219002
09/01/16 00:03:00,61.743,37.027996,58.404999,56.234997,56.520998,55.559999,45.843,46.452999,47.48,46.778004,47.610001,39.269005,64.224995,57.494997,37.770004,42.810997
09/01/16 00:04:00,61.743,57.027996,58.508999,56.234997,56.520998,55.676996,45.843,46.452999,47.48,46.778004,47.610001,39.269005,64.224995,57.494997,35.599998,41.198002
09/01/16 00:05:00,61.860001,57.132,58.508999,56.234997,56.520998,55.676996,45.843,46.452999,47.48,46.778004,47.610001,39.269005,64.224995,57.585997,36.960995,35.764
09/01/16 00:06:00,61.860001,57.132,58.508999,56.234997,56.520998,55.676996,45.843,46.452999,47.48,46.778004,47.610001,39.269005,64.302998,57.598997,35.014,36.401997
09/01/16 00:07:00,61.860001,57.132,58.508999,56.234997,56.624998,55.676996,45.843,46.452999,47.48,46.778004,47.610001,39.269005,64.328999,57.598997,36.185997,33.353001
09/01/16 00:08:00,61.860001,57.132,58.508999,56.234997,56.624998,55.676996,45.843,46.452999,47.48,46.778004,47.610001,39.269005,64.328999,57.598997,36.977001,33.562
09/01/16 00:09:00,61.860001,57.132,58.508999,56.234997,56.624998,55.676996,45.843,46.452999,47.48,46.778004,47.610001,39.269005,64.225002,57.598997,37.085995,32.710999
09/01/16 00:10:00,61.860001,57.132,58.508999,56.234997,56.624998,55.676996,45.843,46.452999,47.48,46.778004,47.610001,39.269005,64.225002,57.598997,35.574001,32.542
09/01/16 00:11:00,61.860001,57.132,58.508999,56.234997,56.624998,55.676996,45.843,46.452999,47.48,46.778004,47.610001,39.269005,64.225002,57.598997,35.414005,29.472
09/01/16 00:12:00,61.860001,57.132,58.508999,56.286995,56.624998,55.676996,45.843,46.452999,47.48,46.778004,47.610001,39.269005,64.225002,57.598997,35.321999,28.153001
09/01/16 00:13:00,61.860001,57.028,58.235001,56.234997,56.676996,45.843,46.452999,47.48,46.778004,47.610001,39.269005,64.225002,57.598997,35.137001,27.777
09/01/16 00:14:00,61.860001,57.028,58.508999,56.338997,56.624998,55.676996,45.843,46.452999,47.48,46.778004,47.610001,39.269005,64.225002,57.560003,35.396,29.416
09/01/16 00:15:00,61.860001,57.028,58.508999,56.338997,56.624998,55.676996,45.843,46.452999,47.48,46.778004,47.610001,39.269005,64.225002,57.598997,35.150002,31.155998
09/01/16 00:16:00,61.834,57.028,58.508999,56.338997,56.624998,55.676996,45.843,46.452999,47.48,46.778004,47.610001,39.269005,64.225002,57.493001,34.344006,33.505005
09/01/16 00:17:00,61.834,57.028,58.508999,56.338997,56.521002,55.676996,45.843,46.452999,47.48,46.778004,47.610001,39.269005,64.225002,57.493001,37.085995,37.022999
09/01/16 00:18:00,61.834,57.132,58.508999,56.338997,56.521002,55.676996,45.843,46.452999,47.48,46.778004,47.610001,39.269005,64.225002,57.495001,36.137001,40.483002
09/01/16 00:19:00,61.834,57.132,58.508999,56.338997,56.624998,55.676996,45.843,46.452999,47.48,46.778004,47.610001,39.269005,64.225002,57.598997,36.467003,41.018997
09/01/16 00:20:00,61.964001,57.236,58.612999,56.443001,56.624998,55.780996,45.843,46.452999,47.48,46.778004,47.610001,39.269005,64.432995,57.598997,38.085995,40.617001
09/01/16 00:21:00,61.964001,57.236,58.612999,56.443001,56.624998,55.689997,55.780996,45.843,46.452999,47.48,46.778004,47.610001,39.269005,64.432995,57.598997,38.076,40.557999
09/01/16 00:22:00,61.964001,57.352997,58.729996,56.443001,56.728998,55.884996,45.843,46.452999,47.48,46.778004,47.610001,39.269005,64.432995,57.676996,37.84,40.329998
09/01/16 00:23:00,62.068001,57.352997,58.729996,56.547001,56.728998,55.884996,45.843,45.518002,47.48,46.778004,47.610001,39.269005,64.432995,57.702997,38.254002,40.098003
09/01/16 00:24:00,62.068001,57.456997,58.729996,56.547001,56.728998,55.884996,45.843,45.518002,47.48,46.778004,47.610001,39.269005,64.432995,57.702997,38.368001,40.042
09/01/16 00:26:00,62.198002,57.456997,58.859997,56.547001,56.832998,55.988996,45.946997,46.57,47.48,46.778004,47.610001,39.269005,64.549,57.702997,36.387001,39.742004
09/01/16 00:27:00,62.198002,57.456997,58.859997,56.547001,56.832998,55.988996,45.946997,46.57,47.48,46.778004,47.610001,39.269005,64.549,57.702997,37.080002,39.566002
09/01/16 00:28:00,62.198002,57.559998,58.859997,56.547001,56.872,55.988996,45.946997,46.57,47.48,46.778004,47.610001,39.269005,64.549,57.806997,37.776997,39.469997
09/01/16 00:29:00,62.198002,57.559998,58.859997,56.547001,56.936998,55.988996,45.946997,46.57,47.48,46.778004,47.610001,39.269005,64.549,57.806997,37.763996,39.329998
```

FIGURE 9.11   Data from a sensor (DEMO).

Although there are some methods of equipment condition monitoring, there are great limitations in the traditional methods. For example, most device monitoring methods rely on "alarm" mechanisms and lack response time; early warning is highly dependent on the responsibility and experience of the staff, easy to overlook; it is easy for "Window" monitoring model to ignore structural changes, and so on.

It can help to greatly reduce the cost of modeling and analysis of complex massive data and to realize early warning of abnormal status of monitored objects by using artificial intelligence, big data analysis and other technologies, which are especially suitable for the analysis, processing and utilization of dynamic, massive, high concurrency, nonlinear, strong noise and heterogeneous data generated by industrial equipment and processes. The data intelligence process can be illustrated by the Fig. 9.12.

For example, an enterprise successfully realized 3 minutes in advance of the detection of the generator set equipment fault warning, to avoid accidents.

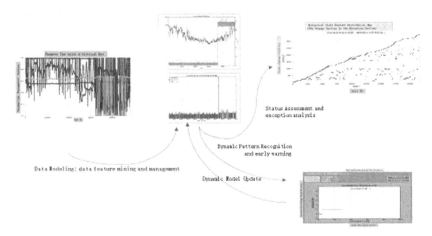

FIGURE 9.12   Data intelligence process diagram.

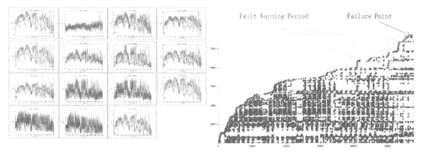

Distribution of operating state of generator sets

(3 minutes before the failure of the generator set was found t have obvious abnormal status)

FIGURE 9.13   **Early warning of equipment failure.**

The method is as follows: (1) collect all the equipment sensor parameters of the generator set according to the time series (more than 50,000 pieces of data in the three weeks before the equipment failure), and extract the state characteristics of the equipment operation efficiently by using AI technology; (2) model the device state characteristics and find out the regular distribution of the device running state; (3) monitor real-time device sensor data based on the model, so as to provide real-time early warning based on fault precursor features. Therefore, this method can evaluate the status of the equipment in real time and capture the fault symptoms. The Fig. 9.13 is a schematic diagram of the fault warning period and the actual fault point clearly displayed on the obtained distribution diagram of the operating state of the equipment.

9.5.2   Equipment fault recognition and diagnosis

The deep learning algorithm is used to construct the fuzzy diagnosis model library of all kinds of equipment faults, which provides a new method for the final judgment of equipment faults. The knowledge mapping technology is applied to realize the correlation mapping between equipment failure and expert experience, so as to provide auxiliary reference for the timely and accurate diagnosis and treatment of various equipment failure.

Through the efficient mining of historical data and the analysis of abnormal state, the technology can quickly locate the period of obvious abnormal state in the history of the equipment. For example, a company conducted AI technology data mining for 3.6 million pieces of historical data of a key pump in the past seven years, and located the abnormal state period in the pump's history, including process adjustment, failure symptoms, failure and shutdown, within 5 minutes. The method is as follows: (1) collect the historical data of the pump according to the time series, and then use AI technology to extract the characteristics of the running state of the pump efficiently; (2) model the pump state characteristics and find out the running state and regular distribution of the pump; and (3) locate and analyze the abnormal period and abnormal state characteristics of the pump in its operation history, as shown in the following Fig. 9.14.

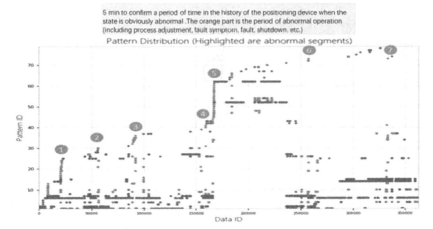

FIGURE 9.14   Distribution of historical running state of a key pump.

If there are some abnormal changes with the sensor values (in the initial stage of the abnormal period, there are usually obvious changes in some parameters), then based on the historical big data of the sensor, the operating state character-istics of the device can be extracted efficiently with artificial intelligence tech-nology, and the device can be conducted the abnormal state analysis. Normally, the initial abnormal changes of sensor parameters with will lead to abnormal changes in the values of other sensors, so it is possible to locate the fault source point through such abnormal analysis, so as to find out the cause of the fault. The following figure is a schematic diagram of the equipment fault diagnosis analysis of a compressor unit in a refinery enterprise. The above figure shows the param-eters of one of the sensors in the compressor that have significantly changed dur-ing the initial phase of the abnormal period. The figure below (Fig. 9.15) shows

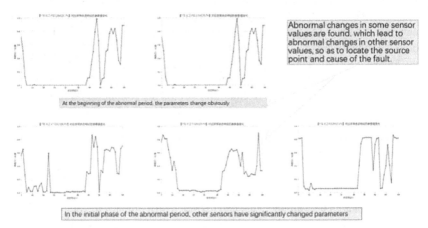

FIGURE 9.15   Parameter variation of compressor during abnormal state period.

the abnormal changes of values of other sensors in the same time series caused by the change of sensor parameters in the figure above. Such abnormal changes in the value of one sensor lead to abnormal changes in the value of other sensors to locate the cause and source of the fault.

Therefore, the equipment running state diagnosis and fault cause analysis, focusing on the state characteristics before and at the time of the failure. It means to find out what are the signs before the failure, when did these signs occur, when the fault occurs, which sensor parameters should be found abnormal. By paying attention to and focusing on the abnormal features of the equipment and establishing the corresponding model, the early warning features of the equipment can be found through intelligent analysis methods and means, so as to prevent the occurrence of the problem.

9.5.3 Equipment health status, residual life prediction and other management

Usually, there are more than one or a variety of different principles of monitoring sensors to monitor monitoring in one (set) of equipment, such as temperature, vibration, displacement, differential pressure, and so on. Even if the measurement of temperature, there may be a variety of measurement techniques. For example, there are 12 sensors in a rotating device, and there are 12 sequence diagrams of a local time period displayed on DCS, which makes it difficult for operators and engineers to be clear. Therefore, by using AI technology to find out the characteristic distribution of the running state, it is possible to display images of 12 sensors in a single screen, and to full grasp the long-term state evolution of a device. Fig. 9.16 shows a schematic diagram of this process.

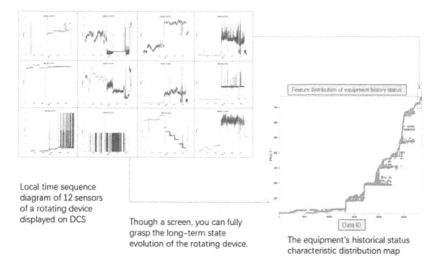

Local time sequence diagram of 12 sensors of a rotating device displayed on DCS

Though a screen, you can fully grasp the long-term state evolution of the rotating device.

The equipment's historical status characteristic distribution map

FIGURE 9.16 The long-term state evolution of a device.

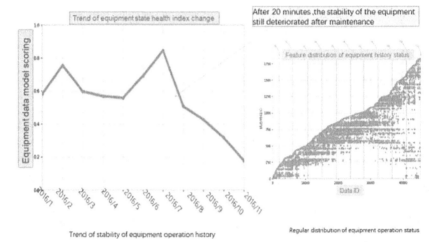

FIGURE 9.17 The stability of equipment in historical operation.

As mentioned earlier, the use of big data of equipment sensor and artificial intelligence technology can extract equipment running state characteristics. Then it is possible to automatically, multi-angle analyze the state of the equipment characteristics, to calculate the relative health of equipment operation stage, to achieve the change trend of equipment health index, to provides the basis for the preventive maintenance for equipment maintenance. The figure is the analysis result of the operation stability trend of a certain unit. Within 20 minutes, the indication that the stability of the equipment is still deteriorating after maintenance is obtained, shown in Fig. 9.17.

The fuzzy evaluation model of corrosion failure is established by support vector machine, artificial neural network and other technologies through learning process data and testing data independently, so as to predict the corrosion rate and residual life of the equipment, to provide decision basis for the enterprise's corrosion protection plan and avoid safety accidents caused by corrosion.

References

Boman, K. (2015). Big data, internet of things transforming oil and gas operations. *Rigzone Oil and Gas News.*

Martinotti, S., Nolten, J., & Steinsb, J.A. (2014). Digitizing oil and gas production. Available from: http://www.mckinsey.com/industries/oil-and-gas/our-insights/digitizing-oil-and-gas-production.

MR Brulé Group IBMS (2015). The Data Reservoir: How Big Data Technologies Advance Data Management and Analytics in E&P Introduction—General Data Reservoir Concepts Data, Reservoir for E&P.

Chapter 10

Soft Sensors for NOx Emissions

Patrick Bangert

Artificial Intelligence Team, Samsung SDSA, San Jose, CA, United States; algorithmica technologies GmbH, Küchlerstrasse 7, Bad Nauheim, Germany

10.1 Introduction to soft sensing

A *sensor* is a physical device that measures a quantity due to a physical reaction to a change in the environment of the device. For example, a temperature sensor usually consists of a material called a thermistor that changes its electrical resistance in response to changes in temperature. The electrical resistance is measured using ohmmeter. During the calibration phase, a researcher would establish a conversion formula that can convert ohmmeter readings into temperature readings. This formula is usually a simple relationship, but it is instructive to note that even simple and regular sensor require a model to convert the physical effect of note (here the resistance) to the effect being measured (here the temperature).

A *soft sensor* is a formula that converts various inputs from simple sensors and combines them to mimic the output of a more complex sensor. In this chapter, we will discuss a sensor for the nitrogen oxides (NOx) emissions of a co-generation power plant. Directly measuring the NOx is possible but the sensor is expensive and fragile. If it were possible, which it is, to substitute it with a formula that is based on a set of much cheaper and robust sensors (such as temperatures, pressures, flow rates, and so on), then we would have several benefits from this:

1. The value from the soft sensor is cheaper both initially and over the lifetime of the plant as no device needs to be purchased or maintained.
2. It is always available as the soft sensor will never fail or need to be removed for recalibration.
3. It is available in real-time as it is just a calculation and we do not have to wait for some physical reaction to take place.
4. It is scalable over many assets or locations without investment or difficulty.

It makes sense to substitute expensive and fragile sensors with soft sensors. Many plants must rely on laboratories to perform certain measurements that are

Machine Learning and Data Science in the Oil and Gas Industry.
http://dx.doi.org/10.1016/B978-0-12-820714-7.00010-8

211

too difficult to perform in the stream of the plant. This necessitates the taking of a sample that is usually done by a person. The sample is taken to the laboratory that measures the value in question. The values are usually made available by manually typing the value into a spreadsheet, which becomes available several hours after the sample was originally taken. In this way, we can expect to receive one or two values per day, at most. This process is expensive, slow and error prone. A soft sensor can eradicate the cumbersome nature of this process and make a continuous value available to the plant. In turn, this value then can be used in advanced process control applications that can improve the workings of the plant. Therefore, a soft sensor can directly contribute to the bottom line of a plant by enabling hitherto impossible control strategies.

Measuring pollutants like NOx and SOx, gas chromatography in chemical applications, or multiphase flow in upstream oil and gas applications are some examples where this has been successfully done on an industrial scale.

The idea of a soft sensor is the same as the idea of the thermistor temperature sensor. We are measuring something simple such as the electrical resistance that we then convert computationally into what we really want to know such as the temperature. The only two differences are that we now have multiple simple quantities that flow into the computation and, usually, a non-linear relationship between these quantities to get to the final output. These form our dual challenge.

First, we must collect as much domain knowledge about the system as we can to determine the fundamental question: Which simple measurements do we need in order to be able to compute the quantity of interest? If we leave out something important, the quality of the model may be poor. If we include something that is not connected to the output, then this may disturb the calculation.

This idea is subtle and so let us look at an everyday example of this. The sale of ice cream is correlated to the number of drownings in swimming pools. While this is true, the relationship is obviously not causal. It is the higher ambient temperature that results in higher ice cream sales and more pools visits, which in turn increase the drownings. It would be better to predict the ice cream sales using temperature and not the number of drownings. This is obvious to a human being who knows what is going on—the domain expert—and not at all obvious to an automated computer analysis method that just looks at correlations.

Second, we must establish the non-linear relationship between all the variables to get the best formula to calculate our desired output. This is the realm of machine learning.

10.2 NOx and SOx emissions

NOx is an umbrella term combining several different nitrogen oxides. Most notably among them are nitric oxide (NO) and nitrogen dioxide (NO_2). These gases contribute to smog, acid rain, and have various detrimental health effects. The term also includes nitrous oxide (N_2O), which is a major greenhouse gas.

SOx is an umbrella term combining several different sulfur oxides. Most notably among these are sulfur dioxide (SO_2), which is a toxic gas, and sulfur trioxide (SO_3), which is the main component of acid rain. The combination of NO_2 with SOx will form sulfuric acid, which is harmful.

VOC is an umbrella term for volatile organic compounds, which are a variety of chemicals that have detrimental health effects.

The combustion of fossil fuels produces all these gases. They must be removed by so-called scrubbers, which spray a reactant into a chamber with the gas that reacts with the unwanted chemicals and converts them into a chemical that is harmless and can be captured. For example, SO_2 is converted by limestone into gypsum, which is a material used by the construction and agricultural industries. While most of the produced pollutants is captured, some amount is released into the atmosphere causing several unwanted effects such as the greenhouse effect. How much of each gas is released is the question at hand.

The recognition that these gasses are harmful has led to governmental regulations capping the amount that may be released. In the case of another greenhouse gas, CO_2, it has even led to the creation of a carbon credit economy where permissions to release the gas are traded on the financial markets. There is significant interest on the part of the general public, governments and regulators to keep the released gasses to a minimum. The amount of these gasses that are produced by combusting a specific amount of the fossil fuel is determined by the fuel itself—even though it varies significantly depending on where the fossil fuel was mined—the critical element in this effort is the scrubber system. Knowing exactly how much is being released and knowing it in real-time allows advanced process control in order to properly operate the scrubber and therefore minimize the emitted gasses.

The regulators require the emission to be determined with accompanying documentation to make sure that permitted limits are adhered to (California Energy Commission, 2015). The standard method of doing this is to measure these gases with a sensor array and to record the values in a data historian.

Three problems are encountered in practice with this approach. First, the sensor array is usually quite expensive running into several hundred thousand dollars for each installation. Second, the sensor array is fragile especially in the harsh environments typical of the oil and gas industry, which in turn leads to significant maintenance costs and monitoring efforts that distract from the process equipment. Third, whenever the sensor array does not provide values—due to a failure or temporary problem—the regulator may conclude that permitted limits have just been exceeded and charge a fine. Fines are costly but may even lead to limits being reduced or public perception being affected.

If the released amounts could be calculated from process measurements, rather than measured directly, then this would solve all the above problems. A calculation is cheap, robust, and does not break down.

Due to all these factors, the soft sensing of emissions is directly tied to the minimization of emissions and the adherence to regulations.

10.3   Combined heat and power (CHP)

In our application, we consider the process of a *combined heat and power* (CHP) generation station, also known as *cogeneration*. This type of power plant uses a heat engine, such as a gas turbine, to generate both electricity and heat. The combustion of fossil fuels produces high-temperature gas that drives the turbine to produce electricity and the low-temperature waste heat is then provided as the heat output, often in the form of steam.

In applications where both electricity and heat are needed, cogeneration is significantly more efficient (about 90%) than generating electricity and heat separately (about 55%). As such, this is a more environmentally friendly way to generate power.

See Fig. 10.1 for an overview of the cogeneration setup. In the beginning, air is mixed with natural gas to be combusted in a gas turbine. The hot gas causes the turbine to rotate that rotates the generator that makes electricity. The exhaust gas is treated in the heat recovery steam generator (HRSG). This uses the hot gas to heat water in order to generate high-pressure and high-temperature steam. The steam is passed through a steam turbine connected to a second generator that produces further electricity that would have been lost if we had not recovered it. The remaining steam at this point can be extracted to provide heat to some application. The rest is cooled and can be recycled in the steam cycle.

As a last step, just before the stack, the flue gas is scrubbed in a scrubber, see Fig. 10.2. The flue gas enters at the bottom and the scrubbing liquid (e.g., pulverized limestone in water) enters at the top and flows down through various levels. The levels provide the gas multiple opportunities to react with the liquid before it exits at the top. Such scrubbers generally remove about 95% of the pollutants in the gas. For extra purity, a second scrubber can be added.

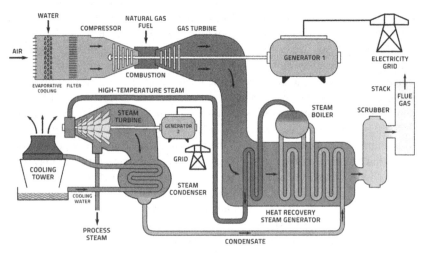

FIGURE 10.1   The general setup of a cogeneration plant.

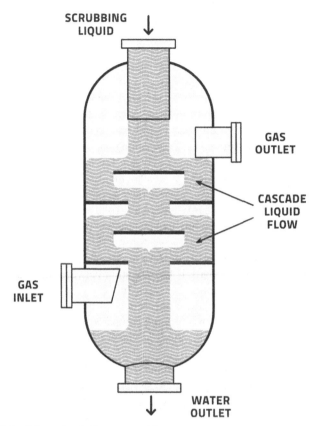

FIGURE 10.2   Schematic of a flue gas scrubber.

10.4   Soft sensing and machine learning

A soft sensor is a formula that calculates a quantity, y say, from measured quantities, say, so that the quantity of interest does not have to be measured. This formula can take any form. In many cases, we have an explicit formula that is known from physics or chemistry. In some cases, however, we do not have a known formula and so the formula must be determined empirically from available data.

Determining a formula from data is the purpose of machine learning (Goodfellow, Bengio, & Courville, 2016). Usually the function takes one of a few standard forms such as a linear regression (Hastie, Tibshirani, & Friedman, 2016), feed-forward neural network (Hagan, Demuth, & Beale, 2002), recurrent neural network (Yu, Si, Hu, & Zhang, 2019), random forest (Parmar, Katariya, & Patel, 2018), and others (Bangert, 2012). The choice of the form is made by the data scientist and machine learning determines the parameters that make the chosen form fit best to the data. Having gotten the best parameters, we can try the formula on some more empirical data to see how it performs. If the accuracy of the formula on this new data is better than the required threshold, then the soft sensor is finished. In this sense, machine learning a soft sensor is comparable to calibrating a physical sensor.

As the physical sensor is the standard, we must compare any soft sensor to it on the same principles that a physical sensor is deemed fit for its purpose. This is *calibration*. According to the International Bureau of Weights and Measures (BIPM), calibration is an "operation that, under specified conditions, in a first step, establishes a relation between the quantity values with measurement uncertainties provided by measurement standards and corresponding indications with associated measurement uncertainties (of the calibrated instrument or secondary standard) and, in a second step, uses this information to establish a relation for obtaining a measurement result from an indication." (JCGM, 2008)

In this sense, calibration is the process of comparing the output of the sensor in question to some standard. If the deviation between the two is less than a certain amount, then calibration is successful. The deviation must always be seen as relative to the measurement uncertainty of both the sensor in question as well as the standard to which it is being compared. The usual assessments of machine learning with its root-mean-square-errors and distribution of residuals is exactly this kind of assessment.

There are various standards around the world that require sensors to be calibrated, and re-calibrated at regular intervals. These are, for example, ISO-9001, ISO 17025, ANSI/NCSL Z540, and MIL-STD-45662A. However, these standards do not define the precise requirement that the sensor has to satisfy in order to be deemed calibrated.

Whether a sensor is calibrated or not is usually decided based on the maximum deviation between the sensor in question and the standard as compared to the measurement uncertainty of the sensor in question. If the deviation is always less than ¼ of the measurement uncertainty, the sensor is considered calibrated. This ratio of ¼ is, of course, a somewhat arbitrary choice that emerged historically due to the practices of the military of the United States (Jabloński and Březina, 2012; Department of Defense, 1984). Since this is an emergent practice and not a legally defined requirement, it is often up to the regulator to define this ratio and it can be larger, such as ½. The other variable is the measurement uncertainty. If the sensor in question is outfitted with an appropriately large uncertainty, it will always meet these criteria. So, the question is not whether a sensor is calibrated or not but rather (1) at what ratio and (2) with what uncertainty the sensor is calibrated.

In conclusion, if we have a good model in the sense of machine learning, it is *ipso facto* a calibrated model. Providing this assessment and documentation to a regulator will allow them to accept a soft sensor in lieu of a physical sensor relative to the same standards.

10.5   Setting up a soft sensor

To begin with, all available empirical data is examined by domain experts to extract those sensors that have the most relevant information content to produce NOx or SOx. It is important to present all the information that we have and not to present any distracting data that may only serve to distract the model. Machine learning is fundamentally unable to differentiate between correlation and causation, which makes the initial selection of sensors vital to the success and this is best performed by an experienced expert (Guyon and Elisseeff, 2003).

The full dataset is now split into a training dataset that will be used to make the model and a testing dataset that will be used to assess the goodness-of-fit of the model. It is typical to randomly select 75% of the data for training and the rest for testing.

Two basic problems plague machine learning and they are underfitting and overfitting (Bishop, 2006). Underfitting results from having so few parameters, that the model cannot possibly represent the phenomenon. Overfitting results from having so many parameters, that the model can memorize the data without learning the underlying dynamics. Clearly, both possibilities will produce a poor model in the sense that a novel datapoint will obtain an unsatisfactory computational result. To avoid both, we must find a medium number of parameters and this is usually problem dependent. A very rough rule of thumb is to obtain at least 10 data points for every one parameter in the model. This is a rough rule as the information content is not proportional to the number of points in a dataset; for instance, if a point is duplicated, then a point has been added but no new information was added. A solution typically recommended in machine learning textbooks—getting more data—is often infeasible especially in an industrial context where data acquisition entails costs of money and time. The available data therefore puts an *a priori* limit on the possible complexity of the model.

In order to reduce the necessary parameter-count in the model while keeping the information content in the dataset the same, we usually employ a dimensionality reduction method. Such a method takes the dataset and projects it into a space with fewer dimensions without merely deleting any dimensions. There are many such methods but the most popular is principal component analysis (Guyon and Elisseeff, 2003).

For the cogeneration plant that we are dealing with here, the following variables were selected as suitable inputs to a soft sensor:

1. Heat recovery steam generator (HRSG)
   a. Burner: temperatures, fuel pressure, flame strengths
   b. Feedwater temperature rise (FWTR): temperature, pressure, flowrate
   c. Control valves: opening
   d. Ammonia: leak detector
   e. General: inlet temperature, tube skin temperatures, steam pressure, flowrates, turbine exhaust heat rate, position feedback, heat rate, differential

pressure across selective catalyst reduction (SCR), heat recovery, water flowrate, fuel flowrate, quality trim factor, consumed heat water
2. Gas turbine
   a. General: fuel pressures, fuel volume rates, fuel heat rates, ammonia mass injection
   b. Engine: thermocouple, variable frequency drive frequency, compensator temperature
3. Weather
   a. General: wind direction, rain rate, air pressure, humidity

As some of these variables are available multiple times, we had 31 variables available that are the vector. They are sampled once every 10 min. The NOx and SOx output as measured by the sensor array is also provided and that is the known *y* that we are going to use to train our soft sensor using machine learning. We perform principal component analysis to reduce the dimensionality from 31 to 26. This retains 99% of the variance of the original dataset. A deep feed-forward neural network is trained on data from one year.

In order to demonstrate the model capability for diversity, we choose to model the NOx concentration in parts per million and the SOx flowrate in pounds per hour.

10.6 Assessing the model

In training the model, we split the data into three parts. The training data is used to adjust the model parameters so that the model is as close to these training data as possible. The testing data is used to evaluate when this training is making no more significant progress and can be considered finished. The validation is not used for training at all. The finished model is executed on the validation data so that we can judge how well the model performs on data it has never seen before. This is the true test of how well the model performs. This would be the basis for any calibration documentation provided to a regulator.

We thus compare the computed result of the model with the known result of the measurement, *y*, using all three datasets. For every datapoint , we compute the residual . If the residual is very close to zero, the model is doing well and if it is far away from zero, the model is doing poorly. In order to study many datapoints it is helpful to plot the probability distribution of residuals, see Fig. 10.3.

This distribution is arrived at by counting how often a particular residual occurs over a large dataset. We then plot the residual on the horizontal axis and the frequency of occurrence on the vertical axis. Where the curve is highest is then the most often occurring residual.

We expect that this distribution is a bell-shaped curve, centered on zero, symmetrical about zero, exponentially decaying to either side, and having a small width. What makes the width "small" depends on the use case. Some uses need a more accurate model than others and this is the most important success criterion to be set for the model. It is usually unrealistic to expect this

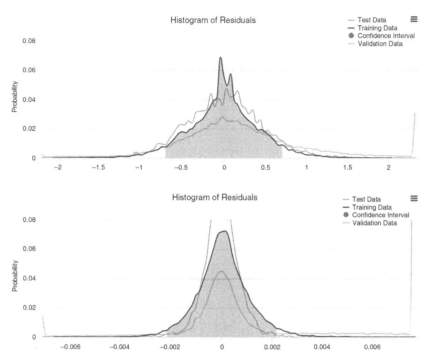

FIGURE 10.3   The probability distribution of residuals of NOx and SOx for training data, testing data, and validation data. All three distributions are bell-shaped curves centered on zero and so are what we expect to see.

distribution to be normal as the normal distribution usually decays faster in its tails than the distributions we encounter in practice. This is due to a variety of factors some of which are artefacts of empirical data analysis such as a limited number of datapoints. Other factors are due to the measurement process itself, i.e. there are causal factors that are not measured and cannot be included in the dataset at all (Tsai, Cai, & Wu, 1998).

Loosely speaking, the standard deviation of the residual distribution is the accuracy we can expect of the model and it is the final conclusion to be presented upon calibration.

In our case, the distribution for the residuals for training, testing, and validation adhere to these principles. We see that the residuals are larger for testing and validation than for training. This behavior is to be expected as a model usually performs better on data that was used to make the model. This is probably also due to the system slightly changing its behavior over time, that is, aging. There are two factors to the aging process in industrial applications. First, there is genuine mechanical degradation over time of all manner of physical components of the system. Second, there is sensor drift that changes the numerical values recorded for any sensor. It is due to sensor drift that sensors occasionally need re-calibration.

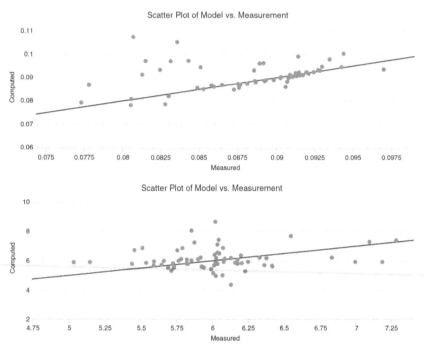

FIGURE 10.4 **The computed value plotted against the measured value for NOx and SOx.** Ideally, this data would form a *straight line,* displayed in black for reference.

In addition to checking the plot of residuals, we may plot the measured against the modeled value directly, see Fig. 10.4. Here, we can see that the data points are distributed closely around the ideal straight line where the computation would equal the measurement.

Finally, we may plot a timeline of the measurement and the computation to assess the true values against each other over time, see Fig. 10.5. This is the output that an operator would see.

In assessing these figures, we must keep in mind that no sensor is precise in its measurement. Every measurement is uncertain to some degree and has an inherent uncertainty. The inherent uncertainty is best assessed by the standard deviation of the variable. In our case, this is 2.43 ppmc for NOx and 0.02 lb/h for SOx. Comparing this to the standard deviation of 0.75 ppmc for NOx and 0.01 lb/h for SOx in the results of the physical measurement, we may conclude that the soft sensor is roughly as accurate as the physical sensor array.

10.7 Conclusion

We find that the model for NOx and SOx is comparable in accuracy to the physical measurement. This makes the models suitable for use in real CHP plants instead of the physical sensor array. This use saves significant cost in

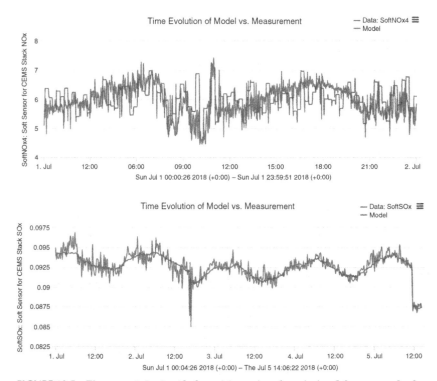

FIGURE 10.5   The computed value *(dark gray)* *(green in web version)* and the measured value *(light gray)* *(red in web version)* for NOx and SOx.

the acquisition and maintenance of the sensors as well as in the fines associated with sensor malfunction and failure.

As the models were made using machine learning based on empirical data, they did not consume much human time and effort. However, they benefitted significantly from human expertise mainly due to the careful selection of input data. Since the models include the important human expertise and that they were carefully examined against measurement data, we can conclude that the model output is reliable.

References

Bangert, P. (2012). *Optimization for industrial problems*. Springer.

Bishop, C. M. (2006). *Pattern recognition and machine learning*. Springer.

California Energy Commission. (2015). Guidelines for certification of combined heat and power systems pursuant to the waste heat and carbon emissions reduction act. CEC-200-2015-001-CMF.

Department of Defense. (1984). *Military handbook: evaluation of contractor's calibration system*. MIL-HDBK-52A.

Goodfellow, I., Bengio, Y., & Courville, A. (2016). *Deep learning*. MIT Press.

Guyon, I., & Elisseeff, A. (2003). An introduction to variable and feature selection. *Journal of Machine Learning Research, 3,* 1157–1182.

Hagan, M. T., Demuth, H. B., & Beale, M. (2002). *Neural network design.* PWS Publishing.

Hastie, T., Tibshirani, R., & Friedman, J. (2016). *The elements of statistical learning: data mining, inference, and prediction* (2nd ed.). Springer.

Jabłoński, R., & Březina, T. (2012). *Mechatronics.* Springer.

JCGM (2008). *International vocabulary of metrology—basic and general concepts and associated terms (VIM).* BIPM.

Parmar, A., Katariya, R., & Patel, V. (2018). *A review on random forest: an ensemble classifier. Proceedings of ICICI 2018: International Conference on Intelligent Data Communication Technologies and Internet of Things (ICICI).* Springer.

Tsai, C. L., Cai, Z., & Wu, X. (1998). The examination of residual plots. *Statistica Sinica, 8,* 445–465.

Yu, Y., Si, X., Hu, C., & Zhang, J. (2019). A review of recurrent neural networks: LSTM cells and network architectures. *Neural Computation, 31,* 1235–1270.

Chapter 11

Detecting Electric Submersible Pump Failures

Long Peng, Guoqing Han, Arnold Landjobo Pagou and Jin Shu
China University of Petroleum, Beijing, China

11.1 Introduction

Electric Submersible Pumps are currently widely employed to help enhance the production for nonlinear flowing well with high production, high water cut, and offshore oil wells due to its simple structure and high efficiency (Ratcliff, Gomez, Cetkovic, & Madogwe, 2013). Among all the oil artificial lift systems, ESP is preferred because it can produce high volumes in higher temperatures and reach deeper depths. However, it is often observed that the ESP system reaches the point of service interruption. The breakage of the pump shaft is a severe issue for the operating company as it generates an estimated loss of hundreds of millions of oil barrels. Should the pump shaft break, the motor current would drop suddenly, and the production would be interrupted. The reason of the pump shaft breakage is bad pump assembly or pump aging.

The development of sensors and data acquisition systems make it possible for ESP systems to continuously record the intake pressure, intake temperature, pump head, discharge pressure, discharge temperature, motor temperature, motor current, leakage current, vibration and so on. Those data would be recorded at regular intervals and transmitted to surface Remote Terminal Units (RTUs) (Bates et al., 2004). For those broken shaft wells, statistics, or machine learning algorithms can be used for failure analysis and health monitoring.

The objective of this paper is to evaluate principal component analysis (PCA) as a monitoring tool to forecast the breakage of the ESP shaft.

11.2 ESP data analytics

ESP operation system has developed over the years and is considered as an effective means to lift crude oil under wellbore conditions. There is increased attention on ESP systems due to the fast development of ESP sensors in the oil industry. The ESP sensors gather a vast amount of data, including dynamic data, static data, and historical data (Abdelaziz, Lastra, & Xiao, 2017), as shown in

Machine Learning and Data Science in the Oil and Gas Industry.
http://dx.doi.org/10.1016/B978-0-12-820714-7.00011-X

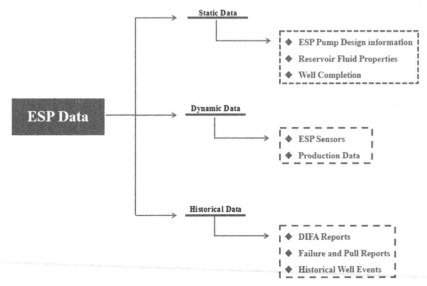

FIGURE 11.1 ESP data are available.

Fig. 11.1. In the past few years, wells using ESP technology have been tracked by physical well site visits, which require enormous human resources to make necessary adjustments to the well operating system. The oil and gas industry has begun applicating data analytic to improve production efficiency (Stone et al., 2007; Takacs, 2017) proposed the ammeter charts as the primary diagnostic method to monitor ESP performance.

Furthermore, the emergence of the Supervisory Control and Data Acquisition (SCADA) system has provided convenience for field personnel to monitor and control the ESP well behavior (Ratcliff et al., 2013). The SCADA system can achieve the full realization of continuously recording ESP production data in real-time.

In recent years, "Big Data" collected on ESP sensors is framing the key point of how to extract the most essential information to evaluate the ESP operating system. Data-driven models coupled with machine learning, have been used to judge and optimize well production. Bravo, Rodriguez, Saputelli, & Rivas Echevarria (2014) described data analytic as an integral process of collecting and analyzing big data.

Pump shaft fracture is a common fault in the ESP operating system. Based on operation data acquired by ESP ground and downhole sensors, data-driven model analytics will play an important role in monitoring the breakage of the pump shaft. There is a need to evolve from a supervised method toward fault diagnosis to an approach according to data-driven models for predictive maintenance. PCA is widely considered as a pre-processing method for dimensionality reduction, eigenvalue extraction and data visualization (Abdelaziz et al., 2017).

PCA can be used as an unsupervised machine learning technique to analyze the reason of the pump shaft breakage.

11.3   Principal Component Analysis

Jackson (2005) defines PCA as an unsupervised dimensionality reduction means, which transfers data linearly and creates a new set of parameters called "principal component". What is known to us is that ESP data are generally highly correlated; i.e. an increase of the wellhead pressure will lead to an increase of intake and discharge pressure, which will finally cause an increase of the motor temperature. PCA makes use of the interdependence of original data to build a PCA model. This results in the reduction of the production parameters' dimension by taking advantage of the linear combinations and by creating a new Principal Component space (PCs). This PCs can evaluate the ESP system only by several principal components, making the process much easier.

For those wells with pump shaft fracture, a PCA model can be established to analyze a few months of production data before the pump breaks. Once the robust PCA model is built, the cause of the breakage of the pump shaft will be monitored and diagnosed. The basic PCA model can be represented as follows:

$$X = TP^T + E \qquad (11.1)$$

where X is the input matrix ($n \times p$); it represents original parameters; P is the loading matrix ($p \times k$); it represents the contribution of original parameters; T is the score matrix ($n \times k$); it represents the relationship between original parameters; E is the residual matrix ($n \times p$); it represents the uncaptured variance; n is the number of time steps (Gupta, Saputelli, & Nikolaou, 2016); p is the number of original parameters; k is the number of principal components.

The first principal component contains the highest variance, implying that the first principal component contains the most information. The second principal component will capture the next highest variance, which has already removed the information of the first principal component. By this means, the third, fourth, …, k th principal component can be constructed to evaluate the original system. Fig. 11.2 summarizes the comments above.

As is often the case, the PCA model finds k principal component to construct the PCs, which retains most of the information that belongs to the initial system. The kth principal component is denoted in the following equation taking PC1 as an example.

$$PC_1 = a_{11} \times P_{(intake\ pressure)} + a_{21} \times P_{(discharge\ pressure)} + a_{31}$$
$$\times P_{(intake\ temperature)} + a_{41} \times P_{(motor\ current)} + \cdots + a_{p1} \times P_{(vibration)} \qquad (11.2)$$

When working with two dimensions, Ionita and Schiopu (2010) depicted this case using Fig. 11.3. PCA is used to find patterns in high-dimensional data

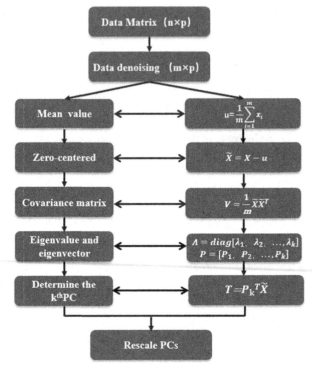

FIGURE 11.2   The architecture of PCA.

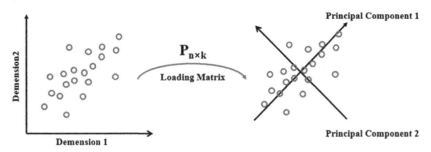

FIGURE 11.3   The geometric meaning of PCA with two dimensions.

and to transform the stable region data, which is usually characterized as a tight cluster or cloud data sets. In this way, anomalies in the ESP operation system will be detected by establishing a PCA model with a normal production dataset. The first two PCs have the highest variance, explaining most of the information in the original parameters visualized only by the first two PCs.

11.4 PCA diagnostic model

PCA diagnostic model is applied to identify the cause and the time of pump shaft fracture. The Hotelling Tsquare statistic (T^2) and Squared Prediction Error (SPE) are used to numerically visualize scalar statistics (Yue and Qin, 2001). T^2 is a univariate statistic that plays an important role in multivariate hypothesis testing, SPE is frequently used in multivariate statistical process control. T^2 and SPE are applied to analyzing whether the decision variable is satisfactory with the requirement of stable running. With this method, the contribution of each decision variable towards failure can be determined. The failure issue will ultimately be diagnosed due to the ranking of the contribution of each of the decision variables. The potential anomalies are related to the correlative higher ranking or highest decision variables.

Yue and Qin (2001) described T^2 and SPE at time step as follows:

$$T^2 = \Lambda^{-\frac{1}{2}} P^T x(t)^2 = x^T(t) P \Lambda^{-1} x(t) \leq \delta_T^2 \qquad (11.3)$$

$$SPE^2 = \left(I - PP^T\right) x(t)^2 = P_e^T x(t)^2 \leq \delta_{SPE}^2 \qquad (11.4)$$

where $x(t)$ is the tth timestep of the input matrix; Λ^{-1} is the inverse of the covariance matrix (Westerhuis, Gurden, & Smilde, 2000); P is the eigenvectors of the covariance matrix; P_e is the residual loading matrix; I is the identity matrix;

$$\delta_T^2 = \frac{(N-1)(N+1)k}{N(N-k)} F_\alpha(k, N-K) \qquad (11.5)$$

$$\delta_{SPE}^2 = \theta_1 \left[\frac{C_\alpha h_0 \sqrt{2\theta_2}}{\theta_1} + 1 + \frac{\theta_2 h_0 (h_0 - 1)}{\theta_1^2} \right]^{\frac{1}{h_0}} \qquad (11.6)$$

where δ_T^2 and δ_{SPE}^2 donate the confidence limits for T^2 and SPE. T^2 follows an F-distribution, the F-distribution is a right-skewed distribution for studying population variances. Where α represents the boundary that the Cumulative Distribution Function of the possible distribution is 0.99. Once it exceeds the control limit δ_T^2, T^2 is regarded as a potential anomaly. According to Jackson and Mudholkar (1979), SPE follows a Gaussian distribution, Gaussian distribution is symmetric about the mean, showing that data near the mean are more frequent in occurrence than data far from the mean. Where C_α represents the boundary and α is equal to 0.99. SPE is considered abnormal when it exceeds the control limit δ_{SPE}^2.

Cho, Lee, Choi, Lee, & Lee (2005) proposed the following equation defining the contribution of each decision variable P based on T^2 and SPE

$$cont_i^{T^2} = x^T P^T \Lambda^{-1} P \xi_i \xi_i^T x, \; i = 1, \cdots, m \qquad (11.7)$$

$$cont_i^{SPE} = \left(\xi_i^T \left(I - PP^T \right) x \right)^2, \; i = 1, \cdots, m \qquad (11.8)$$

where ξ_i represents the i column of identity matrix I. The higher the contribution of each decision variable, the greater the possibility of potential anomalies.

11.5 Case study: diagnosis of the ESP broken shaft

11.5.1 Selection of the ESP broken shaft variables

Production data of ten ESP wells with broken shaft are recorded at a frequency of 20 minutes by ESP downhole and ground sensors. The ESP downhole and ground sensors start to collect the production data since ten ESP wells are put into production and come to an end when the breakage of the pump shaft occurs in these wells. These ESP broken shaft wells, including E52ST1, C06ST1, B50ST2, E20ST2, A11ST1, B03ST1, B48ST1, E21ST1, E47ST1, and E42ST1 are from the Penglai block of Bohai Oilfield in China. Two different types of datasets are collected. They include:

- Data records containing input variable parameters of casing choke, casing line pressure, casing pressure, casing gas rate, ESP intake pressure, ESP discharge pressure, flowline pressure, flowline temperature, intake temperature, motor current, motor leak current, motor power, motor temperature, motor torque current, motor vibration, motor voltage, tubing choke, and VFD frequency.
- Data records containing information on the time when the breakage of the pump shaft occurs in each well.

11.5.2 Score of principle components

A PCA model is constructed based on the input variables obtained. Different principal components are ranked according to the decreasing order of variance captured. Taking well E52ST1, for example, it is observed that eight principal components capture more than 99% of the variance of the original input parameters, as shown in Fig. 11.4.

The first two principal components have the highest variance and capture probably 70% variance in the original data. A two-dimensional plot of scores of Principal Component 1 and Principal Component 2 is used to observe different clusters during the stable, unstable or failure periods. The ESP operates normally, and all the input variable parameters are in normal working range during the stable periods. When it comes to the unstable periods, some of the input variable parameters are obviously abnormal, but the ESP is still operating.

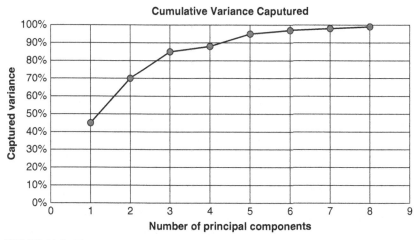

FIGURE 11.4   The captured variance of well E52ST1 by Principal Components.

Furthermore, the breakage of the pump shaft occurs, and the ESP breaks down during the failure periods.

Fig. 11.5 represents the score plot of Principal Component 1 and Principal Component 2 of the historical data from well E52ST1 with pump shaft fracture. During this time frame, as the time step increases, the result clearly shows three different clusters for the stable region, unstable region and failure region. In the beginning, the ESP is put into production. It is observed that the normal operating input variable parameters form a stable region cluster. After working long

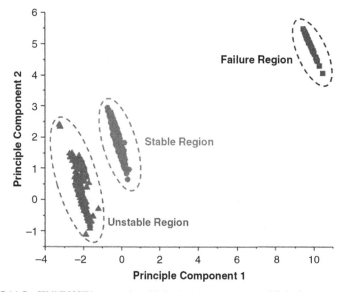

FIGURE 11.5   Well E52ST1 scores plot of Principal Component 1 and Principal Component 2.

hours, some of the input variable parameters start to deviate from the normal working range, but the ESP is still operating. When this abnormal behavior takes place, an unstable region deviating away from the stable region will form. The ESP continues to work for a period of unstable time. Finally, the breakage of the pump shaft occurs, and the ESP breaks down. From the plot below, we can observe that the black failure region is far away from the stable region. This two-dimensional plot of scores of Principal Component 1 and Principal Component 2 can be of great significance to monitor ESP performance in real-time against the previously normal operating zone to forewarn field engineers of potential failure if the cluster starts deviating away from the stable region.

11.5.3 Pump broken shaft identification

Production data from the stable region is normalized and used as the input matrix (Xtraining) to construct the robust PCA model. Besides, historical data corresponding to an unstable or a failure region period is selected as a testing dataset (Xtesting) fed to the PCA model. This process can be repeated for the historical broken shaft events leading to a failure. Also, the PCA diagnostic model is built to predict the time at which the breakage of the pump shaft occurs, and to determine the decision variable most responsible for the pump shaft fracture.

The contribution of each of the decision variables can be calculated by the PCA diagnostic model based on Eqs. (11.7) and (11.8). The decision variables with higher ranking or highest contribution are more related to the pump shaft fracture. The decision variables are ranked based on their contribution. Taking well E52ST1 as an example, it is shown in Fig. 11.6 that the motor torque

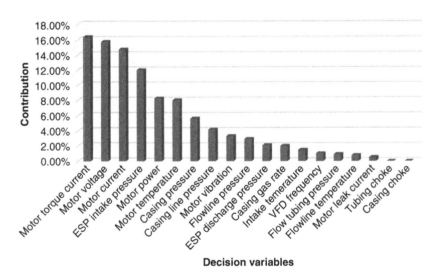

Decision variables

FIGURE 11.6   Well E52ST1's ESP diagnostic dashboard.

current has the highest contribution for the failure region. Therefore, this kind of contribution chart can be used to diagnose the decision variables most responsible for the breakage of the pump shaft in a real-time monitoring platform if the cluster starts deviating away from the stable region.

Once a robust PCA model is constructed, the time at which the fracture of the pump shaft occurs could be forecasted. T^2 and SPE equations are applied to determine the time of potential anomalies preceding the breakage of the pump shaft. Dunia and Joe Qin (1998) suggested four possible detection results under the PCA diagnostic model, those results are as follows: (1) both the T^2 and SPE indices exceed the control limits; (2) neither T^2 nor SPE indices exceed the control limits; (3) the T^2 index exceeds the control limit, but SPE does not; (4) the SPE index exceeds the control limit, but T^2 does not.

As is often the case, detection results (1) and (4) are usually regarded as potential breakage of the pump shaft. This paper contains information regarding the data of the time at which the rupture of ESP shaft occurred in each well. A detailed comparison is made between the predicted ESP's shaft breakage time by the PCA diagnostic model and the actual ESP's shaft breaking time.

Taking wells E52ST1, CO6ST1 and B50ST1 as examples, T^2 and SPE indices are computed to predict the anomaly of ESP production shown in Figs. 11.7–11.9. When the ESP shaft breaks, T^2 and SPE are used to determine the breakage time by employing the detection results (1) and (4).

Table 11.1 shows the comparison of the PCA diagnostic model prediction time of the pump shaft fracture and the actual ESP shaft braking time. An analysis of Table 11.1 reveals that the breakage time predicted by the PCA diagnostic model is a little earlier than the actual pump shaft breakage time. Consequently, the PCA diagnostic model has excellent accuracy in predicting the breakage time for ESP broken shaft wells and learning technique to predict the ESP broken shaft in real-time. Moreover, the PCA technique can be applied as the foundation for the development of better tools to predict ESP failure.

11.6   Conclusions

This paper presents a big data-driven analytical model to predict impending broken shaft in the ESP operation system. The big-data model depends on real-time data collected by ESP downhole and surface sensors. It can be concluded that PCA has the potential to be used as a recognition technique to predict dynamic changes and therefore identify the impending breakage of the ESP shaft. Key conclusions from this study can be summarized as follows:

1. A two-dimensional plot of scores of Principal Component 1 and Principal Component 2 can be used to identify different clusters of the stable region, unstable region, and failure region. From this two-dimensional plot, field engineers will be reminded of potential ESP shaft fracture if the cluster is far away from the stable region.

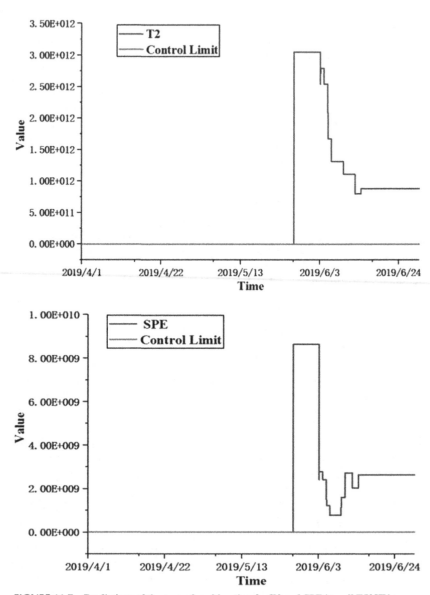

FIGURE 11.7 Predictions of the pump breaking time by T2 and SPE in well E52ST1.

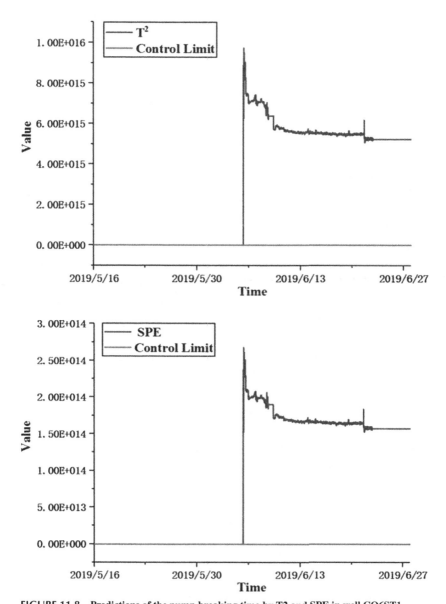

FIGURE 11.8   Predictions of the pump breaking time by T2 and SPE in well CO6ST1.

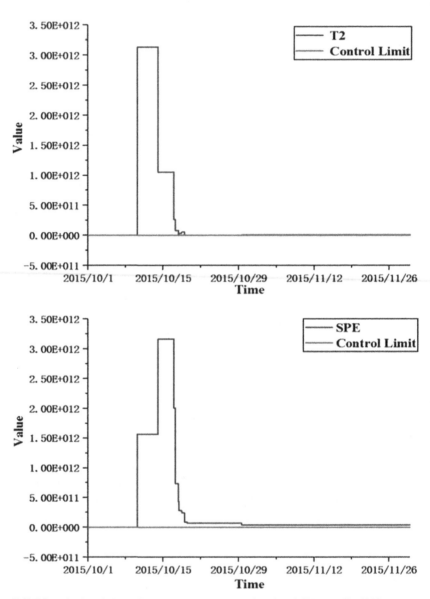

FIGURE 11.9 Predictions of the pump breaking time by T2 and SPE in well B48ST1.

2. Once a robust PCA diagnostic model is built, it is of great importance that the decision variables most responsible for the breakage of the ESP shaft will be determined to explain the deviation of the cluster from the stable region.

3. By implementing T^2 and SPE equations, the PCA diagnostic model has excellent accuracy in predicting the ESP shaft breakage time.

TABLE 11.1 Comparing of the PCA diagnostic model prediction time of the ESP shaft fracture to the actual breakage time.

Wells no.	PCA model predictions	Actual breakage time
E52ST1	2019-5-26 13:40	2019-5-26 16:00
CO6ST1	2019-5-27 22:20	2019-5-28 6:40
B48ST1	2015-10-7 21:40	2015-10-8 16:20
E20ST2	2019-9-12 10:20	2019-9-12 12:40
B50ST2	2019-6-15 8:00	2019-6-16 15:40
A11ST1	2015-5-10 4:40	2015-5-10 10:00
B03ST1	2015-8-30 15:40	2015-8-31 1:00
E21ST1	2015-8-26 10:00	2015-8-26 23:40
E47ST1	2018-1-14 8:40	2018-1-15 9:20
E42ST1	2018-4-24 22:40	2018-4-25 2:00

4. PCA can be used as an important pre-processing method and as an unsupervised machine learning technique to predict the developing ESP failures.

References

Abdelaziz, M., Lastra, R., Xiao, J., et al. (2017). Esp data analytics: Predicting failures for improved production performance. *Abu Dhabi International Petroleum Exhibition & Conference*. Society of Petroleum Engineers.

Bates, R., Cosad, C., Fielder, L., Kosmala, A., Hudson, S., Romero, G., & Shanmugam, V. (2004). *Taking the pulse of production wells-esp surveillance*. Oilfield Review.

Bravo, C., Rodriguez, J., Saputelli, L., Rivas Echevarria, F., et al. (2014). Applying analytics to production workflows: Transforming integrated operations into intelligent operations. *SPE Intelligent Energy Conference & Exhibition*. Society of Petroleum Engineers.

Cho, J. H., Lee, J. M., Choi, S. W., Lee, D., & Lee, I. B. (2005). Fault identification for process monitoring using kernel principal component analysis. *Chemical Engineering Science, 60*, 279–288.

Dunia, R., & Joe Qin, S. (1998). Subspace approach to multidimensional fault identification and reconstruction. *AICHE Journal, 44*, 1813–1831.

Gupta, S., Saputelli, L., Nikolaou, M., et al. (2016). Big data analytics workflow to safeguard esp operations in real-time. *SPE North America Artificial Lift Conference and Exhibition*. Society of Petroleum Engineers.

Ionita, I., & Schiopu, D. (2010). Using principal component analysis in loan granting. *Bulletin of PG University of Ploiesti, Series Mathematics, Informatics, Physics, 62*, 88.

Jackson, J. E. (2005). In *A user's guide to principal components* (vol. 587). John Wiley & Sons.

Jackson, J. E., & Mudholkar, G. S. (1979). Control procedures for residuals associated with principal component analysis. *Technometrics, 21*, 341–349.

Ratcliff, D., Gomez, C., Cetkovic, I., Madogwe, O., et al. (2013). Maximizing oil production and increasing esp run life in a brownfield using real-time esp monitoring and optimization software: Rockies field case study. *SPE Annual Technical Conference and Exhibition*. Society of Petroleum Engineers.

Stone, P., et al. (2007). Introducing predictive analytics: Opportunities. *Digital Energy Conference and Exhibition*. Society of Petroleum Engineers.

Takacs, G. (2017). *Electrical submersible pumps manual: design, operations, and maintenance*. Gulf professional publishing.

Westerhuis, J. A., Gurden, S. P., & Smilde, A. K. (2000). Generalized contribution plots in multivariate statistical process monitoring. *Chemometrics and Intelligent Laboratory Systems, 51,* 95–114.

Yue, H. H., & Qin, S. J. (2001). Reconstruction-based fault identification using a combined index. *Industrial & Engineering Chemistry Research, 40,* 4403–4414.

Further reading

Eriksson, L., Byrne, T., Johansson, E., Trygg, J., & Vikström, C. (2013). *Multi-and megavariate data analysis basic principles and applications* (vol. 1). Umetrics Academy.

Chapter 12

Predictive and Diagnostic Maintenance for Rod Pumps

Patrick Bangert

Artificial Intelligence Team, Samsung SDSA, San Jose, CA, United States; algorithmica technologies GmbH, Küchlerstrasse 7, Bad Nauheim, Germany

Approximately 20% of all oil wells in the world use a beam pump to raise crude oil to the surface. The proper maintenance of these pumps is thus an important issue in oilfield operations. We wish to know, preferably before the failure, what is wrong with the pump. Maintenance issues on the downhole part of a beam pump can be reliably diagnosed from a plot of the displacement and load on the traveling valve; a diagram known as a *dynamometer card*. This chapter shows that this analysis can be fully automated using machine learning techniques that teach themselves to recognize various classes of damage in advance of the failure. We use a dataset of 35292 sample cards drawn from 299 beam pumps in the Bahrain oilfield. We can detect 11 different damage classes from each other and from the normal class with an accuracy of 99.9%. This high accuracy makes it possible to automatically diagnose beam pumps in real-time and for the maintenance crew to focus on fixing pumps instead of monitoring them, which increases overall oil yield and decreases environmental impact.

12.1 Introduction

12.1.1 Beam pumps

Of all oil wells worldwide, approximately 50% have some form of artificial lift system installed. Of those, approximately 40% make use of the *beam pump*, also known as the *rod pump* or sucker-rod pump. That accounts for approximately 500,000 beam pumps in use worldwide (Takacs, 2015). The beam pump is comprised of a standing valve at the bottom of the well, and a traveling valve attached to a rod that moves up and down the well driven by a motorized horsehead assembly on the surface, see Fig. 12.1.

The journey of the traveling valve from the top of the well to the bottom and back up again is called a *stroke*. As the pump returns to the same configuration at the start of every stroke (unless the pump breaks), the motion is inherently

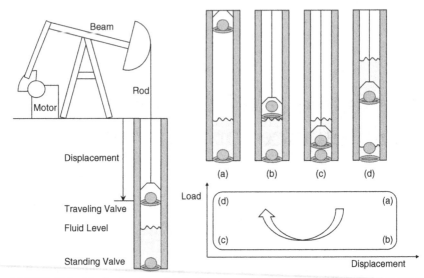

FIGURE 12.1 The basic schematic of a beam pump on the left with the evolution of the stroke in four stages on the right. Plotting of displacement and load against each other over the stroke produces the dynamometer card on the bottom right.

periodic. When the rod starts its downward journey, the standing valve closes; Fig. 12.1(a). The traveling valve opens as soon as it encounters fluid in the well and allows it to pass through; Fig. 12.1(b). At the bottom of the stroke, the traveling valve closes, and the journey is reversed at which point the standing valve opens again allowing fluid to enter the well; Fig. 12.1(c). As the closed traveling valve moves up, it transports the fluid it collected during the downstroke to the surface; Fig. 12.1(d). We can measure both the load, that is, the weight of the fluid above the traveling valve, and the displacement from the surface during one full stroke. If we graph these two variables against each other, we get a diagram known as a dynamometer card; see Fig. 12.1 bottom-right and compare with Fig. 12.2 for a real example.

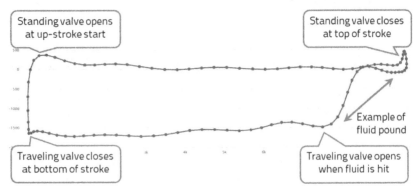

FIGURE 12.2 An example of a dynamometer card with the four major points labeled, see the introductory text for an explanation.

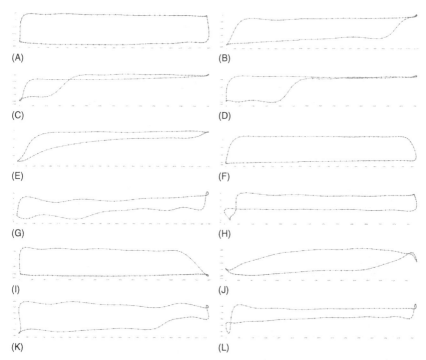

FIGURE 12.3   **Some cards, obtained in the Bahrain oilfield, which illustrate that the diverse conditions of a beam pump can be visually identified.** Please see the text for a complete list of conditions. (A) Normal; (B) Fluid pound (Slight); (C) Fluid pound (Severe); (D) Pumped off; (E) Gas interference (Severe); (F) Traveling valve or plunger leak (G) Standing valve, traveling valve leak, or gas interference (H) Pump hitting down; (I) Hole in barrel or plunger pulling out of barrel (J) Inoperative pump; (K) Pump hitting up and down; (L) Inoperative pump, hitting down.

12.1.2   Beam pump problems

It was discovered by Walton Gilbert in 1936 that the shape of the dynamometer card allows an experienced person to precisely diagnose any of the typical problems that a beam pump can have downhole (Gilbert, 1936). Please see Fig. 12.3 for some examples of the various categories that are encountered. A list of typical down-hole beam pump problems follows here.

1. Normal
2. Full Pump (Fluid Friction)
3. Full Pump (Fluid Acceleration)
4. Fluid Pound (Slight)
5. Fluid Pound (Severe)
6. Pumped Off
7. Parted Tubing
8. Barrel Bent or Sticking
9. Barrel Worn or Split
10. Gas Locked

11. Gas Interference (Slight)
12. Gas Interference (Severe)
13. Traveling Valve or Plunger Leak
14. Traveling Valve Leak and Unanchored Tubing
15. Traveling and Standing Valve Balls Split in Half
16. Standing Valve Leak
17. Standing Valve Leak and Gas Interference
18. Standing Valve, Traveling Valve Leak, or Gas Interference
19. Pump Hitting Down
20. Pump Hitting Up
21. Pump Sanded Up
22. Pump Worn (Slightly)
23. Pump Worn (Severe)
24. Pump Plunger Sticking on the Upstroke
25. Pump Incomplete Fillage
26. Tubing Anchor Malfunction
27. Choked Pump
28. Hole in Barrel or Plunger Pulling out of Barrel
29. Inoperative Pump
30. Pump Hitting Up and Down
31. Inoperative Pump, Hitting Down

It is difficult to measure the load on the moving rod directly and so we measure it at the top of the rod and infer the downhole conditions by solving the wave equation (Bastian, 1989; Gibbs, 1963). Based on this, we can calculate several physical quantities such as a pump intake pressure without measuring them (Gibbs & Neely, 1966) and use the computed downhole card effectively to diagnose problems with the beam pump (Eickmeier, 1967). This method of determining the dynamometer card (measure at the top of the well and compute the downhole conditions) is now industry standard and relies on accurately approximating the friction laws that the moving parts experience on their journey (DaCunha and Gibbs, 2007).

12.1.3 Problem statement

The question arises: Can this diagnosis be done by a computer? If it could, then the computer would automate the diagnosis of every dynamometer card in real-time as it is measured. Depending on the stroke duration and amount of idle time, a single beam pump may produce several cards per minute. An oilfield will have hundreds or thousands of pumps. It is practically impossible for a human operator to look at them all. This automation frees up the maintenance staff to focus on fixing the pumps that need attention instead of determining which pumps need attention. It therefore decreases the environmental impact of the pump (spills, pollution, spare parts, waste, and so on) while increasing its availability and, in turn, production volume (Bangert, Tan, Zhang, & Liu, 2010).

We will demonstrate in this chapter that it is possible to reliably perform the diagnosis of a beam pump using an automatic method from machine learning that teaches itself how to distinguish the various classes of damage.

12.2 Feature engineering

Every dynamometer card in our case, has 100 load and displacement measurements over the duration of the stroke. Thus, 200 variables characterize each card and would need to be fed into whatever classification method we choose to use.

One of the first questions in every machine learning project is to determine how the available input data (in our case the 100 loads and displacements) can be transformed in order to (1) lower the number of input variables without losing any important information, and (2) bring out any quantity that we know, by expert domain knowledge, to be helpful in deciding the question at hand. This process is known as feature engineering and the resulting transformed variables are called features. Sometimes, the data obtained in the field is provided to the machine learning as it is (Sharaf, Bangert, Fardan, & Alqassab, 2019), or some of the available input variables are simply removed. This is a quite simple version of feature engineering. It is at the point of feature engineering that we have the greatest opportunity to inject domain knowledge into the machine learning task.

A model obtained by machine learning often performs significantly better on the training data for the simple reason that this data is known to it while the model is being constructed. If the model performs significantly worse on testing data, then the model cannot generalize beyond its experience to new situations and we speak of overfitting. If the model performs poorly even on training data, we speak of underfitting. Both situations are undesirable, and we look for a model that performs well on both data sets so that we may be assured that the model has learned the task that we wanted it to learn. Such a model can then be applied to new data and its conclusions become useful (Bangert, 2011).

For the problem of characterizing a dynamometer card, several papers have been published in the literature that make suggestions for suitable features. We briefly review some of these papers here and state, wherever possible, how accurately the methods performed both on the data used to construct the method (training data) and the data used to assess the method (testing data). We summarize our literature review of features in Table 12.1.

12.2.1 Library-based methods

A prominent idea in the older literature is that of a library. A library is a set of cards with known classification. The idea is then to compute a distance function, called a metric, between the card that we are interested in and each member of the library. The library card with the smallest distance is selected and its classification becomes the classification of the new card. This approach relies on selecting representative examples of classes for the library. It is limited by

TABLE 12.1 A summary of the features suggested in the literature.

	Summary	Type	Cards to learn	Cards to test	Test error	References
1.	Metric is sum of differences in both dimensions	Library	—	—	—	Keating et al. (1991)
2.	Fourier series or gray level	Library	—	—	—	Dickinson and Jennings (1990)
3.	Geometric moments	Library	—	—	2.6%	Abello et al. (1993)
4.	Fourier series and geometric characteristics	Library	—	1500	13.4%	de Lima et al. (2012)
5.	Extremal points	Library	?	2166	5%	Schnitman et al. (2003)
6.	Average over segments	Model	2400	3701	2.2%	Bezerra Marco et al. (2009); Souza et al. (2009)
7.	Centroid and geometric characteristics	Model	230	100	11%	Gao et al. (2015)
8.	Fourier series	Model	102	?	5%	de Lima et al. (2009)
9.	Line angles	Segment	6132	?	24%	Reges Galdir et al. (2015)
10.	Statistical moments	Segment	88	40	2%	Li et al. (2013a)
11.	Geometric characteristics	Segment	—	—	—	Li et al. (2013a)

the fact that the library cannot be large if this procedure is to be carried out in real-time. The most difficult element is choosing the metric.

In one case, the library contains 37 cards and the metric is the sum of differences between the loads and displacements of the two cards being compared (Keating, Laine, & Jennings, 1991).

Another paper suggests measuring the difference between coefficients in the Fourier series representation of the card or to use a form of gray-level analysis on the image produced by a plot of the card (Dickinson and Jennings, 1990). Here the library contained 28 cards. While the authors claim that the methods performed well, we do not know the number of test cases.

Note that it makes sense to describe a dynamometer card by a Fourier series, as it is inherently periodic. If we take each measurement of displacement $x(s)$ and load $y(s)$ where s is a parametric variable that indexes the various measurements around one card. We can then form the complex variable $u(s) = x(s) + iy(s)$ and expand it in a Fourier series

$$u(s) = \sum_{n=-M}^{M} a_n exp\left(\frac{2\pi ins}{S}\right)$$

where S is the total card perimeter and M is the number of moments. The Fourier coefficients a_n are

$$a_n = \frac{1}{S}\int_0^S u(s)exp\left(\frac{-2\pi ins}{S}\right)ds$$

One may also describe the cards by geometric moments and threshold the differences to determine the most similar library member (Abello, Houang, & Russell, 1993). This approach yields a 2.6% test error on synthetically generated cards.

Geometric characteristics and Fourier series were evaluated in another study that found Fourier series to be more representative (de Lima, Guedes, Affonso, & Silva Diego, 2012). This study used 1500 cards and obtained an error of 13.4%.

An interesting approach is the attempt to extract from a card some few important points that act as local extrema and to compare them to similar points on the library cards (Schnitman, Albuquerque, Correa, Lepikson, & Bitencourt, 2003). This approach was evaluated using 2166 cards and seems to achieve an error of 5% but this is hard to quantify based on the description.

12.2.2 Model-based methods

Some features are very simplified. For instance (Bezerra Marco, Schnitman, Barreto Filho, & de Souza, 2009; Souza, Felippe, Bezerra Marco, Schnitman, & Barreto Filho, 2009) breaks the card into 32 segments and takes an average load over each segment. They had 8 different classes with 300 training examples each, 6101 manually classified cards in total and achieved a test error of 2.2%.

Another idea is to locate the centroid of the card and to compute the total area above and below the card in an extremal rectangle drawn around the card (Gao, Sun, & Liu, 2015). They then used extreme learning machines, a form of single-layer perceptron neural network, to train on 230 cards and test on 100 cards to obtain an error of 11% where we cannot be sure whether this is a total error or a test error. The centroid is

$$(\bar{x}, \bar{y}) = \left(\frac{1}{N}\sum_{i=1}^{N} x_i, \frac{1}{N}\sum_{i=1}^{N} y_i\right)$$

244 Machine Learning and Data Science in the Oil and Gas Industry

Decomposing the card into a Fourier series and using the coefficients of the first moments is another idea to lower the number of dimensions. Using this, one paper reports a 5% error on 102 known cards (de Lima, Guedes, & Silva, 2009). However, it is not clear what method they used to classify the cards and how many of the cards were used for training and testing.

12.2.3 Segment-based methods

Some proposals depend upon the ability to detect the four points on a card at which the standing and traveling valves open and close, see Fig. 12.2. There are two methods to determine those points. One relies on a pure geometric interpretation called the chain code (Reges Galdir, Schnitman, & Reis, 2014) and another on a physically motivated heuristic (Hua and Xunming, 2011). Supposing that we can determine these opening and closing points, we may then proceed to compute some quantities for each of the four segments of the card.

One paper suggests three geometric quantities based on the angle between the straight lines between any two neighboring pairs of observations (Reges Galdir, Schnitman, Reis, & Mota, 2015). They then build a fuzzy logic classifier for each one of 11 classes. Having 6132 manually classified cards in total, it appears from the unclear description that the error was approximately 24%. As the paper does not break its data into training and test, the test error would be higher than that.

Another approach requiring the opening and closing points is to compute the first seven statistical moments of the data in each segment (Li, Gao Xian, Yang Wei, Dai Ying, & Tian, 2013a). They use support vector machines to learn from 88 training cards and 40 test cards to obtain a test error of 2%. Another paper uses geometric characteristics of the four segments and claims effective performance using a simple look-up table of which class has or does not have which characteristic, but the authors provide no numerical results (Li, Gao Xian, Tian & Qiu, 2013b).

12.2.4 Other methods

A completely different approach is to try to extract knowledge and experience of human experts into the form of rules, usually known as an expert system. These rules would have to be formulated in precise numerical terms, so they can be evaluated on a card in an automated computer system. This has been done for this task and the management process is discussed of how one obtains these rules (Alegre, de Rocha, & Morooka, 1993). The authors do not report on the accuracy of the system.

Notably, one paper claims to have achieved perfect classification results on 6113 manually classified cards (Reges, Schnitman, & Reis, 2013). As they do not describe their features, their classification method, or in what manner, if any, they divided the data set into a training and testing data set, it is difficult to evaluate this result.

12.2.5   Selection of features

We have attempted each of the above referenced feature methodologies. Our conclusion is that the automatic detection of the four-valve opening and closing points is numerically unstable in general and particularly for non-normal dynamometer cards that are, of course, of particular interest to this application. Thus, we have decided to exclude all the features that require this basis.

In selecting features, we must make a trade-off between bias and variance (Dong and Liu, 2018; Zheng and Casari, 2018). A high bias means that the test error is high, and a high variance means that the variance of the test error is high across several runs of the algorithm with different training and testing data sets. A smaller number of features will generally lower the test error but increase the variance while a small number of features will remove essential information from the problem and thus increase the test error. We must therefore select the right number of features to characterize the situation with enough (bias) but not too much (variance) detail.

Having several candidate features available for selection, we would hope to rank them and take the first few features that contribute most to the task at hand. Unfortunately, this simple approach is generally doomed to failure. A central result in the field of feature engineering is that a feature that is insignificant on its own can provide significant performance increases together with other features, and two features that are both insignificant individually may be significant together. Therefore, we must not select features individually but rather in groups. This is problematic in the case, as it is here, that we have many features to choose from and the number of combinations of features is too large to try in a realistic amount of computation time. For each combination, we must train the model several times as we need to know not only the performance but the variance of any selected set of features. The normal way to solve this problem is with so-called wrapper methods that train a simple model for each combination to save overall computation time. The selected feature set is then used to train the complex model (Guyon and Elisseeff, 2003).

For example, the method of representing the dynamometer card by its Fourier series leads to many numerical experiments as the number of moments in the series must be chosen. To illustrate the trade-off, we plot the bias and variance in Fig. 12.4. The plot is a standard box-and-whisker-plot for various scenarios. The vertical axis represents the number of test errors obtained with a model trained on 30000 cards and tested on 5292 cards. Each time a model is trained, the training and test examples are chosen at random. All models that went into the figure exhibited zero training errors. The vertical position of the box on the plot, the average number of test errors over 5 models, is the bias. The vertical size of the box is a representation of the variance. The first 13 boxes correspond to representing the card by the Fourier series of that many moments, starting with zero moments. The

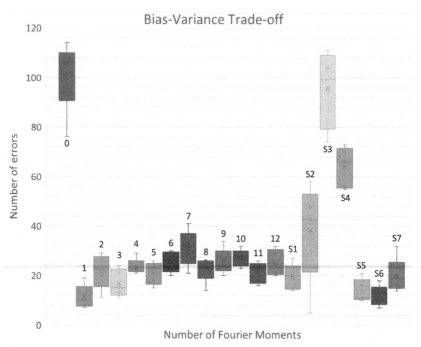

FIGURE 12.4   An illustration of the bias-variance trade-off in choosing the number of moments in a Fourier series representation of a dynamometer card. See text for an explanation.

seven boxes after that, labeled S1 through S7, correspond to various other combinations,

- S1: Fourier series with 1 moment and all 5 geometrical features of (Gao et al., 2015).
- S2: The 5 geometrical features of (Gao et al., 2015).
- S3: The centroid coordinates of the card.
- S4: The centroid coordinates and the average line length.
- S5: Fourier series with 1 moment, the centroid coordinates, and the average line length.
- S6: Fourier series with 1 moment and the centroid coordinates.
- S7: Fourier series with 1 moment, the centroid coordinates, and the two area integrals.

From these experiments, we can conclude that a Fourier series of 4 moments has the least variance but this is achieved at a significant cost in bias. The Fourier series of 1 moment has the least bias but this is achieved at a cost in variance. The Fourier series of 1 moment in combination with the centroid coordinates (option S6) offers the best compromise with 12 test errors on average with a standard deviation of 4. This implies a test error rate of 0.0023 ± 0.0008 or an accuracy rate of 99.8 ± 0.1%. The best single model encountered, overall and with the S6 feature combination, only exhibited 7 test errors (accuracy of $(5292 - 7)/5292$ x $100\% = 99.9\%$). It is this model that we have chosen to put into production.

In conclusion, our investigation of feature engineering yields that the best features are the Fourier series coefficients to the first moment and the coordinates of the centroid of the card. We therefore have five features. While this may be surprising due to its simplicity, it does lead to very good classification results. In comparison to the papers cited above, this study uses a far larger dataset both to train the model as well as to test its performance.

12.3   Project method to validate our model

12.3.1   Data collection

We must measure some cards in the field. This was done on 299 of the approximately 750 different beam pumps distributed in the Bahrain oilfield all of which are instrumented, and the data is relayed by a digital canopy to a central data facility where the cards can be obtained in real-time and saved to a database. In total 5,380,163 cards were collected for a first study during the fall of 2018.

12.3.2   Generation of training data

Some training data must be generated. The operating company of the Bahrain oilfield, Tatweer Petroleum, has four experts on its staff who are responsible for determining maintenance activities based on dynamometer cards. These experts were asked to look at the cards measured and to classify them into the classes mentioned above. Over a few weeks, they classified 35,292 dynamometer cards. These cards make up the dataset that can be used to both train the machine learning model and to determine its effectiveness.

12.3.3   Feature engineering

We computed the features for each dynamometer card; see the discussion of feature engineering earlier. We also divided the 35,292 known cards into two groups. 85% of the cards were used to train the model and 15% were used to assess the model after training. These two data sets are usually called the training data and the testing data.

12.3.4   Machine learning

We presented the training data set to several machine learning algorithms. Each algorithm has parameters that a human expert must tune experimentally to the task. We performed parameter tuning in each case to get the best performance out of each algorithm type. We attempted a single-layer perceptron neural network (Goodfellow, Bengio, & Courville, 2016), multiple-layer perceptron

TABLE 12.2 Classification performance of the best model on the test data set.

Category	Training samples	Testing samples	Incorrect
Normal	8557	1529	0
Fluid pound (Slight)	5347	955	0
Fluid pound (Severe)	93	15	0
Inoperative pump	1981	379	0
Pump hitting down	1740	303	2
Pump hitting up and down	2258	407	2
Inoperative pump, hitting down	9045	1626	1
Traveling valve or plunger leak	98	15	1
Standing valve, traveling valve leak, or gas interference	345	62	0
Pumped off	234	39	1
Hole in barrel or plunger pulling out of barrel	132	20	0
Gas interference (Severe)	101	11	0
Total	29931	5361	7

For each class, we specify how many training and testing samples were used. The model performed perfectly on all training samples but made a few errors on testing samples, as specified.

neural network (Goodfellow et al., 2016), extreme learning (Gao et al., 2015), and decision trees (Breiman, Friedman, Stone Charles, & Olshen, 1984).

The stochastic gradient-boosted decision tree is the most effective method for this task. It shows perfect performance on the training data set and 99.9% accuracy on the testing data set. The errors that it makes on the testing data set are roughly evenly distributed among the various classes, which is another desirable fact, see Table 12.2. Some other algorithms were competitive on the error rate but clustered its errors in one or other of the classes, which showed a systematic problem with detecting that class accurately.

12.3.5 Summary of methodology

The entire procedure of generating and using the classification model is summarized in Fig. 12.5. The workflow on the bottom of the figure describes the production situation in which a dynamometer card is measured, its features extracted, and its classification computed by the existing classification model. The model itself is a formula that takes a card's features as input and produces a

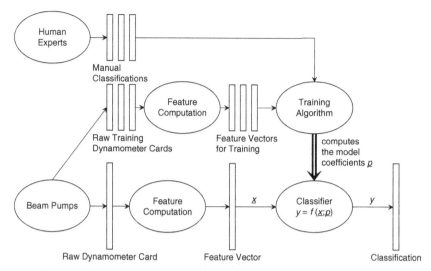

FIGURE 12.5 Elements of the diagnosis procedure.

class number as output using several parameters that have been learned. The workflow at the top of the figure produces the model, which is nothing more than computing the parameters of the formula that will be the model in the end. This learning procedure uses card features and the known classification, produced by human experts, of those cards. This schematic workflow is always the same for the general problem of classification, no matter what type of raw data, feature engineering, or model type we choose to use.

12.4   Results

12.4.1   Summary and review

Based on 35292 manually classified cards, we performed feature engineering to determine that the best set of features to balance between bias and variance of the model is to take the Fourier series representation of the card to the first moment and to take the centroid coordinates of the card. We found that computationally detecting the four-valve opening and closing points of the traveling and standing valves as proposed in the literature is numerically unstable and so the features depending on this basis were not investigated. Using 85% of the data for training and 15% for testing, we obtained a model that makes no training errors and 7 test errors. This performance can be reproduced over multiple training runs, where each training run chooses the training and testing samples randomly, with a variance of 4 test errors. The distribution of test errors across the different classes is relatively even, so that we seem not to be making a systematic error, see Table 12.2.

12.4.2 Conclusion

We conclude that the classification accuracy is high enough for the model to be used practically to identify the various beam pump problems in a real oilfield. It can do this in real-time for many beam pumps simultaneously and thus offer a high degree of automation in the detection of problems. If a card is identified as non-normal, the method will release an automated alert with its diagnosis and thus generate a maintenance measure. This procedure frees up human experts from the job of monitoring and diagnosing beam pumps to the more important task of fixing them. It also alerts the experts sooner than they would have discovered the problems by themselves as the algorithm can diagnose every dynamometer card as it is measured; a volume of analysis that would be impossible for a human team of realistic size.

We note that this work can classify cards into one of 12 classes. We have identified 31 classes in principle. The existing model will not work for the remaining 19 classes, as we did not have any training samples for them. When these problems occur on real beam pumps in the field, the relevant cards should be manually classified and then the training process can be repeated to expand the model's classification ability beyond its current domain. We estimate that about 100 samples per class are necessary to achieve reasonable results for that class. It is however clear that more data will always lead to either a more accurate or a more robust model. No matter how mature the model, feedback by manual classification is always of value.

References

Abello, J., Houang, A., & Russell, J. (1993). *A hierarchy of pattern recognition algorithms for the diagnosis of sucker rod pumped wells. Proceedings of ICCI'93: 5th International Conference on Computing and Information.* Canada: Sudbury, Ontario. 27–29 May 1993.

Alegre, L., de Rocha, A. F., & Morooka, C. K. (1993). *Intelligent approach of rod pumping problems.* SPE. 26253.

Bangert, P. (2011). *Optimization for industrial problems.* Springer.

Bangert, P., Tan, C., Zhang, J., & Liu, B. (2010). Mathematical model using machine learning boosts output offshore China. *World Oil, 11,* 37–40.

Bastian, M.J. (1989). A two equation method for calculating downhole dynamometer cards with a study of damping effects. MS Thesis. Texas A&M University.

Bezerra Marco, A. D., Schnitman, L., Barreto Filho, M. de A., & de Souza, J. A. M. F. (2009). *Pattern Recognition for Downhole Dynamometer Card in Oil Rod Pump System using Artificial Neural Networks. ICEIS 2009 - Proceedings of the 11th International Conference on Enterprise Information Systems.* Milan, Italy: Volume AIDSS. May 6–10, 2009.

Breiman, Leo, Friedman, Jerome, Stone Charles, J., & Olshen, R. A. (1984). *Classification and regression trees.* Chapman and Hall.

DaCunha, J. J., & Gibbs, S. G. (2007). *Modeling a finite-length sucker rod using the semi-infinite wave equation and a proof to Gibbs' conjecture.* SPE. 108762.

de Lima, F. S., Guedes, L. A. H., & Silva, D. R. (2009). *Application of Fourier descriptors and person correlation for fault detection in sucker rod pumping system.* Mallorca, Spain: IEEE Conference on Emerging Technologies & Factory Automation. 22–25 Sept. 2009.

de Lima, F. S., Guedes, L., Affonso, H., & Silva Diego, R. (2012). Comparison of border descriptors and pattern recognition techniques applied to detection and diagnosis of faults on sucker rod pumping systems. In S. G. Stanciu (Ed.), *Digital image processing.* IntechOpen.

Dickinson, R. R., & Jennings, J. W. (1990). Use of pattern-recognition techniques in analyzing downhole dynamometer cards. SPE 17313. *SPE Production Engineering, 5,* 187–192.

Dong, G., & Liu, H. (2018). *Feature engineering for machine learning and data analytics.* CRC Press.

Eickmeier, J. R. (1967). Diagnostic analysis of dynamometer cards. SPE 1643. *Journal of Petroleum Technology, 19,* 97–106.

Gao, Q., Sun, S., & Liu, J. (2015). *Fault diagnosis of suck rod pumping system via extreme learning machines.* Shenyang, China: Proceedings of the 5th IEEE International Conference on Cyber Technology in Automation, Control and Intelligent Systems, June 8–12 2015. 503-507.

Gibbs, S. G. (1963). *Predicting the behavior of sucker-rod pumping systems.* SPE 588. SPE Rocky Mountain Regional Meeting, Denver. May 27–28, 1963.

Gibbs, S. G., & Neely, A. B. (1966). Computer diagnosis of down-hole conditions in sucker rod pumping wells. SPE 1165. *Journal of Petroleum Technology,* 93–98.

Gilbert, W. E. (1936). An oil-well pump dynagraph. *Drilling and Production Practice, 4,* 84–115.

Goodfellow, I., Bengio, Y., & Courville, A. (2016). *Deep learning.* MIT Press.

Guyon, Isabelle, & Elisseeff, André (2003). An introduction to variable and feature selection. *Journal of Machine Learning Research, 3,* 1157–1182.

Hua, Liang, & Xunming, Li (2011). Accurate extraction of valve opening and closing points based on the physical meaning of surface dynamometer card. *Petroleum Exploration and Development, 38,* 109–115.

Keating, J. F., Laine, R. E., & Jennings, J. W. (1991). *Application of a pattern-matching expert system to sucker-rod, dynamometer-card pattern recognition.* SPE. 21666.

Li, K., Gao Xian, W., Yang Wei, B., Dai Ying, L., & Tian, Z. D. (2013a). Multiple fault diagnosis of down-hole conditions of sucker rod pumping wells based on freman chain code and DCA. *Petroleum Science, 10,* 347–360.

Li, K., Gao Xian, W., Tian, Z. D., & Qiu, Z. (2013b). Using the curve moment and the PSO-SVM method to diagnose downhole conditions of a sucker rod pumping unit. *Petroleum Science, 10,* 73–80.

Reges, G. D., Schnitman, L., & Reis, R. A. (2013). *Application of curvature-based descriptors for fault diagnosis in sucker rod oil pumping systems.* Brazil: 22nd International Conference on Production Research.

Reges Galdir, D., Schnitman, Leizer, & Reis, Ricardo (2014). Identification of valve opening and closing points in downhole dynamometer cards from sucker rod pumping systems based on polygonal approximation and chain code. *Rio Oil and Gas Expo and Conference.* Sept. 15–18, 2014.

Reges Galdir, D., Schnitman, L., Reis, R., & Mota, F. (2015). *A new approach to diagnosis of sucker rod pump systems by analyzing segments of downhole dynamometer cards.* SPE. 173964.

Schnitman, L., Albuquerque, G. S., Correa, J. F., Lepikson, H., & Bitencourt, A. C. P. (2003). *Modeling and implementation of a system for sucker rod downhole dynamometer card pattern recognition.* SPE. 84140.

Sharaf, S. A., Bangert, P., Fardan, M., & Alqassab, K. (2019). Beam pump dynamometer card classification using machine learning. *SPE 194949. Middle East Oil & Gas Show and Conference (MEOS), Bahrain.* March 18–21, 2019.

Souza, de, Felippe, J. A. M., Bezerra Marco, A. D., Schnitman, L., & Barreto Filho, M. de A. (2009). *Artificial neural networks for pattern recognition in oil rod pump system anomalies.* Las Vegas Nevada, USA: Proceedings of the 2009 International Conference on Artificial Intelligence. ICAI 2009, July 13–16, 2009,.

Takacs, G. (2015). *Sucker-rod pumping handbook.* Gulf Professional Publishing.

Zheng, A., & Casari, A. (2018). *Feature engineering for machine learning.* O'Reilly Media.

Chapter 13

Forecasting Slugging in Gas Lift Wells

Peter Kronberger
Wintershall Dea Norge AS, Stavanger, Norway

Digital transformation of existing processes using technologies such as big data, advanced data analytics, machine learning, automation, and cloud computing will enable continuous performance improvements within the operational sphere. The application of the technology will link the physical and digital world, providing a digital model of physical assets and processes. It will represent the evergreen, wholly integrated digital asset model-from reservoir to export pipeline.

The Brage operations team identified processes with the highest potential for digital transformations during an initial opportunity-framing workshop. Based on pain points and business needs clear emphasis is in the areas of production and well performance optimization, live-data implementation, and handling, as well as the entire flow of information from database integration up to dashboarding.

Strong reliance on an improved data infrastructure and IT/OT performance will require close cooperation with the IT/OT team. New solutions will be aligned with and integrated into already identified global solutions. The improved acquisition of data, together with the integration of already existing "Digital Twin" technologies will enable the first redesign of work processes. Prioritization of processes to be transformed will be mandated by their business impact as well as possibility to scale to other Wintershall Dea assets too.

Focus areas, such as slugging, digital production engineer, intelligent maintenance, dashboarding, and planning and scheduling were defined as more and more ideas evolved. Individual processes such as water injection and scaling surveillance were made more time-efficient and transparent. The vision is to bring all processes together into a customizable and collaborative dashboarding solution.

Machine Learning and Data Science in the Oil and Gas Industry.
http://dx.doi.org/10.1016/B978-0-12-820714-7.00013-3

13.1 Introduction

The digital transformation is revolutionizing the oil and gas industry, together with the fast amount of data collected throughout the lifetime of the platform, it is now possible for production technology and operators to utilize this to lower operational expenditure and improve production efficiency.

The Brage field consists of four reservoirs, Fensfjord, Statfjord, Sognefjord, and Brent where in each of the mentioned reservoirs gas lifting is applied. Water breakthrough already happened a couple of years ago. Last year the Brage field celebrated its 25th anniversary, which means production optimization is key. Many parameters such as maximum water and gas lift capacity, slugging, sand production, and scaling issues show that it is not a straightforward process to do. That challenges are not something new and to better understand, solve and mitigate them digitalization is key.

The motivation to tackle the digitalization journey with Brage in a brown field environment is fueled by production increase due to even more efficient production operations, cost savings as possible predictive maintenance and automated processes can proactively mark possible failures and routine tasks can be done by the system. This, eventually, results in a more efficient way of working and ultimately in a lifetime extension of the platform.

Digitalization in brown versus green field environments, are two different challenges due to various reasons. In green field developments, one can design the system and data environment from the very beginning. Digitalization is a topic from the beginning and sensors, systems, and processes can be designed and set up accordingly.

In brown field environments, however, the challenge is a different one as the underlying systems are usually old and as the platform at its end of planned life. Often data systems need to first get exchanged or upgraded to have the basis for starting to implement digitalization processes. Another important aspect is the data availability and quality. In brown field environments a lot of data was collected of the lifetime of the platform, however, that often cannot be used, as the quality is too low. Moreover, data sensors might be missing, as they were not installed at the start of the platform.

What are the key benefits it will give to Brage? Is the underlying question for starting the digitalization journey?

1. To connect humans, devices, and systems across one entire value chain to have all information available in real-time whenever needed.
2. Turning data fast into better-informed decision do work more efficient and optimized to lower the operation costs, increase production, and increase the field lifetime.
3. Fostering partner and stakeholder management.
4. Improving HSEQ performance by faster and better diagnosing deficiencies and causalities to correct and reveres negative trends for more safely work.

13.2 Methodology

Starting with an internal ideation phase to map the pain points and possible work processes to improve in the daily working environment. In the next stage external providers assisted in structuring these findings and mapping the key areas by, for example, design thinking methodologies.

These key areas were then steadily refined and structured and business cases were created, where eventually five focus projects were defined based on their areas of impact and economic values.

This process, from the first get together and ideation phase until the definition of the business cases and mapping of the focus projects took several months, whereas some high value projects did already start earlier.

The approach was therefore driven by analysis of pain points and opportunities, rather than digitalizing the entire platform. Individual processes and their business cases were defined and selected. How would digital technologies change the way we work and create efficiencies or cost savings? Answering this question helped developing the current Brage Digital twin project portfolio, of which all address three steps of data process:

Data Liberation: Generally said, putting operational data into a data lake infrastructure, which is for a 25-year-old platform a complex, as hardware components need to be updated, yet very important step. The ultimate vision of that step is to have one single source of data, available for anyone at any time when needed.

Data Analytics: Once the data is available, the next step is to optimize and improve the processes and workflows by getting the most out of the data. Collaborations with different specialized vendors for applying advanced analytics that help us to operate more safely and efficiently were established.

Data Visualization: Insightful visualization of data is then the key for communicating these insights to the individual engineers and operators. Live dashboards will support in decision-making processes and notifications of unfavorable conditions. The focus is on customizability and collaboration. Often that is already the solution.

Digitalization and digitization are two different, yet sometimes exchanged expressions. Digitization is the process or task to make physical items, for example, a written letter, digital to make it for instance usable for processing on a PC. Digitalization is then the process to use digital technologies or digitized tasks to improve processes.

This is important to understand as the Brage Digital Twin project is a digitalization project, which therefore utilizes existing and proven technologies such as machine learning to improve and optimize the daily operational work to improve efficiency, increase production, and decrease costs.

After the ideation phase and structuring the ideas different topics were identified and explained in more detail, whereas a first estimation of business cases was estimated. The business cases were based production increase, more efficient work and therefore saved hours which could be used for other more important tasks rather than routines tasks and cost reduction. The first estimation was very rough and simply done to get an understanding of the magnitude comparability of the different topics. Over time those estimations got more and more matured and therefore refined in more detail.

The initiatives were ranked based on their estimated business cases and probability of success; however, it was and is still a living process and changes on the priority might be made based on new information and adjustments in business cases or resent findings in increasing the probability of success.

The project is mainly be executed in an agile project management manner. The scrum framework is applied as much as possible, which is explained in detail in Chapter 7 of this book.

13.3 Focus projects

The five focus projects selected are defined as the following:

1. Dashboarding Landscape: The key aspect within this idea lays on transparency and collaboration. To unbreak the chain of data silos and spreadsheet-based workflows.
2. Digital Production Engineer (DPE): The DPE consist of several sub-projects as Production Optimization, Injection Optimization, Scale Treatment Optimization, and Sand Choke Optimization.
3. Slugging: Due to its early kick off, importance, and high
4. Intelligent Maintenance
5. Planning/Scheduling

These projects and sub projects were assessed to have the highest impact with the highest probability of success. However, there is also a side project going on with regards to the already installed Narrowband technology on Brage. The goal of this project is to have a cheap and fast proof of concept for IoT technology for a more efficient cooler monitoring on the platform. Benefits of IoT sensors are amongst others its cheap prices and installation, long battery life, and easy maintenance.

13.3.1 Dashboarding landscape/architecture

Data visualization is a major aspect in spotting erroneous behavior and possibly prevents failures or costly interventions. Often data is not easily accessible, and it takes time to access and extract data for analysis. Transparency and accessibility of data is therefore an important topic.

A specialized vendor called Eigen was selected for a dashboarding solution pilot (Fig. 13.1). For testing purposes, an Eigen Ingenuity framework was installed in the Azure cloud. This included a VPN link to the historian AspenTech IP.21, allowing process and production data to be displayed a manipulated in various ways.

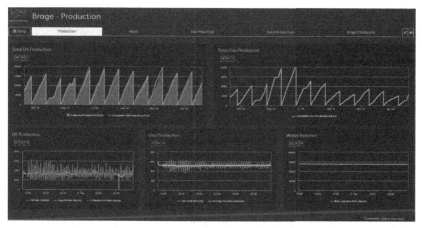

FIGURE 13.1   First implementation of production dashboard in pilot phase.

This stage took a lot of time as getting access to production data is a complex task due to various restrictions and security barriers. A clear reference on how to establish such a connection was not available as it was not done before.

After successfully setting up the live connection, the first tasks were defined for a 3 months testing period, this is important to test the functionality and integration of the solution in the daily operations. These task are (1) creating a new dashboard overview for more transparency and faster spotting unwanted behavior, (2) creating an overview and trends of all wells and parameters including analysis and notifications for engineers being on duty for an efficient follow up, (3) automating and improving the process and its underlaying tasks for creating the weekly production plan, and (4) automating the well rate allocation process to have live rates implemented in the Eigen dashboard environment.

The intent of the pilot project is to test the analytics, trending and dashboarding capability of the Eigen Ingenuity platform for your operations and optimize or re-define some of our work processes. The work focuses on the replacements of systems used in the technical engineering management and monitoring of the Brage field, more specifically the field's production and injection wells.

13.3.2 Slugging

The life of an oil production well can often be very challenging and differs from well to well. Some wells are drilled and produce high oil rates from the beginning without any further operational issues; however, other wells start to have challenges from early on. These production challenges, such as slugging, stress the facilities to a point that production engineers are forced to operate the well intermitted or even shut in the well permanently, which will result in production losses in every case. Besides production losses slugs can damage the processing equipment and accelerate erosion effects, which result in higher failure rates and maintenance costs. This all together leads to significant economic losses.

Slugging can be induced by different underlying causes. The main types of slugs are severe, or riser induced slugs, which are caused by offshore pipeline rises, where liquid can accumulate at the bottom of the riser and block the gas flow. However, pressure will build up upstream the liquid accumulation and at some point, will push the liquid up. Another type is the pigging induced slug. A pig is a pipeline-cleaning device, which is send through the pipeline to clean possible precipitations. This device then creates a liquid slug. A third type is the hydrodynamic slugging, which is often caused in pipelines when liquid and gas flow at different speeds. The difference in speed can, at a certain point, cause flow instabilities-in a horizontal well-small waves in the liquid phase, which ultimately leads to a slug. Hydrodynamic slugging mostly, however, dependent on the phase rates and the topography, has a high frequency.

The fourth type, and for our use case the most interesting one, is terrain slugging. As the name says terrain slugging is induced by the topography of a pipeline or horizontal well. Often occurring in the late life of wells with long horizontal sections when the reservoir pressure has declined. Topographical differences can occur because of geo-steering, which might result in a water lag where liquids are likely to accumulate at a certain point in time. The force upstream coming from the produced gas is not high enough to push the liquid phase through the liquid lag further up the wellbore and fluids start accumulating. The gas phase builds up pressure and at some point it will be high enough to push through the water leg and pull the liquid as a slug along up the well until the hydrostatic pressure of the liquid is bigger than the gas pressure again, when the next volume of liquid starts accumulating in the leg.

Mitigating these kinds of slugs is particularly challenging as the trajectory of the well is given and sometimes, even if known it will cause challenges, cannot be changed due to geological and drilling reasons. Therefore, it is of up most importance for a production engineer to be prepared and equipped with mitigation tools.

Over the years a lot of data was collected from each individual well such as wellhead pressures and temperatures. Often it was not utilized for its full potential to optimize the production of each well. This big amount of data per well combined with data science knowledge can be used to predict and mitigate production issues. This is particularly interesting as some wells on the Brage field are struggling with slugging. Major production losses are due to slugging therefore wells often cannot even be put on production.

A slug flow is characteristic for oil wells during their tail end production, where a volume of liquid followed by a larger volume of gas causes intermitted liquid flow and therefore instabilities in the production behavior. These instabilities are undesirable as the consequences in the facilities can be severe: from insufficient separation of the fluids, increased flaring to tripping the whole field in the worst case. The intention of the slugging project has been to develop an algorithm, which can predict the state of the well some time in advance, identify a possible upcoming slug to have the possibility to intervene timely.

The first question to ask is how to model the slug flow physically. If there would be a physical model, which is tuned to the well it might be possible to optimize the flow. That approach exists with industry proven simulators. However, these purely physical models need constant tuning and adjustment. Moreover, slug frequencies are often very short, but a calculation cycle is performance intensive and takes time. Although the model might be well maintained, and the performance is good, the predictability is often mediocre.

Seeing the value that can be generated by mitigating slugging and guaranteeing stable flow makes the problem especially important to solve. Understanding the restrictions and downsides of a purely physical model approach, such as its tuning and maintenance time, the value of trying a data-driven approach became attractive.

A data driven approach must provide the key behaviors of needing low to no maintenance with a low uncertainty in predicting upcoming slugs and an even more important than a long and uncertain prediction horizon is an accurate optimization.

Besides the above-mentioned points physical restrictions also play a role, such as the reaction time and the acting speed of the choke. If a well would suffer from high frequency slugging, but the reaction time and closing speed of the choke is too slow, the best algorithm will not be able to help mitigating the slugging.

Concluding that it is important understand the technical feasibility of the slugging problem throughout the whole solution process, starting from the data. What data is available (which measurements) in which frequency and quality? Understanding the kind and characteristics of the slugging and the key parameters that is causing it. Investigate and understand the equipment restrictions, such as what is needed if an optimization technique is in place. Is the choke able to be automatically changed? Is it necessary to make modifications topside to get the solution running?

These questions need to be answered beforehand, as they are important milestones in the maturation process to not get unpleasant surprises in the end.

After understanding these unknowns, some key milestones and phases for the project where outlined:

1. Technical feasibility: Can a data-driven algorithm predict an upcoming slug at least a required amount of time so that the choke can be adjusted?
2. Is the algorithm reliable and trustworthy? Put the algorithm on a screen so that it creates visibility and trust. Visually show that the predictions are good. Try to manually adjust the choke as the optimization suggests and observe the reaction.
3. Operationalization: Starting to let the optimization algorithm control the choke first for a restricted amount of time and after a while more and more extent the time until the trained model is fully controlling the choking in order to mitigate the slugging.

A vendor with experience and expertise in data science and the oil and gas industry named algorithmica technologies was selected for a pilot project. An important aspect already early in the phase was to get a proof of concept, whether it is possible to identify a slug, at least within the choke reaction time, to possibly interact proactively. Therefore, the project was divided into 4 main milestones. The first to answer the question of proof of concept to model the well state, the second to verify it is possible to indicate the slug proactively. The third phase is to manually test different choke settings. This step is important and necessary, as the algorithm needs to be trained on how to optimize the well to mitigate slugging automatically in the end. The fourth step, after verifying that the optimization would theoretically work, is a modification project to upgrade the choke actuators to have the possibility for a faster and automatic reaction.

To start with the proof of concept in time, a selected dataset was exported from the historian for one slugging well. However, in that process it was realized that it is of upmost importance to start as early as possible finding a solution for an IT environment so that a possible solution can run real-time. Questions such as will the trained model run as an application in the environment of the company and how will that look like, or will the real-time data stream transferred to algorithmica technologies and the calculations will take place there? The decision was made that these questions will be answered as soon as the proof of concept is successful.

Two major pathways for modeling the well state were initiated:

1. Explicitly feeding a limited history into an unchanging model, for example, multi-layer perceptron; however, it is (1) difficult to determine the length of this limited history and other parameters and (2) this model results in lower performance $\underline{x}_{t+\tau} = \mathcal{L}\left(\underline{x}_t, \underline{x}_{t-1}, \cdots, \underline{x}_{t-h}\right)$
2. Feeding the current state to a model with internal memory that retains a processed version of the entire past, for example, LSTM (Long Short-Term Memory). Characteristics of this approach are (1) easier to parametrize and (2) better performance; however, the training time is longer $\underline{x}_{t+\tau} = \mathcal{L}\left(\underline{x}_t\right)$

An LSTM ingests and forecasts not just one variable but multiple ones. For our purposes, all the quantities measured and indicated in Fig. 13.2 were provided, including the production choke setting. The model therefore forecasts the slugging behavior relative to the other variables. As the production choke setting is a variable under our influence, we may ask whether we can mitigate the upcoming slug by changing the choke. The model can answer this question.

An optimization algorithm is required to compute what change to the choke will accomplish in the best reaction of slugging and this is a guided trial and error calculation, which is why it is necessary to have a model that can be computed quickly in real-time. The method of simulated annealing was used because this is a general-purpose method that works very quickly, converges to the global optimum, requires little working memory, and easily incorporates even complex boundary conditions.

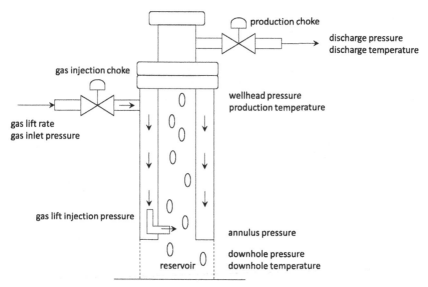

FIGURE 13.2   Schematic of a gas lift well.

An LSTM network is composed of several nested LSTM cells, see Fig. 13.3. Each cell has two different kinds of parameters. Some parameters are model coefficients that are tuned using the reference dataset and remain constant thereafter. The other parameters are internal states that change over time as each new data point is presented for evaluation. These state parameters are the "memory" of the network. They capture and store the information contained in

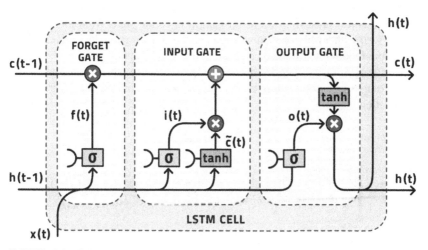

FIGURE 13.3   Schematic of a LSTM network. x, input signal; h, hidden signal; A, cell with steps in each cell to determine the state, forget and update functionalities. (*Olah, 2015*).

the observation made now, which will be useful in outputting the result at some later time. This is how the LSTM can forecast events in the future (Hochreiter and Schmidhuber, 1997).

It was discovered that a forecast horizon of 300 minutes was achievable given the empirical data. This forecast has a Pearson correlation coefficient R^2 of 0.95 with the empirical observation and so provides a forecast of acceptable accuracy. The next challenge was to define a slugging indicator so that the system can recognize a certain behavior as a slug.

Defining an indication of what a slug is for the system was a challenge on its own. The peaks of a slug as can be seen in Fig. 13.4 would be too late for a reaction for the choke. Therefore, as can be seen the indicator (orange | vertical lines) is going to 1 (representing a slug) before the peak and going back to zero (no slug).

These findings-being able to forecast the wellhead pressure changes of a well up to 300 minutes, which gives enough time for the choke to react and being able to isolate, meaning detecting slugs automatically.

Although these results seem very promising and positively surprising as up to 300 minutes is definitely enough to interact in some matter to mitigate a slug, the point is definitely "in some matter." Forecasting the behavior of the well is one concept that needed to be proven, however, forecasting alone is fascinating to see, but does not give any added value, as the how to intervene is important. Should the wellhead choke be changed, if yes, how much? Or should the gaslift choke be changed? If yes, to which degree?

For the model to know how to mitigate the slugging optimally it needs data from when the variable parameters are changed-in this case the wellhead choke and possibly the gaslift choke. One might think that in the life of a well are certain times with different choke positions, however, as many wells are infill wells that is often is not the case. The wells are often long horizontal wells with a low reservoir pressure, as a result gas lift injection starts from the early life of the well with a fully opened wellhead choke. Training the data-driven model therefore required to produce variation in the data by choking the well. For this task, a program was designed and carried out offshore which captured all relevant ranges of combinations in an efficient way, as choking results in certain production losses.

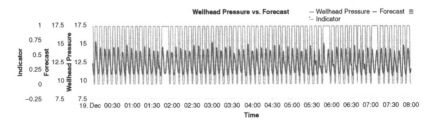

FIGURE 13.4 Comparison of observed WHP data with modelled data, almost on top of each other, including a defined slugging indicator *(vertical lines)*.

After finishing the program, the data was extracted again and sent to algorithmica technologies for training the model to that newly gained data, which took several days. The results looked good and the data-driven model gave suggestions to optimize the well.

In the meantime, training the model to the newly gained choking data, the well became less and less economical and the decision was made to shut it in. So, on the one hand there was a trained and basically ready to implement data-driven model to mitigate slugging, but on the other hand it was decided to shut in the well as the production gain was far too low.

One of the big learnings was that timing is an important factor. It takes time to train a data-driven model and it takes even more time to fully implement such a solution, therefore the selected might not be within economical margins anymore.

It was decided to train the model again on anther well, as after the proof of concept, the confidence was in place that it would work on any other well too. The chosen well did not fully slug yet, but already showed some indications. Meaning that during normal production the well shows stable flow, however, if the well was shut in and then started up again it slugged for some time. This behavior made it the perfect candidate.

A direct connection between the data historian and algorithmica technologies was also set up in the meantime, which helped a lot to start training the new well optimally. The scope for this attempt was slightly adjusted. In addition to be able to mitigate slugging, it should be first used as a surveillance tool so that the model can give an alarm when the well might start slugging more heavily in before though stable conditions, as the slugging itself is not a great problem yet. It needs to be mentioned that experience shows that the selected wells are due to the formation they are producing from and shape of well trajectory very likely to start slugging at some point after startup as can be seen in Fig. 13.5.

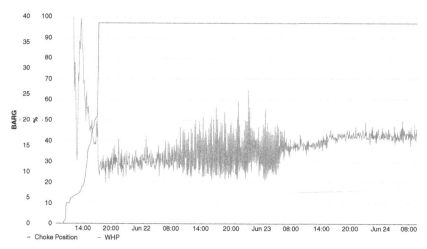

FIGURE 13.5 Second attempt to surveil a slugging prone well and get a notification if the slugging will become more severe.

After the gained confidence from of proofing the concept from the originally selected well, it was started to train the model to the new well with the adjusted scope. Interestingly, it seemed to take longer than before to train the model to that data, which might already give an indication of a higher complexity. That assumption was then verified after the model was trained. The forecasting horizon showed to be shorter than the one from the first well trained with 300 minutes. However, that turned out not to be an obstacle as it was still enough for the tasks of surveillance and possible mitigation later.

By now there are two well behaviors trained with a data driven model, and one also for optimization purposes including an offshore test. It can be said that forecasting the state of the well is possible with only having the historian available. The next step is to put the calculations in real-time, so that engineers can judge the forecasts compared to the actual behavior the well will then show. Assuming that this is successful it will trigger a new project to update the choke actuators on the wells suffering from slugging so that the model can automatically make changes. This last step is exciting, as a data-driven model will-on a long-term perspective-monitor and optimize a well autonomously. It might then reduce then on the one hand the amount of time an engineer needs to spend on such a problem well and on the other hand also increase the value of that well. Besides that, it might also reduce maintenance costs as the flow in the flowlines and into the separator is more stable and therefore induces less stress on the equipment.

The mitigation of slugging by automatically choking the well is not a trivial task and requires a throughout planning of the phases and milestones until a model will automatically optimize a well on stable flow. However, the proof of concept for two wells has shown that it is possible to forecast the well state and to get recommendations on optimal choking. Besides that, it needs to be mentioned that it was not yet tried in practice, which is still one major step to do. This might give also further insights as how stable the forecast is on changing the choke and therefore answering the question how it works in operations. Furthermore, field testing will also show on how precise the choking must be to get the wanted positive effect. What would happen for example if the choke reacts with a 5 minutes delay and slightly differently compared to the suggestion of the model? Will the output be the same or will the slugging even get worse? Generally, how sensitive or stable will be model and the suggestions be?

13.4 Data structure

A solid underlying data structure is key for a successful digital transformation. It is a complex process as different kind of information are saved in different databases and structures and security features on top make it a great challenge to build a solid data lake architecture.

As of this importance Wintershall Dea started its own company-wide project called data integration project with the goal of building a data lake infrastructure.

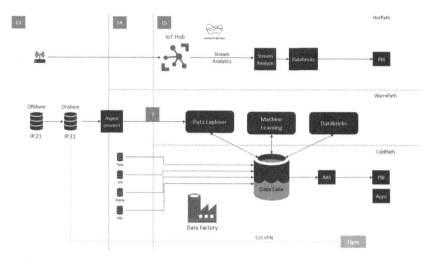

FIGURE 13.6 Outline of a possible data lake solution for Brage.

As a result of the project there should be "one single source of data truth" to connect.

A possible outline of such an infrastructure was created for Brage (Fig. 13.6) where the Data Lake would be the center of information connected via the data explorer with the historian AspenTech IP.21 databases. Since January 2019, a narrowband infrastructure is available at Brage to install IoT sensors, that technology can be integrated by streaming data from an IoT hub into the system.

All data sources, currently saved in different databases such as SAP, Tieto or SQL can be pushed to the data lake where it can be accessed via different applications through one connection. Applications then can be selected based on their need and functionality as a service (AAS-application as a service), rather than its connectivity to different data sources.

As by today this infrastructure does not exist yet, as an example the pilot for the dashboarding solution is for the proof of concept phase currently directly connected to the historian database.

13.5 Outlook

Digitalization is a key aspect for the future to optimize processes and reduce costs by utilizing the available data combined with the right know how. In the short and mid-term, the importance lay with manifesting the solutions within the Brage operations to daily work will transform in being more efficient and effective. In a long-term perspective, it should be able to upscale solutions based on their needs to all business units in the Wintershall Dea environment.

13.6 Conclusion

This chapter showed the journey from the ideation phase until the first implementations of a digital transformation of a brownfield platform. It explained why it makes sense to digitalize a brownfield platform. The origin of many challenges was often culturally rather than technologically.

A consequential finding already in the ideation phase was that it is important that the external facilitator has great experience in the oil and gas industry with a deep understanding of the uniqueness of the business otherwise it was experience that the progress in identifying and developing ideas is slow and the engagement of people is less strong.

Creating an understanding of scrum and its benefits in product development and creating an understanding of ownership for the initiatives. A challenge was and still is in some cases to have product ownership.

Data availability, connectivity, liberation, and a clear data foundation is still a vision rather than reality. This, however, is a particularly important topic as it gives the foundation for all following analysis and data science work. Currently data connects are established on request and often takes weeks or months. Data availability, not only process data, but also getting data out of software tools and in the right format is complex and often a bottleneck in executing digitalization initiatives.

Beside the described focus projects and the deeper dive into two of them, so called "low hanging fruits," easy and quick implementations, often with a great time saving benefit where executed as side initiatives, such as the water injection surveillance. Previously the surveillance was performed with various spreadsheets; a quick MVP was designed and then implemented in the Eigen Ingenuity platform, which provides now a compact dashboard overview with automatic alarms if threshold values were exceeded. This implementation did not take more than three days of active work.

Further reading

Olah, C. (2015). Understanding LSTM networks. colah.github.io/posts/2015-08-Understanding-LSTMs/.

Hochreiter, S., & Schmidhuber, J. (1997). Long short-term memory. *Neural Computation, 9*, 1735–1780.

Index

Note: Page numbers followed by "f" indicate figures, "t" indicate tables.

Printed in the United States
By Bookmasters

TOPICS IN RECREATIONAL
MATHEMATICS

TOPICS IN RECREATIONAL
MATHEMATICS

BY J.H.CADWELL

CAMBRIDGE UNIVERSITY PRESS

CAMBRIDGE

LONDON   NEW YORK   NEW ROCHELLE

MELBOURNE   SYDNEY

CAMBRIDGE UNIVERSITY PRESS
Cambridge, New York, Melbourne, Madrid, Cape Town, Singapore, São Paulo, Delhi

Cambridge University Press
The Edinburgh Building, Cambridge CB2 8RU, UK

Published in the United States of America by Cambridge University Press, New York

www.cambridge.org
Information on this title: www.cambridge.org/9780521044097

First published 1966
Reprinted 1970, 1977, 1980
This digitally printed version 2008

A catalogue record for this publication is available from the British Library

Library of Congress Catalogue Card Number: 67–10013

ISBN 978-0-521-04409-7 hardback
ISBN 978-0-521-09620-1 paperback

Contents

Preface

The topics discussed in this book have formed the subject of talks given by the author to mathematical and scientific audiences. The aim was to show something of the fascination and beauty of mathematics, as well as its enormous extent.

These talks convinced the author that most scientists and many mathematical specialists are unfamiliar with the problems and ideas discussed here. This fact, coupled with the belief that such material can provide great enjoyment, led him to write this book.

The mathematical background assumed does not often go beyond that of G.C.E. at Advanced Level, but it is rather more extensive than that required for many of the excellent accounts of recreational mathematics available. Like most of these accounts this book requires the persistence needed to follow a chain of reasoning of moderate length and complexity. It can be argued that the exercise of this facility is the main source of the pleasure to be found in this sort of reading. However, it is hoped that, where a formal proof has to be skipped, the reader will still be able to appreciate and enjoy the result being discussed. The various chapters are virtually independent and can be read in any order.

I would like to record my indebtedness to my colleagues Dr S. H. Hollingdale, Mr J. B. J. Thorpe and Mr D. E. Williams for their encouragement and interest in this project. In addition Dr Hollingdale has read the text and made many valuable suggestions for its improvement. I would also like to thank the Staff of the Cambridge University Press for their care and skill in planning and producing this book.

J.H.C.

Mathematics Department,
Royal Aircraft Establishment,
Farnborough

Note

Superior figures in the text refer to the lists of references at the end of each chapter.

1

Regular polyhedra

1. The Platonic solids

A polyhedron is a solid bounded by plane faces, the three-dimensional analogue of a plane polygon. Such a plane figure is said to be regular if all sides and angles are equal; there is evidently an infinite number of regular polygons. The best-known regular polyhedron is probably the cube, distinguished by its 6 regular polygonal faces, each of 4 sides. In addition, at each vertex 3 edges meet symmetrically. A section of the cube near a vertex, by a plane perpendicular to the diagonal through that vertex, is an equilateral triangle. We therefore describe the cube by the number pair $(4, 3)$, a valuable symbolism due to Schläfli. The first number indicates that each face contains 4 edges, the second that 3 edges meet at each vertex.

 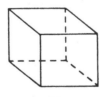 

Fig. 1

In Figure 1 we have 2 other regular polyhedra. The first is a regular tetrahedron with 4 equilateral triangular faces, occurring 3 at a vertex; its Schläfli symbol is $(3, 3)$. The third of the group is the regular octahedron, called $(3, 4)$, since there are 4 faces, each of 3 sides, at any vertex.

In Figure 2 the remaining Platonic solids are depicted. The regular dodecahedron $(5, 3)$ has 12 pentagonal faces; while the regular icosahedron, with 20 triangular faces, is $(3, 5)$. Table 1 summarises this information.

In this table, and throughout the rest of this chapter, the word regular is understood to apply to any polyhedron mentioned, unless the contrary is indicated.

1

The first three occur in nature as crystals, and while a dodecahedral crystal with irregular pentagonal faces exists, crystallographic theory shows that neither of the last two can form a crystal and retain its regularity. All have long been known, and dodecahedral charms date from Etruscan times. A pair of icosahedral dice may be seen in one of the Egyptian rooms of the British Museum. The Greeks were much interested in these solids, hence the name Platonic. Euclid in his *Elements* discusses their properties.

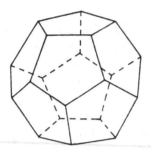 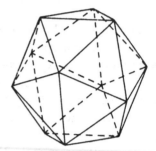

Fig. 2

Table 1. *The Platonic solids*

Name	No. of faces	No. of vertices	No. of edges	Schläfli symbol
Tetrahedron	4	4	6	(3, 3)
Hexahedron (cube)	6	8	12	(4, 3)
Octahedron	8	6	12	(3, 4)
Dodecahedron	12	20	30	(5, 3)
Icosahedron	20	12	30	(3, 5)

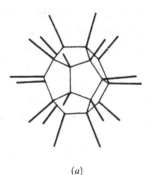 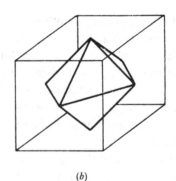

(a) (b)

Fig. 3

Surprisingly, they occur in the skeletons of tiny sea creatures called radiolarians. These are made of silica, and measure only a fraction of a millimeter in diameter. Figure 3(a) illustrates the dodecahedral version, a regular framework carrying 20 symmetrically disposed spikes. Further illustrations are given by Weyl.[6]

Figure 3(b) illustrates the fact that the centres of the 6 faces of a cube are vertices of an octahedron. A similar construction applied to the octahedron yields the cube, and these solids are said to be dual to each other. The dodecahedron and icosahedron also form a dual pair, while the tetrahedron is self-dual. The Schläfli symbol (p, q) indicating p-gonal faces, arranged q at a vertex, is dual to the symbol (q, p).

2. Proofs of completeness

The need for proof that the above list is complete was appreciated by the mathematically mature Greek geometers, and they found such a proof. We give an alternative demonstration, based on Euler's relation for convex polyhedra. Such a solid has F faces, V vertices and E edges; these numbers are linked by the equation

$$F + V = E + 2.$$

A proof will be found in Chapter 8. Each edge belongs to 2 faces and joins 2 vertices. If there are p edges to a face, and q edges at a vertex, we count edges in 2 ways to get

$$pF = qV = 2E.$$

Combining these results with Euler's, we get

$$E = \frac{2pq}{4 - (p-2)(q-2)}.$$

It is necessary for E to be a positive integer, hence neither p nor q can exceed 5. Thus only a small number of possibilities need be examined, and just 5 are found to give a possible E. Each corresponds to one of the Platonic solids, and, within our assumptions, no others can arise.

We next outline another proof. In Figure 2 we can imagine the dodecahedron divided into 2 congruent pieces, each consisting of a pentagon bounded by 5 others. The edge of either of these cup-shaped pieces is a skew polygon of 10 sides. For the polyhedron (p, q) there is

3

a similar skew polygon of h sides, called its Petrie polygon. Coxeter[1] shows that

$$\cos^2 \frac{\pi}{p} + \cos^2 \frac{\pi}{q} = \cos^2 \frac{\pi}{h}.$$

The possible sets of rational values of p, q and h satisfying this relationship have been determined. They are 9 in number, 5 of them being

p	q	h
3	3	4
3	4	6
3	5	10
5/2	3	10/3
5/2	5	6

Interchange of p and q leads to the other 4. Of these sets, 5 correspond to the Platonic solids, while the other 4 involve fractional values of either p or q.

3. The Kepler–Poinsot polyhedra

Before dismissing the fractional solutions obtained above as meaningless, we look for analogous results with plane polygons.

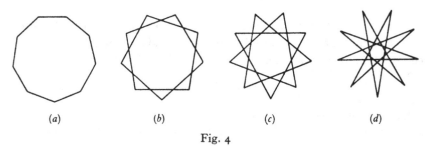

(a)　　　　　(b)　　　　　(c)　　　　　(d)

Fig. 4

In Figure 4 the first diagram is of a simple regular nonagon or 9-sided figure. Each side subtends an angle of $\frac{2}{9}\pi$ at its centre. The second diagram can also be called regular, but each side now subtends an angle of $2\pi/(9/2)$ at the centre. We thus choose to regard it as a figure with 9/2 sides. Figure 4(d) can likewise be called a regular figure with 9/4 sides, it is easy to see that a 9/5-sided figure is of the same form. Figure 4(c) consists of 3 symmetrically related equilateral triangles.

Kepler decided to include crossed polygons in the search for regularity, and discovered the first 2 solids shown in the plate (facing p. 6).

4

Their Schläfli symbols are (5/2, 3) and (5/2, 5), in each case faces being regular star pentagons of 5/2 sides. Being unaware of the principle of duality he left the discovery of the next pair of solids to Poinsot. These have simple polygonal faces, but in each case the vertex section is a star pentagon, their symbols are (3, 5/2) and (5, 5/2). For each of the 4 Kepler–Poinsot polyhedra one face of the model shown in the plate has been painted in a lighter colour than the rest.

In our first proof of completeness, we disregarded the possibility of fractional p or q; indeed Euler's relation may cease to hold in such cases. The trigonometric relation of the second method depends essentially on the angle subtended at the centre of a regular polygon, and it remains valid for fractional p, q or h. Consequently it supplies a completeness proof for our list of generalized regular polyhedra.

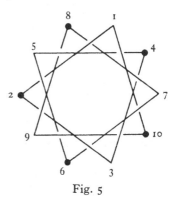

Fig. 5

The horizontal projection of the Petrie polygon for (3, 5/2) appears in Figure 5. It is a regular polygon of 10/3 sides. The Petrie polygon itself is skew, with odd-numbered vertices lying in one plane, and even numbered ones lying in a parallel plane. The side (1, 2) intersects both (4, 5) and (8, 9) and so on.

4. The regular compounds

We return to Figure 4(c), and ask if a polyhedral analogue can be found.

The stella octangula of Figure 6, discovered by Kepler, is such an analogue. It consists of a symmetrical compound of 2 tetrahedra; they can be regarded as inscribed in a cube. Dually they circumscribe an octahedron defined by the volume common to the pair.

In Figure 7 we have a cube inscribed in a dodecahedron, and the dual situation, an octahedron circumscribing an icosahedron. The cube and the icosahedron are assumed to be opaque in the diagrams, and the visible faces of the icosahedron that lie in faces of the octahedron have been shaded. Each construction can be carried out in five ways, with the same dodecahedron and icosahedron respectively. The resulting symmetrical compounds of 5 cubes and 5 octahedra are illustrated by the

5

fifth and sixth solids of the plate. In each set, one of the 5 has been painted in a darker colour than the rest.

There is a further development: we can inscribe a stella octangula in each of the 5 cubes. This gives the seventh solid in the plate; it is a symmetrical combination of 10 tetrahedra. Finally, we can omit 5 of these to arrive at the last regular compound, one consisting of 5 tetrahedra. This can be done in two essentially different ways, only one being shown. The resulting solids, although not identical, are related as are

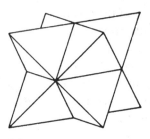

Fig. 6

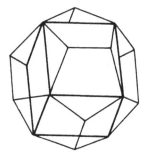
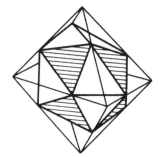

Fig. 7

object and image in a mirror, and are said to constitute an enantio-morphous pair. It has been proved that the stella octangula, together with the 4 regular compounds just discussed, exhaust all possibilities of this kind.

5. *N-dimensional space*

In Figure 8 the triangle, a two-dimensional object, is regular, but we have to put up with irregularity in our two-dimensional picture of the regular three-dimensional tetrahedron. The faces are assumed to be transparent, so all edges are drawn as full lines.

6

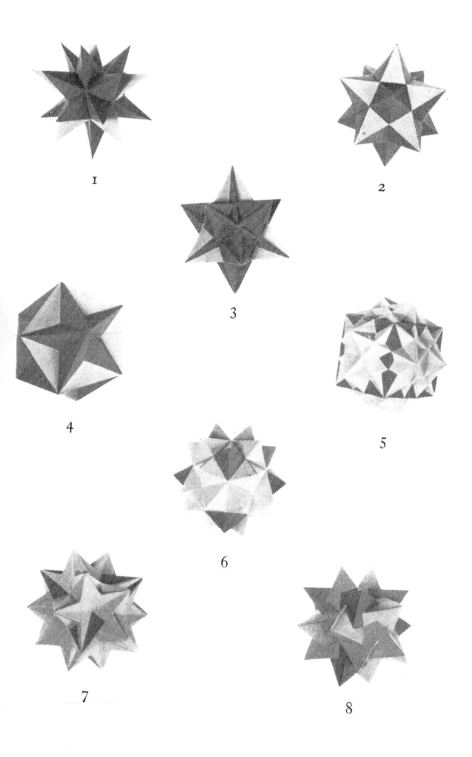

1

2

3

4

5

6

7

8

By introducing another point P, and joining it to the other 4, we obtain a two-dimensional presentation of the regular simplex in four dimensions. A three-dimensional presentation of this configuration can be constructed, but must also exhibit some irregularity. We see that the four-dimensional simplex contains 5 tetrahedra. The regular simplex in n dimensions has $(n+1)$ vertices, all pairs of which are equidistant; these pairs define its edges. It contains $(n+1)$ simplexes each of dimensions $(n-1)$.

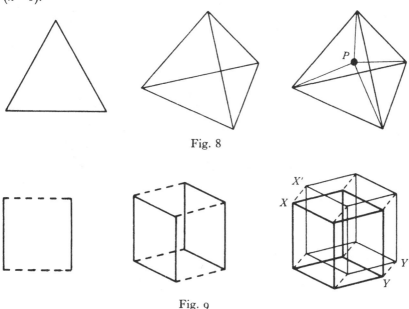

Fig. 8

Fig. 9

Figure 9 shows the measure polytopes in 2, 3 and 4 dimensions respectively. The three-dimensional version is obtained by displacing a square in a direction perpendicular to both its sides, just as a square is obtained by shifting a line in a direction perpendicular to itself. The four-dimensional version arises by moving a cube in a direction perpendicular to the 3 edges meeting at a vertex. The familiar facts that a plane can be covered with equal squares, and that 3-space can be filled with equal cubes, gives rise to the adjective measure. The word polytope describes the two-dimensional polygon, the three-dimensional polyhedron and all analogues in space of higher dimensions. We see that the four-dimensional version contains 8 cubes. These are the original cube XY, the displaced cube $X'Y'$ and the 6 cubes generated on the faces of XY.

7

The series of cross-polytopes is illustrated in Figure 10. In two dimensions 4 points at unit distance from the origin, and lying on a pair of perpendicular axes, are joined to form a square. In three dimensions 6 points lying on 3 mutually perpendicular lines form an octahedron. In four dimensions we can introduce a further axis PQ perpendicular to all three, and through their point of intersection. Joining P and Q to the previous 6 points the four-dimensional cross-polytope is formed. It is easy to verify that it contains altogether 16 regular tetrahedra, 8 with one vertex at P and 8 with one at Q.

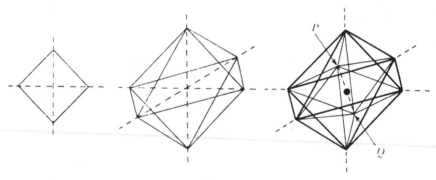

Fig. 10

We summarise the results discussed above in Table 2.

Table 2. *Structure of regular polytopes*

Type	Two-dimensional	Three-dimensional	Four-dimensional	n-dimensional
Simplex S_n	3 lines	4 triangles	5 tetrahedra	$(n+1)$ figures S_{n-1}
Measure polytope M_n	4 lines	6 squares	8 cubes	$2n$ figures M_{n-1}
Cross-polytope X_n	4 lines	8 triangles	16 tetrahedra	2^n figures S_{n-1}

We have seen how the simple Platonic solids of Figure 1 give rise to n-dimensional analogues. Other extensions will be found in Coxeter[1], where two-dimensional representations of some of them are given. Table 3 enumerates the possibilities.

There are just 3 convex cases and no stellated forms for all dimensions greater than 5. No regular compounds exist in five or six dimensions, but some are known in seven.

8

Table 3. *Numbers of regular polytopes*

Type	Two-dimensional	Three-dimensional	Four-dimensional	Five-dimensional
Convex	∞	5	6	3
Stellated	∞	4	10	0
Compound	∞	5	30	0

6. Some further developments

If the mid-points of edges of a cube are joined, as in Figure 11 (*a*), and the 8 corner pyramids removed, a solid known as a cuboctahedron results. It is described as a semi-regular polyhedron. It has two types of regular face, and only one type of vertex. This is no longer a regular

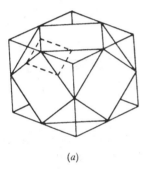
(*a*)

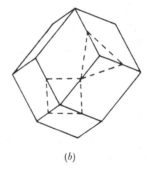
(*b*)

Fig. 11

vertex, for the section shown dotted in Figure 11 (*a*) is a rectangle rather than a square. This is one of the 13 Archimedean solids, discussed by Rouse Ball[4], and Cundy and Rollett[5]. If we allow stellation as well as semi-regularity there are 53 further known solids. These are beautifully drawn and photographed in Coxeter, Longuet-Higgins and Miller.[2] Had the cuts been further from the centre of the cube in Figure 11 (*a*), they would have marked out octagons on the cube faces; for one particular distance these would have been regular. The resulting facially regular solid is called the truncated cube. The truncated tetrahedron and the truncated octahedron, two other Archimedean solids, are discussed in Chapter 9.

The centres of the faces of an Archimedean solid define the dual figure. This has regular vertices of more than one type, and just one kind of non-regular face. The rhombic dodecahedron, shown in

9

Figure 11(b), is dual to the cuboctahedron, having 12 congruent rhombic faces. At some vertices 3 faces meet symmetrically, and at others 4. The dotted figures indicate equilateral triangular and square sections at these two types of vertex.

If we start with a sufficiently large block of wood, and make twelve plane cuts suitably disposed, we can form a regular dodecahedron. However, there will also be other finite solids; apart from those including faces of the original block, and potentially of infinite extent.

These pieces are illustrated in Figure 12. There are 12 pentagonal pyramids, 30 (irregular) tetrahedra and 20 triangular bi-pyramids. If the 12 pyramids are stuck to the faces of the dodecahedron we obtain the

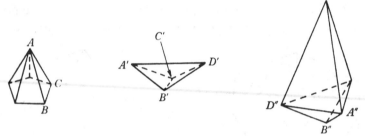

Fig. 12

solid (5/2, 5). Now the 30 tetrahedra are used to fill up gaps between pyramids. In this process the face ABC will be in contact with a face $A'B'C'$, and so on. The result is now (5, 5/2), and finally the 20 bi-pyramids are inserted in its dimples, faces like $A'B'D'$ and $A''B''D''$ being placed in contact. This gives (5/2, 3), and the process, referred to as stellation, has produced 3 of the 4 Kepler–Poinsot polyhedra.

If we generate the icosahedron in the same way, there are 472 additional pieces. From these no less than 59 solids can be formed, each having the rotational symmetry of the regular icosahedron. Very fine drawings of these will be found in Coxeter, DuVal, Flather and Petrie[3]. We have already met a few, thus (3, 5/2) and the regular compounds of 5 tetrahedra, 5 octahedra and 10 tetrahedra are all stellations of the icosahedron. Of the 59, 27 are like the compound of 5 tetrahedra, having mirror images not directly congruent to themselves.

References

1. H. S. M. Coxeter. *Regular Polytopes*, 2nd edition (Macmillan, 1963.)
2. H. S. M. Coxeter, M. S. Longuet-Higgins and J. C. P. Miller. 'Uniform Polyhedra'. *Phil. Trans.* A, **246**, 401 (1954).
3. H. S. M. Coxeter, P. DuVal, H. T. Flather and J. F. Petrie. *The Fifty-Nine Icosahedra* (University of Toronto Press, 1938).
4. W. W. Rouse Ball. *Mathematical Recreations and Essays*, 11th edition (Macmillan, 1940).
5. H. Martyn Cundy and A. P. Rollett. *Mathematical Models*, 2nd edition (Oxford University Press, 1961).
6. H. Weyl. *Symmetry* (Princeton University Press, 1952).

11

2

The Fibonacci sequence

1. *Defining the sequence*

Leonardo of Pisa, also known as Fibonacci (son of good-nature) was the author of a text on algebra called *Liber Abaci*. This appeared in 1202, and contained the following problem.

Rabbits are assumed to breed as follows. A pair of rabbits in its first month of life does not produce young. During the second and each ensuing month they produce a new pair. Starting with a single pair, how many pairs will there be at the beginning of months 1, 2, 3, ...? Deaths are supposed not to occur in the period considered.

The numbers sought form the Fibonacci sequence:

$$1 \quad 1 \quad 2 \quad 3 \quad 5 \quad 8 \quad 13 \quad ....$$

Each entry is the sum of the preceding pair of terms, since, at the beginning of month $(n+1)$ there will be

(i) The u_n pairs alive at the beginning of month n.

(ii) Pairs born during the nth month. These came from pairs over 1 month old, i.e. from the u_{n-1} pairs alive at the beginning of the $(n-1)$th month.

Thus we define the F-sequence by

$$u_{n+1} = u_n + u_{n-1}, \quad u_2 = u_1 = 1. \tag{1}$$

Another more plausible biological origin can be framed concerning branching habits of trees. A tree is assumed to have branches that grow indefinitely, producing no new branches during their early stages. All branches send forth a new branch at the end of their second and each succeeding year of life. The original branch $ABCDE...$ puts forth new branches at $B, C, D, E, ...$ The branch starting at B is $BC_1D_1E_1 ...$ and its first subbranch appears at D_1 and so on. At the beginning of the fifth year,

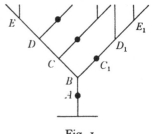

Fig. 1

12

just above the D level, there are 5 branches. Just above level E there are 8, the sixth F-number.

We shall mention other links between the F-sequence and biological growth later.

2. *Algebraic relationships*

As might be expected definition (1) leads to various algebraic results; only a few are discussed here; Coxeter [1] and Vorobyov [3] give many others. Verification of such results often depends on using (1), together with the method of mathematical induction. As an illustration we consider

$$u_{n+1}^2 - u_{n+2}u_n = (-1)^n. \tag{2}$$

Assuming this to be true for n, and using (1) with n replaced by $(n+1)$, we eliminate u_n to get

$$u_{n+1}^2 - u_{n+2}(u_{n+2} - u_{n+1}) = (-1)^n.$$

Collecting up terms, and again using (1), gives

$$u_{n+2}^2 - u_{n+1}(u_{n+1} + u_{n+2}) = (-1)^{n+1},$$

i.e.
$$u_{n+2}^2 - u_{n+3}u_{n+1} = (-1)^{n+1}.$$

Thus, if the formula holds for n, it holds for $(n+1)$. It is true for $n = 1$, hence it holds for $n = 2, 3, \ldots$.

It can be rewritten in the form

$$\frac{u_{n+1}}{u_n} - \frac{u_{n+2}}{u_{n+1}} = \frac{(-1)^n}{u_n u_{n+1}}.$$

Since the right-hand side tends to 0 as n increases, we see that ratios of successive F-numbers get closer and closer to equality. This suggests that the ratio tends to a limit, a fact noted by Kepler.

We now derive a formula, due to Binet, for the nth F-number. Starting with the assumption that

$$u_n = Ax^n,$$

and substituting in (1), we get on cancelling Ax^{n-1}

$$x^2 = x + 1,$$

leading to two values for x

$$g = \frac{1 + \sqrt{5}}{2} = 1 \cdot 62\ldots, \quad h = \frac{1 - \sqrt{5}}{2} = -0 \cdot 62\ldots.$$

13

Considering the expression

$$u_n = Ag^n + Bh^n,$$

since each term on the right-hand side satisfies (1), then the whole formula must do so. The constants A and B are now chosen to make u_1 and u_2 equal to 1; the result is

$$u_n = \frac{1}{\sqrt{5}} \left\{ \left(\frac{1+\sqrt{5}}{2} \right)^n - \left(\frac{1-\sqrt{5}}{2} \right)^n \right\}.$$

As h is numerically less than 1, its nth power tends to zero as n increases, so that approximately

$$u_n \simeq \frac{1}{\sqrt{5}} g^n, \quad \frac{u_{n+1}}{u_n} \simeq g.$$

Thus the ratio of successive F-numbers tends to g; this limit is called the golden number, and we return to it later.

Finally, we prove a result discovered by Lagrange. If each term of the F-series is divided by 4, the remainders are

$$1 \ 1 \ 2 \ 3 \ 1 \ 0 \ 1 \ 1 \ 2 \ 3 \ 1 \ 0 \ 1 \ 1 \ \ldots .$$

This sequence is periodic, the first 6 values being repeated indefinitely. We first note that this set of remainders can be generated by (1), used in a slightly modified form. Thus

$$1+1 = 2, \quad 1+2 = 3, \quad 2+3 = 5.$$

As the third number exceeds 4, we take off 4 before proceeding. Now, on reaching the seventh and eighth terms these are seen to be the same as the first two. Consequently the process cannot fail to reproduce the same set from this point, and so on. In other words periodicity depends on the recurrence of a number pair in the sequence. The argument does not depend on the divisor used.

When dividing by k, there are just k possible remainders from 0 up to $(k-1)$. Thus there are just k^2 possible pairs of consecutive values, the first in the above sequence being (1, 1), the second (1, 2) and so on. In view of the finite number of possibilities a pair must eventually appear for the second time, and then the whole sequence is repeated from this duplicated pair onwards. This is an example of Dirichlet's 'pigeonhole' principle. If more than n items are distributed among n pigeonholes, at least one must contain 2 or more items.

We have not ruled out the possibility that, with the general divisor k

14

rather than 4, the first pair to be repeated is not the initial pair (1, 1). It is left as an exercise for the reader to prove that (1, 1) is always the first pair to recur.

3. Continued fractions

The golden number g satisfies

$$g = 1 + \frac{1}{g}.$$

Using this formula to replace the second g we get

$$g = 1 + \frac{1}{1 + (1/g)}.$$

Repeated use of this device leads to an infinite continued fraction for g, it is

$$g = 1 + \cfrac{1}{1 + \cfrac{1}{1 + \cfrac{1}{1 + \ldots}}}.$$

This fraction can be truncated at any point and the resulting finite fraction simplified, thus

$$1 + \cfrac{1}{1 + \cfrac{1}{1 + \cfrac{1}{1}}} = \frac{5}{3}.$$

We call $\frac{5}{3}$ a convergent of the continued fraction; the whole series of convergents is

$$\frac{1}{1} \quad \frac{2}{1} \quad \frac{3}{2} \quad \frac{5}{3} \quad \frac{8}{5} \quad \ldots.$$

This is the series of ratios of successive F-numbers, so continuation is easy. Successive convergents are alternately less than and greater than g.

Such infinite fractions have long been of importance in number theory; they have also arisen in analysis. The result

$$\tan^{-1} x = \cfrac{x}{1 + \cfrac{(x)^2}{3 - x^2 + \cfrac{(3x)^2}{5 - 3x^2 + \cfrac{(5x)^2}{7 - 5x^2 + \ldots}}}},$$

shows that they are not confined to expressions involving integers. Until recently such expressions for transcendental functions were regarded as mere curiosities. However, the advent of digital computers, with the need to compute standard functions like sine, tan and log, has resulted in much work in this field.

All simple periodic C.F.'s with integer elements lead to quadratic surds; the theory is given by Olds[2]. We content ourselves with illustrating the point by

$$x = 1 + \cfrac{1}{1 + \cfrac{1}{2 + \cfrac{1}{1 + \cfrac{1}{2 + \ldots}}}}.$$

It is readily verified that this implies

$$x = 1 + \cfrac{1}{1 + \cfrac{1}{1 + x}},$$

a result leading to $x^2 = 3$, so the C.F. converges to $\sqrt{3}$. The first four convergents are

$$\tfrac{1}{1} \quad \tfrac{2}{1} \quad \tfrac{5}{3} \quad \tfrac{7}{4} \quad \tfrac{19}{11} \quad \tfrac{26}{15} \quad \ldots.$$

The rule for their formation is found to be

$$5 = 1 + 2 \times 2, \quad 7 = 2 + 5, \quad 19 = 5 + 2 \times 7, \quad 26 = 7 + 19, \quad \ldots,$$

for top lines, with an exactly similar form for the denominators. This differs from the simpler rule for the g convergents only in that, at alternate stages, a multiplier 2 appears. In more general cases, such as the C.F. for tan x, there is still a very simple rule for the formation of successive convergents, a fact that accounts for their utility in digital computer applications.

4. Some geometrical relationships

We start with a paradox related to (2). In Figure 2 the 8×8 square is dissected and reassembled to form the 5×13 rectangle! The explanation of the discrepancy in area is that $ABCD$ is not really a straight line, but a narrow parallelogram of unit area. By starting with a 21×21 square, and noting that (2) implies

$$21 \times 21 + 1 = 13 \times 34,$$

a still more convincing figure results.

Geometrical applications to provide models of biological growth were at one time much studied. Today these ideas have lost much of their

16

appeal. Coxeter[1] gives an interesting account of the segmented pattern to be seen on a pineapple or pine cone. We content ourselves here with outlining the phenomenon of phyllotaxis or leaf growth. On an elm twig successive leaves occur at intervals of half a revolution, the twig is said to exhibit 1:2 phyllotaxis. With beech leaves the ratio is 1:3, while

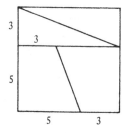 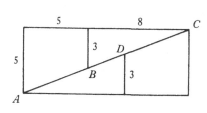

Fig. 2

with oak an advance along the stem of 5 leaves produces just 2 revolutions, i.e. the ratio is 2:5. The theory goes on to allocate 3:8 to poplar and 5:13 to almond. The numbers involved are alternate values from the F-series.

5. Golden section

Euclid posed the question: to find a rectangle such that, when a square is cut from it, as in Figure 3, the remaining smaller rectangle has the same shape as the original. Starting with sides 1 and x, those of the reduced rectangle are $(x-1)$ and 1, so that

$$\frac{x}{1} = \frac{1}{x-1},$$

an equation whose positive root is g, the golden number or golden ratio.

In Figure 3, similar triangles can be used to show that AC, BD are perpendicular, and that $AO:OB = g:1$. Hence, the rectangle with diagonal BD is obtained from rectangle AC by rotation through a quarter-turn about O, together with a size reduction of $1:g$.

We see that BD, CE in the new rectangle have exactly the same relative positions to it as AC, BD in the old. It follows at once that BE is the diagonal of a square, and a further application proves that CF is, and so on. The points A, B, C, ... lie on a logarithmic spiral with its pole at O; Zippin[4] discusses this topic in detail.

17

In the past the golden ratio was made the feature of a whole mathematical theory of aesthetics. While its supposed importance in this field can be disregarded, there is no doubt at all that g appears at many places in elementary geometry.

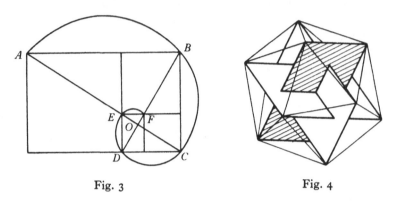

Fig. 3   Fig. 4

As an example we cite Figure 4, where three mutually perpendicular golden rectangles intersect symmetrically to form the basis of a regular icosahedron.

6. *Fibonacci sorting*

In the digital computer field it is often necessary to sort a string of numbers, stored serially on magnetic tape, into ascending or descending order. The need arises mainly in commercial applications, and often the series contains many thousand numbers. Multi-stage sorting processes are essential to deal with such large quantities of data quickly. The basic principle consists in sorting quite small blocks, perhaps of 1,000 numbers each. Then these are merged into sorted blocks of 2,000 each and so on. The merging process continues until there is a single sorted block.

Assume that there are four tape decks A, B, C and D, and that initially the unsorted data is on deck A. The first process consists of reading 1,000 numbers at a time from A into the computer. These are sorted by the machine and then the first block is copied out to deck C; the next block goes to D, and so on. Eventually half the sorted blocks will have been placed on deck C and the rest on D. The presence of an incomplete block, or of one more block on C than on D does not affect the merging process that follows.

Now the first block on C and the first on D are merged by the com-

18

puter into a block of 2,000 and this is placed on A. The process works as follows. The first numbers for C and for D are read into the computer. It compares them, and assuming that ascending order is required, it copies the smaller on to deck A. If this came from C, then the second number on tape C is read, otherwise the second entry of D's block goes in. A fresh comparison is made, with the smaller number going to A, and being replaced in the same way. This process continues until one of the blocks is exhausted. The rest of the other block is then transmitted to deck A. By virtue of the selection process used, and the fact that the two blocks of 1,000 were in ascending order at the start, there is now an ascending block of 2,000 numbers on deck A.

The next block of 2,000 is merged on to B and so on. We end this stage with a number of ordered blocks of 2,000 split evenly between A and B. These now become the input blocks, and merged blocks of 4,000 are built up on C and D. Eventually this doubling process leads to a single ordered block. Fractional blocks do not interfere, since we readily see that merging two blocks does not depend on their being of the same size.

Now tape decks are expensive items, and therefore the following Fibonacci 3-deck sort has come into favour. The initial stage is designed to produce an F-number of blocks; we take 13 for illustrative purposes, but it could be much larger. They are initially stored on deck A as before; 5 of them are at once copied to B, while C is empty. The following list shows successive states of the 3 decks being used.

A	B	C
8	5	0
3	0	5
0	3	2
2	1	0
1	0	1
0	1	0

The second state arises by merging all the blocks of B with the first 5 of A, putting results on to deck C. This leaves A with 3 blocks and B free. Therefore the next merge is on to deck B, 3 blocks coming from A, and leaving it free, while the number on C is reduced to 2. The culmination of the process should now be apparent from the table

We have described the Fibonacci sorting technique in simple form, but a number of more elaborate procedures have been evolved (see Gilstad[6])

19

7. Fibonacci search for a turning point

We conclude with another recent application of the F-sequence. It is sometimes necessary to locate the maximum or minimum of a function $f(x)$ when the derivative $f'(x)$ cannot be found in analytic form. Even if the derivative can be determined, its zeros may be difficult to locate. Consequently, search methods, based on evaluating the function, have been devised. As, in some cases, the evaluation of f itself is a lengthy task, we need an optimum method of conducting the search.

Assume that the function is known to have a unique minimum at an internal point of the interval (a, b). The computing algorithm to be described determines the required value of x to within an uncertainty of $(b-a)/13$. Here 13 is the seventh F-number and 6 evaluations of $f(x)$ are needed. We assume that $a = 0$, $b = 13$; this makes description easier without leading to any loss of generality.

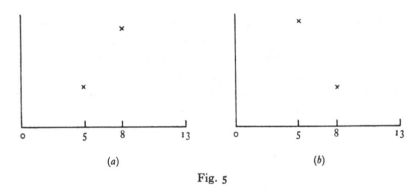

(a)   (b)

Fig. 5

The values $f(5)$ and $f(8)$ are first found, and two possibilities arise. In the first, diagram (a) of Figure 5, $f(5)$ is the smaller, and the minimum sought must lie in the range (0, 8). In case (b), $f(5)$ is the larger, and the required value of x lies in (5, 13). We see that the transformation $X = 13 - x$ turns (b) into a position completely analogous to (a). We assume that (b) holds, and describe the next step. Because of the symmetry just noted, this method, with appropriate modifications, will resolve case (a) equally well.

We next evaluate $f(10)$, and the possible results are illustrated by (a) and (b) of Figure 6. The points 5, 8, 10 and 13 have central symmetry, hence the choice of the value 10. As before, the intervals (5, 10) of (a) and (8, 13) of (*b*) are symmetrical with regard to their information about $f(x)$. We note that, just as $f(5)$ exceeds $f(8)$ in (*a*), so does $f(13)$ exceed $f(10)$ in (*b*), by virtue of the initial assumptions.

Assuming that (*a*) holds, the next stages could be those shown in (*c*). The fourth evaluation is of $f(7)$, this is assumed to be greater than $f(8)$. The fifth evaluation, that of $f(9)$, giving a value less than $f(8)$, shows that the minimum lies in (8, 10). The last evaluation of $f(9 \cdot 01)$ gives the slope of f at $x = 9$ and enables us to decide between the intervals (8, 9) and (9, 10).

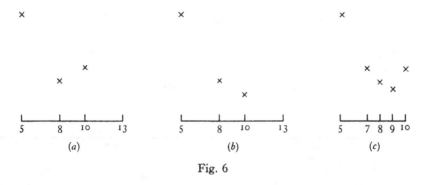

Fig. 6

By specifying more evaluations we could get greater accuracy. As the *F*-numbers are approximately multiplied by the factor g at each step, each extra evaluation divides the uncertainty of the answer by the golden number $1 \cdot 62 \ldots$. Further applications of this type are given by Bellman and Dreyfus[5].

References

1. H. S. M. Coxeter. *Introduction to Geometry* (Wiley, 1961).
2. C. D. Olds. *Continued Fractions* (Random House, New Mathematical Library, 1963).
3. N. N. Vorobyov. *The Fibonacci Numbers* (D. C. Heath and Co., 1963).
4. L. Zippin. *Uses of Infinity* (Random House, New Mathematical Library, 1962).
5. R. E. Bellman and S. E. Dreyfus. *Applied Dynamic Programming* (Princeton, 1962).
6. R. L. Gilstad. 'Read-Backward Polyphase Sorting', *Communications of the A.C.M.* **6**, no. 5, 220 (1963).

21

3

Nested polygons

1. *Sequences of polygons*

We consider a number of cyclic constructions, each of which, when applied to a polygon P of n sides, generates an n-sided polygon P'. Application of the same process to P' leads to P'', and so on. We restrict ourselves for the moment to hexagons, and denote the vertices of the initial polygon in Figure 1(a) by the symbols A_0 to A_5.

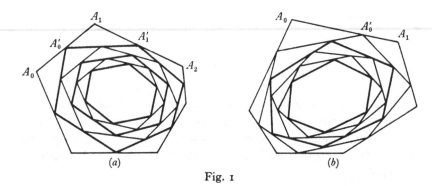

Fig. 1

In the first diagram A_0' is the mid-point of $A_0 A_1$, A_1' of $A_1 A_2$, and so on. After several stages, during which the polygons decrease steadily in size, they appear to be tending to a limiting shape. In this shape opposite pairs of sides are parallel to each other, and to the corresponding diagonals. Moreover, each polygon is approximately similar and similarly situated to the next but one in the sequence.

The trisection of $A_0 A_1$ to give A_0' and so on, leads to Figure 1(b). The same remarks apply here, except that there now appear to be three series of similar and similarly situated hexagons.

Another linear cyclic construction consists in replacing A_1 by the mid-point of $A_0 A_2$, and so on. For convenience we refer to this as 'first diagonal bisection'. This time the limiting shape behaves very differently. In Figure 2 the first diagram shows three stages of the process, in the second a further polygon has been generated from an enlargement of the third stage. The limiting shape now appears to consist of a line counted 6 times. Moreover, it does not shrink to zero size, as did the polygons in previous constructions.

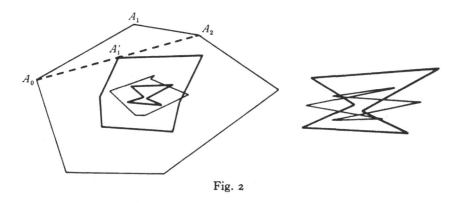

Fig. 2

In the next construction, that of Figure 3, A_0' divides $A_0 A_3$ in the ratio $p:(1-p)$, with $p < 0.5$. Proceeding cyclically we have A_3' dividing the same interval in the ratio $(1-p):p$. Thus the new diagonal has the same mid-point as the old, its length being multiplied by $(1-2p)$. Repeated

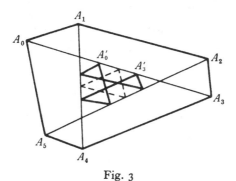

Fig. 3

application leads to a limiting triangle counted twice; it is formed by joining the mid-points of the 3 original diagonals. This is shown dotted in Figure 3 where $p = 0.4$. As diagonals are multiplied by 0.2 at each stage convergence is rapid, and the second stage (not shown) is quite close to the limit.

2. *The Fourier representation of a polygon*

Consider the hexagon with vertices A_r at (x_r, y_r) for $r = 0, 1, \ldots, 5$. The six equations

$$x_r = X + P_1 \cos \tfrac{1}{3}r\pi + Q_1 \sin \tfrac{1}{3}r\pi + P_2 \cos \tfrac{2}{3}r\pi + Q_2 \sin \tfrac{2}{3}r\pi + P_3 \cos r\pi,$$

can be solved for X, the 3 P's and the 2 Q's. By summing them, and noting that the sums of coefficients of each P and Q vanish, we get

$$6X = \sum_{r=0}^{5} x_r,$$

thus X is the x-coordinate of the centre of gravity of the 6 points. Although the results are not used, we note that a similar direct determination of each of the other constants is available, for instance

$$P_1 \sum_{r=0}^{5} \cos^2 \tfrac{1}{3}r\pi = 3P_1 = \sum_{r=0}^{5} x_r \cos \tfrac{1}{3}r\pi.$$

We next observe that, if

$$P_1 \cos \tfrac{1}{3}r\pi + Q_1 \sin \tfrac{1}{3}r\pi = C_1 \cos (\tfrac{1}{3}r\pi + \theta_1),$$

expansion of the right-hand side and comparison of coefficients leads to

$$P_1 = C_1 \cos \theta_1, \quad Q_1 = -C_1 \sin \theta_1,$$

equations easily solved for C_1 and θ_1. Thus we have a typical vertex given by

$$x_r = X + C_1 \cos (\tfrac{1}{3}r\pi + \theta_1) + C_2 \cos (\tfrac{2}{3}r\pi + \theta_2) + C_3 \cos r\pi, \qquad (1)$$

$$y_r = Y + D_1 \cos (\tfrac{1}{3}r\pi + \phi_1) + D_2 \cos (\tfrac{2}{3}r\pi + \phi_2) + D_3 \cos r\pi, \qquad (2)$$

and our analysis of the phenomena is based on this representation.

3. *The case of simple bisection*

The x-coordinates of the first derived hexagon are given by

$$x_r' = \tfrac{1}{2}x_r + \tfrac{1}{2}x_{r+1}.$$

Substituting (1) for the 2 terms on the right, and using the addition formula for cosines gives

$$x_r' = X + \tfrac{1}{2}\sqrt{3}\, C_1 \cos (\tfrac{1}{3}r\pi + \theta_1 + \tfrac{1}{6}\pi) + \tfrac{1}{2}C_2 \cos (\tfrac{2}{3}r\pi + \theta_2 + \tfrac{1}{3}\pi).$$

24

The effect of the process has been to multiply C_1 by $0.866...$, and C_2 by 0.5. The coefficient C_3 vanishes, and additions $\frac{1}{6}\pi$, $\frac{1}{3}\pi$ are made to the 2 angles. Repeated applications will cause both the trigonometric terms to diminish, but the second does so faster than the first. After l steps we neglect this smaller term to get

$$x_r = X + C \cos\left(\tfrac{1}{3}r\pi + \theta\right), \tag{3}$$

$$y_r = Y + D \cos\left(\tfrac{1}{3}r\pi + \phi\right), \tag{4}$$

where C, θ are given by

$$C = \left(\tfrac{1}{2}\sqrt{3}\right)^l C_1, \quad \theta = \theta_1 + \tfrac{1}{6}\pi l,$$

with similar expressions for D and ϕ. For large l, C and D are small, and all vertices lie close to (X, Y), i.e. the hexagon shrinks to zero size.

On expansion (3) and (4) give

$$x_r - X = P \cos \tfrac{1}{3}r\pi + Q \sin \tfrac{1}{3}r\pi, \tag{5}$$

$$y_r - Y = R \cos \tfrac{1}{3}r\pi + S \sin \tfrac{1}{3}r\pi. \tag{6}$$

We write
$$\xi_r = \cos \tfrac{1}{3}r\pi, \quad \eta_r = \sin \tfrac{1}{3}r\pi, \tag{7}$$

and consider the transformation

$$x - X = P\xi + Q\eta, \tag{8}$$

$$y - Y = R\xi + S\eta. \tag{9}$$

If the slope of the line joining points 1 and 2 in the (ξ, η) plane is μ, and for the corresponding transformed points is m, we have

$$m = \frac{y_2 - y_1}{x_2 - x_1} = \frac{R(\xi_2 - \xi_1) + S(\eta_2 - \eta_1)}{P(\xi_2 - \xi_1) + Q(\eta_2 - \eta_1)} = \frac{R + S\mu}{P + Q\mu}.$$

Since m depends only on μ, parallel lines in one system yield parallel lines in the other. If we solve (8) and (9) for (ξ, η) we see that

$$\xi^2 + \eta^2 \to L(x - X)^2 + 2M(x - X)(y - Y) + N(y - Y)^2,$$

so that the unit circle transforms into a conic in the (x, y) plane.

The six values of (7) when r goes from 0 to 5 correspond to vertices of a regular hexagon inscribed in the unit circle as in Figure 4(a). The hexagon, defined by (5) and (6), is obtained from it by the transformation (8), (9). It is therefore inscribed in a conic centre (X, Y), and has its sides parallel to its diagonals, the property noted in Section 1.

Finally, we see that the polygon coming 2 stages after (3), (4) is

$$x_r^* = X + \tfrac{3}{4}C \cos (\tfrac{1}{3}r\pi + \theta + \tfrac{1}{3}\pi),$$

$$y_r^* = Y + \tfrac{3}{4}D \cos (\tfrac{1}{3}r\pi + \phi + \tfrac{1}{3}\pi),$$

here we have introduced the factor $\tfrac{1}{2}\sqrt{3}$ and added the angle $\tfrac{1}{6}\pi$ twice in each coordinate. It follows that

$$x_r^* - X = \tfrac{3}{4}(x_{r+1} - X), \tag{10}$$

$$y_r^* - Y = \tfrac{3}{4}(y_{r+1} - Y). \tag{11}$$

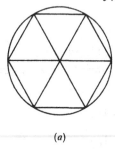 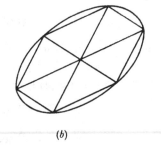

(a)   (b)

Fig. 4

Thus, apart from a renaming of the vertices, polygons occurring two stages apart are similar and similarly situated.

4. The three-dimensional case and trisection

Had we considered a skew hexagon in 3-space, we should have supplemented (5) and (6) by

$$z_r - Z = T \cos \tfrac{1}{3}r\pi + U \sin \tfrac{1}{3}r\pi. \tag{12}$$

Eliminating sine and cosine terms from (5), (6) and (12) gives

$$\begin{vmatrix} x_r - X & P & Q \\ y_r - Y & R & S \\ z_r - Z & T & U \end{vmatrix} = 0.$$

On expansion we get

$$l(x_r - X) + m(y_r - Y) + n(z_r - Z) = 0,$$

a result holding for $r = 0, 1, \ldots, 5$. In other words, the 6 vertices of the limiting form lie in the plane

$$l(x - X) + m(y - Y) + n(z - Z) = 0.$$

26

As this vanishes for $x = X$, etc., the plane goes through the centre of gravity of the original vertices. The limiting polygon is still plane, although arising from an initially skew figure; it will have the same properties as those found in Section 3.

If we look now at the trisection construction

$$x_r' = \tfrac{1}{3}x_r + \tfrac{2}{3}x_{r+1},$$

then (1) gives

$$x_r' = X + \tfrac{1}{3}\sqrt{7}\, C_1 \cos\left(\tfrac{1}{3}r\pi + \theta_1 + \psi_1\right)$$
$$+ (1/\sqrt{3})\, C_2 \cos\left(\tfrac{2}{3}r\pi + \theta_2 + \psi_2\right) - \tfrac{1}{3}C_3 \cos r\pi.$$

As before the C_1 term dominates, and the limiting form is

$$x_r = X + C \cos\left(\tfrac{1}{3}r\pi + \theta\right).$$

Considering the polygon three further on in sequence, we have

$$x_r^* = X + \tfrac{7}{27}\sqrt{7}\, C \cos\left(\tfrac{1}{3}r\pi + \theta + 3\psi_1\right).$$

In addition, the value of ψ_1 leads to

$$3\psi_1 = 122 \cdot 7° \backsimeq \tfrac{2}{3}\pi,$$

and we get the approximate equality

$$x_r^* - X \backsimeq \tfrac{7}{27}\sqrt{7}\,(x_{r+2} - X).$$

So here the property of being similar and similarly situated is only approximately true, even in the limit. The parallelism properties of the limiting hexagon hold as for bisection.

5. The other hexagon constructions

Considering the first diagonal bisection of Figure 2, we have

$$x_r' = \tfrac{1}{2}x_{r-1} + \tfrac{1}{2}x_{r+1},$$

which, together with (1), gives

$$x_r' = X + \tfrac{1}{2}C_1 \cos\left(\tfrac{1}{3}r\pi + \theta_1\right) - \tfrac{1}{2}C_2 \cos\left(\tfrac{2}{3}r\pi + \theta_2\right) - C_3 \cos r\pi.$$

Here the third term is dominant, and the limiting shape is the line joining

$$(X - C_3,\ Y - D_3),\quad (X + C_3,\ Y + D_3),$$

counted 6 times, odd vertices are at one point and even vertices at the other. The multiplier in the dominant term has the numerical value unity, so that the limiting figure is now of finite size.

The diagonal division of Figure 3 leads to

$$x'_r = (1-p)x_r + px_{r+3},$$

and the application of (1) yields

$$x'_r = X + (1-2p)\, C_1 \cos\left(\tfrac{1}{3}r\pi + \theta_1\right)$$
$$+ C_2 \cos\left(\tfrac{2}{3}r\pi + \theta_2\right) + (1-2p)C_3 \cos r\pi.$$

Since $p < 0.5$, the second cosine is dominant, and as r goes from 0 to 5 we get 3 points, each counted twice. Thus the limiting form is a triangle, and again the unit multiplier of the dominant term indicates a non-zero result. With $p = 0.4$, as in Figure 3, the 3 multipliers are 0.2, 1 and 0.2; their relative magnitudes indicate rapid convergence.

6. Some results for pentagons

When considering pentagons our basic representation is

$$x_r = X + C_1 \cos\left(\tfrac{2}{5}r\pi + \theta_1\right) + C_2 \cos\left(\tfrac{4}{5}r\pi + \theta_2\right).$$

For simple bisection of sides this yields

$$x'_r = X + C_1 \cos\tfrac{1}{5}\pi \cos\left(\tfrac{2}{5}r\pi + \theta_1 + \tfrac{1}{5}\pi\right)$$
$$+ C_2 \cos\tfrac{2}{5}\pi \cos\left(\tfrac{4}{5}r\pi + \theta_2 + \tfrac{2}{5}\pi\right),$$

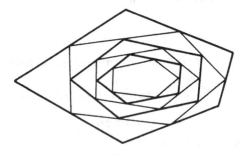

Fig. 5

and the first term dominates. The limit is a transformed regular pentagon, so that in it sides will be parallel to diagonals again. The additive $\tfrac{1}{5}\pi$ again leads to alternate pentagons being similar and similarly placed in the limit. Figure 5 illustrates these results.

28

Turning to first diagonal bisection we get

$$x'_r = X + C_1 \cos \tfrac{2}{5}\pi \cos (\tfrac{2}{5}r\pi + \theta_1) + C_2 \cos \tfrac{4}{5}\pi \cos (\tfrac{4}{5}r\pi + \theta_2). \quad (13)$$

Here the second term dominates, the limiting shape being

$$x = X + C \cos (\tfrac{4}{5}r\pi + \theta).$$

As r goes from 0 to 5, the points of the unit circle are at angles

$$0, \quad \tfrac{4}{5}\pi, \quad \tfrac{8}{5}\pi, \quad \tfrac{12}{5}\pi, \quad \tfrac{16}{5}\pi.$$

They thus form a regular star pentagon, and the limiting shape is crossed. In Figure 6 the first diagram shows two stages of the construction. The result of the second stage is then enlarged and a third polygon, this time a fully crossed pentagon, derived from it.

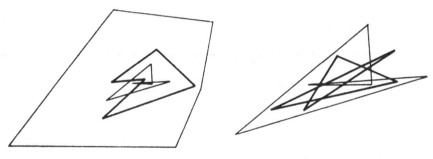

Fig. 6

However, if this construction is applied to a regular convex pentagon, the resulting sequence is one of regular convex figures. The apparent conflict with the above result arises since a regular set of points leads to a zero value of C_2 in (13). Hence C_1 now dominates and the limit is a convex figure.

7. The general theory in matrix form

For an n-sided polygon we consider the cyclic generating relation

$$x_1^{(1)} = \alpha_1 x_1 + \alpha_2 x_2 + \ldots + \alpha_n x_n,$$

with the sum of coefficients α equal to 1. The n results can be summarised by

$$\tilde{x}^{(1)} = A\tilde{x},$$

29

where

$$A = \begin{pmatrix} \alpha_1 & \alpha_2 & \cdots & \alpha_n \\ \alpha_n & \alpha_1 & \cdots & \alpha_{n-1} \\ \vdots & \vdots & & \vdots \\ \alpha_2 & \alpha_3 & \cdots & \alpha_1 \end{pmatrix} \quad \text{and} \quad \tilde{x} = \begin{pmatrix} x_1 \\ x_2 \\ \vdots \\ x_n \end{pmatrix}.$$

When the process is applied k times

$$\tilde{x}^{(k)} = A^k \tilde{x}.$$

The matrix A has n latent roots λ_j, and corresponding vectors $\tilde{u}_j$, such that

$$A\tilde{u}_j = \lambda_j \tilde{u}_j. \tag{14}$$

The vector $\tilde{x}$ can be expressed as follows:

$$\tilde{x} = c_1 \tilde{u}_1 + c_2 \tilde{u}_2 + \ldots + c_n \tilde{u}_n.$$

Hence, by virtue of (14)

$$\tilde{x}^{(1)} = c_1 \lambda_1 \tilde{u}_1 + c_2 \lambda_2 \tilde{u}_2 + \ldots + c_n \lambda_n \tilde{u}_n.$$

If applied k times we have

$$\tilde{x}^{(k)} = c_1 \lambda_1^k \tilde{u}_1 + c_2 \lambda_2^k \tilde{u}_2 + \ldots + c_n \lambda_n^k \tilde{u}_n,$$

and the root λ of greatest numerical value will determine the limiting vector.

There is always a dominant root $\lambda = 1$, with a corresponding latent vector of units. This indicates that, if no other root of modulus as big as 1 occurs, the figure converges to the centre of gravity of the initial set of points. Other roots occur in complex pairs, with a second real root if n is even. Let the pair of largest modulus be

$$\lambda_m \exp\left(\pm i\phi_m\right).$$

It can be proved that the complex roots are given by

$$\lambda_t \exp\left(i\phi_t\right) = \sum_{j=0}^{(n-1)} \alpha_{j+1} \exp\left(\frac{2\pi j i t}{n}\right) \quad (t = 1, 2, \ldots, [\tfrac{1}{2}(n-1)]),$$

so that λ_m is easily determined. The corresponding latent vectors are of the form

$$(1, \omega^{\pm t}, \omega^{\pm 2t}, \ldots, \omega^{\pm(n-1)t}),$$

where ω is a primitive nth root of unity.

30

We leave the matrix theory at this point, but the results will be similar to those obtained earlier by Fourier techniques, terms like

$$\cos\frac{2r\pi}{n} \quad \text{and} \quad \sin\frac{2r\pi}{n}$$

entering via the complex roots of unity.

Reference

J. H. Cadwell. 'A Property of Linear Cyclic Transformations'. *Math. Gaz.* **37**, no. 320, p. 85 (1953).

4

The distribution of prime numbers

1. *The sieve of Eratosthenes*

In order to list the prime numbers less than N we can start by writing out a list of the integers from 2 to N. We ring 2 and cross out all its multiples. The next number 3 is uncrossed, so it is also ringed and its multiples crossed out. Now 4 has been deleted, so 5 is ringed and its multiples removed, and so on. We note that the process of examination need not go beyond M, the square root of N. For any composite number less than N must necessarily have at least one prime factor less than M, and it will have been deleted by the time we reach M. The Greeks regarded this process as a means of listing the primes up to N; Meissel showed that it can be used to count them, provided we have a list up to M.

We define $\pi(N)$ as the number of primes less than or equal to N; it is a quantity of central importance in our topic. For instance,

$$\pi(1,000) = 168,$$

i.e. there are 168 primes among the first 1,000 integers. Meissel's formula is

$$\pi(N) = N - 1 + \pi(\sqrt{N}) - \Sigma \left[\frac{N}{p}\right] + \Sigma \left[\frac{N}{pp'}\right] - \Sigma \left[\frac{N}{pp'p''}\right] + \dots.$$

The square brackets indicate that the integer part of the quantity inside is to be taken. In the first sum all primes p less than M occur, in the second all pairs of different primes p and p', each less than M, are taken, and so on. Signs alternate from sums with an odd number of divisors to sums with an even number, and eventually a point is reached where all further sums vanish because contents of the square brackets are less than 1.

Our proof depends on the sieve of Eratosthenes. From the original $(N-1)$ numbers we subtract the number of numbers deleted because they were multiples of 2, 3, 5, ... As the ringed prime numbers themselves were not deleted, and there are $[N/p]$ multiples of p, this number is

$$\Sigma \left\{ \left[\frac{N}{p} \right] - 1 \right\} = \Sigma \left[\frac{N}{p} \right] - \pi(\sqrt{N}),$$

since the 1 occurs for each prime up to M.

However, numbers twice deleted have been included twice in this formula. To compensate we must add back their number, which is

$$\Sigma \left[\frac{N}{pp'} \right].$$

Still further corrections are needed for numbers deleted 3 or more times. Consider a number like 360, divisible by 2, 3 and 5. It is counted 3 times in the first sum, and 3 times in the second, on account of factors 2×3, 2×5 and 3×5. Thus no allowance for its deletion has been made with the first 2 sums, so we subtract the term

$$\Sigma \left[\frac{N}{pp'p''} \right].$$

Similar corrections for numbers with 4 and more different prime factors are required. A number with k different prime factors is counted

$$\binom{k}{1} - \binom{k}{2} + \binom{k}{3} - \dots + (-1)^{k-1} \binom{k}{k} = 1 - (1-1)^k = 1$$

times, as it should be.

Meissel, using a number of ingenious devices to reduce the labour involved, calculated $\pi(10^8)$, and later Bertelsen found $\pi(10^9)$.

We conclude this section by mentioning the complete tabulation of primes less than 10 million by D. N. Lehmer. Kulik has extended these tables up to 100 million, although his results contain mistakes. Kraitchik found all the prime numbers in the range

$$10^{12} - 10^4 \quad \text{to} \quad 10^{12} + 10^4. \tag{1}$$

We shall see later how this fragmentary tabulation has enabled results in the theory to be tested.

2. *Some elementary results*

Euclid posed the question, are there an infinite number of primes? His proof that there are runs as follows. Suppose that their number is finite, then there will be a largest prime p_n. We construct the larger number

$$2 . 3 . 5 \dots p_n + 1.$$

If divided by any prime up to p_n there is a remainder of 1, therefore its smallest prime factor exceeds p_n, or it is itself a prime. In either case we have constructed a prime number greater than p_n, and the only way out of this contradiction is to abandon the hypothesis that a largest prime exists.

We can write this result in the form

$$p_{n+1} \leqslant 2 \cdot 3 \cdot 5 \cdot \ldots \cdot p_n + 1. \qquad (2)$$

A little experimentation shows that the upper bound it provides is extravagantly large. Rademacher and Toeplitz[4] give a strictly elementary proof for the inequality
$$p_{n+1} < \sqrt[7]{(2 \cdot 3 \cdot \ldots \cdot p_n)},$$

valid for n greater than some fixed value.

We modify Euclid's proof to show that there are infinitely many primes of the form $(4m + 3)$. We first remark that the product of 2 numbers of the form $(4m + 1)$ is of the same form, for

$$(4m + 1)(4m' + 1) = 4(4mm' + m + m') + 1.$$

Next we consider the number

$$2^2 \cdot 5 \cdot \ldots \cdot p_n + 3.$$

This is of the form $(4m + 3)$, and two cases arise; it may be a prime, or it may possess prime factors. These must be of the form $(4m + 1)$ or $(4m + 3)$ with m non-zero. Moreover, there must be at least one of the latter, since a product of factors, all of form $(4m + 1)$, is also of this form. Thus, just as in Euclid's proof, we have discovered a prime $(4m + 3)$ exceeding p_n. Dirichlet proved that the arithmetic progression $(am + b)$ contains an infinity of primes, unless a and b have a common factor, when all its values are composite.

We look now at the spacing of primes. If one excludes the pair 2, 3, the gap between successive primes must be an even number, otherwise one of them would be even. The smallest possible gap of 2 is suspected to occur infinitely often; the successive primes

$$10^{15} + 149341, \quad 10^{15} + 149343$$

are a very large pair of this form. So far the result has defied proof or disproof; we return to it later.

We next ask if there is an upper limit to the gap between successive primes. We can construct the set of $(n - 1)$ consecutive integers

$$n! + 2, \quad n! + 3, \quad \ldots, \quad n! + n.$$

It is easy to see that each number is composite, so that the nearest primes to right and left of the block differ by at least n. As n is arbitrary we can therefore make this gap as large as we please.

We mention one further result, a conjecture of Bertrand's proved by Chebyshev. The theorem states that the interval $(l, 2l)$ contains at least one prime. This can be expressed in the form

$$p_{n+1} < 2p_n, \qquad (3)$$

a much improved version of (2).

3. Numerical evidence on the density of primes

The fact that, by going a long way on in the series of integers, we can be sure of getting a very large gap between successive primes, indicates a decrease in their density as we proceed. Table 1 illustrates the situation.

Table 1. *Counts by blocks of* 100

	0–99	100–199	200–299	300–399	400–499
0	25	21	16	16	17
10^6	6	10	8	8	7
10^{12}	4	6	2	4	2

The first line gives the numbers of primes in the first 5 blocks of 100 numbers. The second gives the corresponding values from 1,000,000 to 1,000,499, and so on. There is a slow fall in density, coupled with an appreciable degree of irregularity. The last line suggests an average gap of about 28 between successive primes. We know that this gap is sometimes as small as 2 in the region of 10^{12}, so the fluctuations are to be expected.

In view of these results it would be optimistic to expect an exact formula for $\pi(N)$ to be available. However, some description of average behaviour is still possible, as we see in the following section.

4. The prime number theorem

Meissel suggested that $\pi(N)$ is close to

$$\frac{N}{\log_e N},$$

35

for large N. Gauss arrived at the same conjecture at the age of 16, later replacing it by the better approximation

$$\pi(N) \sim \int_2^N \frac{dx}{\log x}. \tag{4}$$

Table 2 illustrates just how this formula compares with exact counts. The value of $\pi(10^9)$ was deduced from the Meissel formula as explained earlier.

Table 2. *Counts of primes*

N	$\pi(N)$	Log. integral (4)
10^3	168	178
10^6	78,498	78,628
10^9	50,847,534	50,849,235

It will be seen that the proportional error falls as N increases, and the approximation gives too large a value.

Chebyshev made a serious but unsuccessful attempt to prove the asymptotic equivalence of $\pi(N)$ and the logarithmic integral, and in this work he was led to proving Bertrand's conjecture. It was not until 1896 that a proof appeared, when independent lines of approach by Hadamard and de la Valleé Poussin were successful. Both proofs involved the Riemann ζ function, and the theory of functions of a complex variable. The basis of the method is an identity of Euler's

$$\zeta(s) = \sum_{n=1}^{\infty} \frac{1}{n^s} = \Pi \left(1 - \frac{1}{p^s}\right)^{-1},$$

the product being taken over all primes p. Littlewood later showed that the difference between $\pi(N)$ and its approximation (4) changes sign infinitely often.

The theorem can be restated in other ways. Thus it implies that the interval $(N, N+\Delta N)$ contains about

$$\frac{\Delta N}{\log N}$$

prime numbers. Again this is equivalent to saying that, near the large number N, the average gap between primes is $\log(N)$.

36

5. A heuristic proof of the theorem

Since the original proofs, others of a more elementary nature have appeared, in particular the use of complex variables has been avoided. However, none of the known proofs are at all easy. We present an interesting statistical approach due to the Nobel prizewinner Gustav Hertz, discussed by Courant and Robbins[2]. We assume the existence of a density function $W(x)$ such that there are $W(x)\,dx$ primes between x and $(x+dx)$. This assumption is the sole reason why the proof that follows cannot be made rigorous. It alone of the various steps taken cannot be justified formally.

We shall use Stirling's asymptotic formula

$$n! \sim n^n \sqrt{(2\pi n)}e^{-n}.$$

From this we can obtain the simpler, but less precise, result

$$\log n! \sim n \log n, \tag{5}$$

the error involved being of order n. Next we seek to factorise $n!$, noting that it contains the prime factor p

$$\left[\frac{n}{p}\right]+\left[\frac{n}{p^2}\right]+\left[\frac{n}{p^3}\right]+\dots \tag{6}$$

times. For example 50! is a very large number ending with a string of 12 zeros. It contains the factor 5 once in each of the 10 numbers 5, 10, ..., 50; in addition 25 and 50 contain an extra 5. There are

$$25+12+6+3+1,$$

occurrences of 2, so each occurrence of 5 leads to a zero.

Neglecting the fractional differences involved, we can replace (6) by

$$\frac{n}{p}+\frac{n}{p^2}+\frac{n}{p^3}+\dots = \frac{n}{p-1} \simeq \frac{n}{p}.$$

This leads to the approximation

$$n! \simeq \Pi p^{n/p},$$

the product being taken over all primes less than or equal to n.

Taking logs and using (5) we get

$$\log n! \simeq n \sum_{p \leqslant n} \frac{\log p}{p} \simeq n \log n,$$

37

thus

$$\log n \simeq \sum_{p \leqslant n} \frac{\log p}{p}.$$ (7)

The range from o to n can be divided into intervals Δx, large themselves, but much smaller than n. An interval from x to $(x+\Delta x)$ contains $W(x)\Delta x$ primes of average size $(x+0.5\Delta x)$. Ignoring the relatively small second term in brackets, the primes in this interval contribute about

$$W(x)\frac{\log x}{x}\Delta x,$$

to the sum in (7). We conclude that

$$\log n \simeq \sum \frac{W(x) \log x \, \Delta x}{x} \simeq \int_1^n \frac{W(x) \log x}{x}\,dx,$$

on replacing the sum by an integral. However, we know that

$$\log n = \int_1^n \frac{dx}{x},$$

and comparison of the two integrals leads to the conclusion

$$W(x) \log x \simeq 1.$$

Finally, we have $\quad \pi(N) \simeq \int_2^N W(x)\,dx \simeq \int_2^N \frac{dx}{\log x},$

the prime number theorem.

6. *The average number of factors of n*

A number n may be a prime and possess just one factor. If it is a power of 2 it has the maximum possible number of prime factors, i.e. $\log_2(n)$. We seek the average number of prime factors of n, taken over all values of n less than x. We adopt the definitions

$$n = 2^{k_1}.3^{k_2}.\ldots.p_r^{k_r}, \quad \Omega(n) = k_1+k_2+\ldots+k_r.$$

We next examine the statement

$$\sum_{n \leqslant x} \Omega(n) = \sum_{p \leqslant x}\left[\frac{x}{p}\right] + \sum_{p \leqslant x}\left[\frac{x}{p^2}\right] + \ldots. \quad (8)$$

A particular prime p divides altogether $[x/p]$ numbers less than or equal to x. Of these $[x/p^2]$ will contain this factor twice and so on. Hence, if all the occurrences of p in numbers up to x are counted, there will be

$$\left[\frac{x}{p}\right]+\left[\frac{x}{p^2}\right]+\ldots.$$

Summing this expression for all p less than x gives the total number of factors of all numbers up to x, as in (8).

Euler showed that the sum of the inverses of the prime numbers diverges; Mertens obtained the result

$$\sum_{p \leqslant x} \frac{1}{p} = \log \log x + C.$$

It is interesting to compare this with Euler's result for the natural numbers

$$\sum_{n \leqslant x} \frac{1}{n} \sim \log x + \gamma.$$

The double logarithm in Mertens's formula indicates a much slower divergence of the series when only primes occur. One of the few proved results about prime pairs P, $(P+2)$ is that the sum of reciprocals of all such P converges.

We now observe that

$$\sum_{p \leqslant x} \left[\frac{x}{p}\right] = x \log \log x + xC + k\pi(x),$$

where k is less than 1. This follows since dropping the square brackets on the left-hand side incurs a maximum error of 1 for each term, and there are $\pi(x)$ terms. The other sums in (8) satisfy the inequality

$$\sum_{p \leqslant x} \left[\frac{x}{p^2}\right] + \sum_{p \leqslant x} \left[\frac{x}{p^3}\right] + \ldots < \sum_{\text{All } p} \frac{x}{p^2} + \sum_{\text{All } p} \frac{x}{p^3} + \ldots,$$

and this latter expression can be evaluated giving

$$\sum_{\text{All } p} \frac{x}{p(p-1)} < \sum_{n=2}^{\infty} \frac{x}{n(n-1)} = x.$$

Thus

$$\log \log x + C + k\frac{\pi(x)}{x} < \frac{\sum_{n \leqslant x} \Omega(n)}{x} < \log \log x + C + k\frac{\pi(x)}{x} + 1.$$

Noting that $\pi(x)/x$ behaves like $1/\log x$, and that the central term in the expression is the average number of factors of numbers less than x, we see that this average is asymptotically equal to $\log \log x$. This number increases very slowly with x, numbers up to 10^7 average about 3 prime factors, those up to 10^{80} have a mean of 5. Hardy and Wright[3] treat this and associated topics in greater detail.

7. Some unproved conjectures

The Riemann hypothesis concerning the ζ function is that all its complex zeros have real part $\frac{1}{2}$. It has so far defied all attempts at proof. Its truth would lead to a number of interesting results concerning primes. Thus, as mentioned earlier, Littlewood proved that, for a large enough N, $\pi(N)$ eventually exceeds the integral approximation to it, in spite of the results of Table 2. Skewes proved that, if the Riemann hypothesis holds, this N satisfies

$$\log \log \log N < 79 \text{ or } N < 10^{10^{34}},$$

a very large number.

Lord Cherwell[1] used statistical methods akin to those of Section 5 to investigate the frequency of prime pairs of the form $P, (P+2)$. His result suggests that, for large N, the interval $(N, N+\Delta N)$ contains

$$\frac{1 \cdot 32 \Delta N}{(\log N)^2},$$

such pairs. In the region defined by (1) there are 36 pairs, while this formula gives the number 35. He produced related formulae for prime triplets like 5, 7, 11, and like 13, 17, 19.

Goldbach conjectured that any even number can be expressed as the sum of 2 primes, for instance, $82 = 79 + 3$. Using a generalised Riemann hypothesis, Hardy and Littlewood proved that any odd number could be expressed as the sum of 3 primes. A proof of the latter result, independent of the hypothesis about $\zeta(s)$, was given later by the Russian mathematician Vinogradov, but Goldbach's conjecture still defies proof.

Many fascinating aspects of number theory are brought to light by Hardy and Wright[3], in their discussion of other open and disproved conjectures.

References

1. Lord Cherwell. 'Note on the Distribution of the Intervals between Prime Numbers', *Quart. J. Math.* **17**, no. 65, 46 (1946).
2. R. Courant and H. Robbins. *What is Mathematics?* (Oxford, 1943).
3. G. H. Hardy and E. M. Wright. *An Introduction to the Theory of Numbers*, 4th edition (Oxford, 1960).
4. H. Rademacher and O. Toeplitz. *The Enjoyment of Mathematics* (Princeton, 1957).

5

Inversion in elementary geometry

1. *The inverse transformation*

An inversion is characterised by a centre O and a radius k. Any figure $\mathfrak{F}$ in 2 or 3 dimensions is made up of points typified by P. We join OP and on this line select a point P' so that P and P' lie on the same side of O and

$$OP \cdot OP' = k^2. \tag{1}$$

When applied to each point of $\mathfrak{F}$ this construction will generate the figure $\mathfrak{F}'$ formed by the points like P', and we call $\mathfrak{F}'$ the inverse of $\mathfrak{F}$. The expression (1) is symmetrical with regard to P and P', so the relation between $\mathfrak{F}$ and $\mathfrak{F}'$ is mutual; each is the inverse of the other.

We now present a table showing how various elements behave with regard to inversion.

Table 1. *Pairs of inverse figures*

Line through O	Line through O
Line not through O	Circle through O
General circle	Circle
Plane through O	Plane through O
Plane not through O	Sphere through O
General sphere	Sphere

The first pair needs no comment; for the second we consider Figure 1. If OQ is perpendicular to the line PQ, then (1) leads to

$$OP \cdot OP' = OQ \cdot OQ' = k^2,$$

and hence $PQQ'P'$ is cyclic. Therefore angle $OP'Q$ equals $Q'QP$, a right angle, and the locus of P' as P varies is a circle on OQ' as diameter. We note that we have also answered the question, 'How does a circle through O invert'?

In Figure 2 we consider the inverse of a circle, centre C, with regard to O lying in its plane, but not on it. We draw $P'C_1$ parallel to QC, and have

$$OP \cdot OP' = k^2, \quad OP \cdot OQ = OT^2.$$

These results, with the similar triangles OQC, $OP'C_1$, lead to

$$\frac{P'C_1}{QC} = \frac{OC_1}{OC} = \frac{OP'}{OQ} = \frac{k^2}{OT^2}.$$

The last quantity does not depend on P, so that both $P'C_1$ and OC_1 are fixed. Hence the locus of P' is a circle centre C_1. We note that C_1 is not the inverse of C, i.e. centres do not invert into each other.

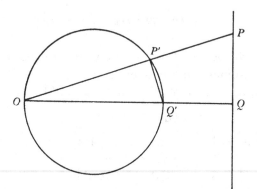

Fig. 1

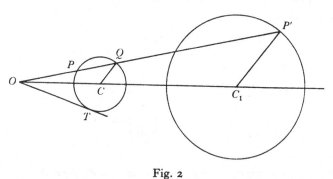

Fig. 2

The three-dimensional statements call for little comment; thus rotation of Figure 2 about OC_1 leads at once to the last result. We have proved the third if O lies in the plane of the circle. If not, the circle can be regarded as the intersection of a pair of spheres in a double infinity of ways. Since the spheres invert into spheres the inverse of their intersection is another circle.

The distinctive property of inversion is the preservation of the angle of intersection of a pair of curves. Thus, if two curves intersect at right angles, their inverses will do the same; if they touch then so do their

42

inverses. In Figure 3, a three-dimensional situation, two curves inter-sect at P, and their inverses cross at P'. The chords PQ, PR at P are short in comparison with the length of OP.

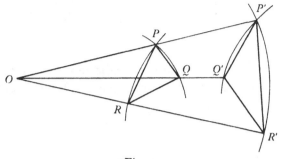

Fig. 3

By property (1) the plane quadrilateral $PP'Q'Q$ is cyclic, and similar triangles OPQ, $OQ'P'$ lead to

$$\frac{P'Q'}{PQ} = \frac{OP'}{OQ} \simeq \frac{OP'}{OP},$$

since PQ is much smaller than OP. By symmetry $P'R'/PR$ is nearly equal to the same ratio. A further application of this argument leads to

$$\frac{P'Q'}{PQ} \simeq \frac{P'R'}{PR} \simeq \frac{Q'R'}{QR}.$$

Hence, in the limit as Q and R approach P, the triangles PQR, $P'Q'R'$ are similar, and the angle between tangents at P equals the angle between tangents at P'.

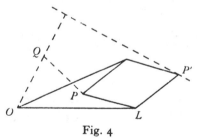

Fig. 4

Figure 4 illustrates the Peaucellier cell, a linkage designed to invert figures mechanically. The point O is fixed, and, as P varies over a figure $\mathfrak{F}$, P' generates the inverse figure $\mathfrak{F}'$. Proof is easy; two appli-cations of Pythagoras lead to

$$OL^2 - (OP + \tfrac{1}{2}PP')^2 = PL^2 - (\tfrac{1}{2}PP')^2,$$

43

or $$OP^2+OP.PP' = OP.OP' = OL^2-PL^2,$$

and the last quantity is fixed. The linkage will generate a straight line motion without linear guides, and was so used in the design of ventilating pumps in the Houses of Parliament in the middle of the last century.

In the diagram Q is fixed and $OQ = PQ$, so that P moves on a circle passing through O. Then P' moves along the dotted straight line. Other straight line linkages are described by Cundy and Rollett[3]. Coxeter[1] and Forder[2] also discuss inversion.

2. Two properties of the Arbelos

The Greeks studied the configuration of the first diagram of Figure 5 in some detail. The three semi-circles with a common diameter form a shape called the Arbelos or shoemaker's knife. We want to prove that PSR, QTR are collinear sets of points, and inversion with regard to O produces the second diagram. The circles on OP, OQ as diameters become straight lines perpendicular to $P'OQ'$, while that on PQ as diameter becomes a circle with $P'Q'$ as diameter. We note in passing

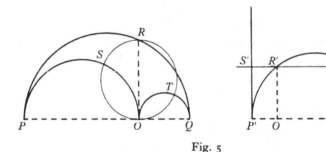

Fig. 5

that there is tangency at P' and Q', just as there was at P and Q. Further, the lines $P'S'$ and $Q'T'$, being inverses of touching semi-circles, are necessarily parallel.

The circle on diameter OR becomes a line parallel to $P'Q'$, since this circle touches PQ. Now $OP'S'R'$ is a rectangle, and therefore cyclic. The inverse of its circumcircle will be a straight line through P, S and R, so these points are collinear. The other collinearity follows in the same way.

The next property is illustrated in Figure 6. A chain of circles is

drawn, the nth having radius r_n and its centre at height h_n above OM. Then we have to prove that
$$h_n = 2nr_n,$$
a result obtained by Pappus without the use of inversion.

Inversion with regard to O turns semi-circles OL, OM into parallel lines, and the circle chain inverts into a new chain, all of the same

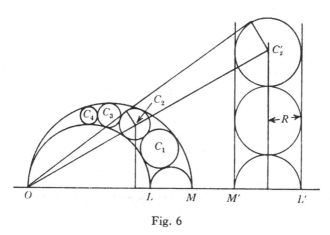

Fig. 6

radius R. The common tangent of the circle with centre C_2 and its inverse, the circle with centre C_2', has been inserted. From two pairs of similar right triangles we see that
$$\frac{r_2}{R} = \frac{OC_2}{OC_2'} \quad \text{and} \quad \frac{h_2}{4R} = \frac{OC_2}{OC_2'}.$$
This proves the result for $n = 2$, the same method applies for any value of n.

3. The problem of Apollonius

We wish to construct a circle to touch three given circles C_1, C_2 and C_3. For the moment the solution of Figure 7, with the three circles external to each other, and with the touching circle external to all three, is considered. If the smallest of the given circles is C_3, of radius r_3, we replace the problem by a new one. Circles C_1 and C_2 are replaced by concentric circles D_1 and D_2 with radii (r_1-r_3) and (r_2-r_3), while C_3 becomes the point D_3. Then the required circle, after expansion by r_3, goes through D_3 and touches D_1 and D_2.

45

Inverting with regard to D_3 this tangent circle becomes a line touching D_1' and D_2'. Four such lines can be found; we need the direct common tangent on the opposite side of the circles to D_3. This is a standard construction, and, once the line is found, its inverse is contracted by r_3 to give the required circle.

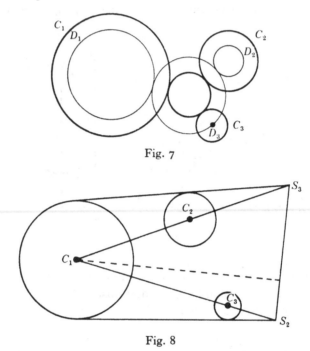

Fig. 7

Fig. 8

There are other possibilities, the tangent circle could include within it 3, 2 or 1 of the given circles. The last two cases each arise in three ways, so that, with the solution found above, there are eight possibilities. The same method will determine each one, but now some of the given circles will expand in the initial step.

The analogous problem in three dimensions is to construct a sphere to touch four given spheres. Again one of these is reduced to a point, and inversion with regard to it generates a simpler problem. We now require the common tangent plane of three spheres. In Figure 8 C_1, C_2 and C_3 are the centres of these spheres, while S_2 and S_3, lying in their plane, arise from common tangents as shown. Any plane through S_2 touching sphere C_1 will also touch C_3; we see this by imagining the plane figure rotated about the line of centres. Similarly, a plane through S_3 touching sphere C_1 also touches C_2.

46

Thus we need only construct a plane through S_2S_3 and touching sphere C_1. A section through the dotted line and perpendicular to the plane of the paper reduces this problem to that of drawing a tangent from a point to a circle. Thus the construction is completed; in general there will be sixteen solutions, as each of the given spheres is either inside or outside the tangent sphere.

4. *Steiner's chain of circles*

In the first diagram of Figure 9 we try to fill the region between circles C_1 and C_2 with a chain of touching circles. To see if this is possible we first choose a centre of inversion in such a way that C_1' and C_2' are concentric. The point O lies on the line of centres of C_1 and C_2 drawn below the circles, and we seek x so that the mid-points of $A_1'A_2'$ and $B_1'B_2'$ coincide. This leads to

$$\frac{k^2}{x} + \frac{k^2}{x + A_1A_2} = \frac{k^2}{x + A_1B_1} + \frac{k^2}{x + A_1B_2},$$

a quadratic giving two values of x.

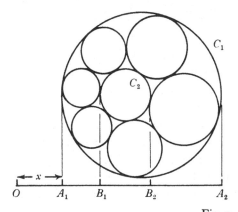
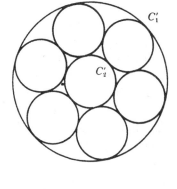

Fig. 9

In the inverse diagram we try to construct a chain of necessarily equal circles. The position of the first circle drawn is of no importance, the closure of the chain depends only on the radii of the base circles. Similarly, in the original picture, if a chain of n circles closes, then there is an infinite number of such chains. The property is said to be poristic; there is either no solution or an infinite number.

Let P_1, P_2, P be the centres and r_1, r_2, r the radii of C_1, C_2, and a

47

circle of the chain. Then we see that

$$PP_1 = r_1 - r, PP_2 = r_2 + r,$$

so that $PP_1 + PP_2$ is the same for all circles of the chain. It follows that the centres P lie on an ellipse with foci at P_1 and P_2. I am indebted to Prof. Ogilvy[4] for this interesting result. He has extended it to the centres of the spheres of Soddy's hexlet.

5. Soddy's hexlet of spheres

Soddy, the discoverer of isotopes, found the following results without using inversion, but this transformation makes their proof almost trivial. We start with spheres A, B and C in contact, and seek a chain of spheres touching all 3. An arbitrary sphere S_1 is chosen to touch A, B and C; then S_2 is constructed to touch S_1, A, B and C, this was shown to be possible in Section 3. Next S_3 touches S_2, A, B and C and so on. Soddy proved that S_6, besides touching S_5, A, B and C, touches S_1, so that a chain of 6 spheres can always be found. As S_1 can be chosen in a single infinity of ways there is an infinity of such hexlets. We can visualise a chain as rotating with continuous change in sizes of its elements, but with all contact conditions always satisfied.

If we consider the case when B and C are large compared with A it is easy to envisage the situation. There will be a gap between A, B and C, and the hexlet surrounds A and threads this gap.

Inversion with regard to the point of contact of B and C greatly simplifies the problem. These spheres become a pair of parallel planes touching sphere A'. There is evidently a chain of 6 spheres equal to A', and touching it as well as the 2 planes. This chain inverts into Soddy's hexlet, it can be rotated, and so can the hexlet. Since the points of contact of the 6 equal spheres with either A' or the planes B' and C' lie on a circle, the same is true for the hexlet, i.e. its 6 points of contact with each sphere lie in a plane.

Moreover, we can find 2 spheres passing through the centre of inversion and touching the 6 equal spheres. Their centres lie on the diameter of A' perpendicular to B' and C'. It follows that the hexlet touching A, B and C is in contact with a pair of planes.

Ogilvy[4] devotes a chapter of his book to Soddy's hexlet.

6. A theorem of Casey

In Figure 10(a) the two circles of radii r_1 and r_2 intersect at A at an angle α. Applying the cosine rule to $C_1 A C_2$ gives

$$\cos(\pi - \alpha) = \frac{r_1^2 + r_2^2 - C_1 C_2^2}{2r_1 r_2}.$$

Using Pythagoras theorem for $C_1 C_2$ we get

$$\cos \alpha = \frac{t_{12}^2 + (r_1 - r_2)^2 - r_1^2 - r_2^2}{2r_1 r_2} = \frac{t_{12}^2}{2r_1 r_2} - 1.$$

Since inversion does not alter α, it follows that the ratio of t_{12}^2 to $r_1 r_2$ is also invariant.

In Figure 10(b) the circles numbered 1 to 4 have a common tangent circle. The direct common tangent of 1 and 2 is of length t_{12} and so on. Casey's theorem states that

$$t_{12} \cdot t_{34} + t_{14} \cdot t_{23} = t_{13} \cdot t_{24}. \tag{2}$$

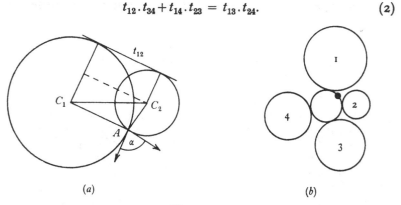

(a) (b)

Fig. 10

We first observe that (2) is equivalent to

$$\frac{t_{12}}{\sqrt{(r_1 r_2)}} \frac{t_{34}}{\sqrt{(r_3 r_4)}} + \frac{t_{14}}{\sqrt{(r_1 r_4)}} \frac{t_{23}}{\sqrt{(r_2 r_3)}} = \frac{t_{13}}{\sqrt{(r_1 r_3)}} \frac{t_{24}}{\sqrt{(r_2 r_4)}}. \tag{3}$$

Since all the ratios are unaltered by inversion, it suffices to prove (3) in a simpler situation obtained as follows.

We invert with regard to a point on the common tangent circle lying between its contacts with 1 and 2. The inverse consists of 4 circles

49

touching a line at points A_1, A_4, A_3 and A_2 taken in that order. We note that

$$A_{12} \cdot A_{43} + A_{14} \cdot A_{32}$$
$$= (A_{13} + A_{32})(A_{42} - A_{32}) + A_{14} \cdot A_{32}$$
$$= A_{13} \cdot A_{42} + A_{32}(A_{42} + A_{14} - A_{13} - A_{32}) = A_{13} \cdot A_{42}.$$

If this result is divided through by the square root of the product of the 4 radii of the new set of circles it gives the expression corresponding to (3) for the inverse figure. Hence (3) holds in the original situation.

7. Coaxal circles and stereographic projection

In Figure 11 we start with a set of concentric circles and a family of lines through their common centre C. Since any circle cuts all the lines at right angles, the same will be true of the inverse of this configuration with regard to a point O. This consists of two sets of coaxal circles; the second, coming from the family of lines, has the pair of points O' and C' in common; and these are point circles belonging to the first set.

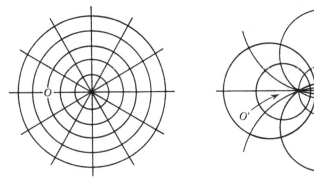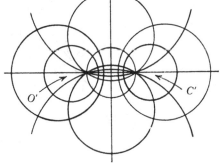

Fig. 11

This situation arises in a number of field problems in physics. An infinite conducting sheet with current entering at an electrode O' and leaving at C' has the set of circles through O' and C' as its lines of flow. The other set constitute the equipotential curves, and is cut at right angles by every line of flow.

50

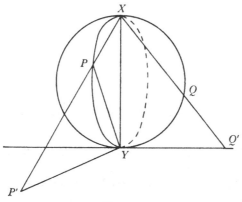

Fig. 12

In Figure 12 the sphere XY is touched by a plane $P'YQ'$ at Y, and we map the sphere on to it by projection from X. Since XPY is a right angle, it follows that

$$XP.XP' = XY^2,$$

hence projection also inverts the sphere into the plane. As inversion preserves angles between curves, stereographic projection does so too. Moreover, any circle on the sphere projects into a circle in the plane.

Assuming that X is the North Pole, then lines of latitude and longitude give concentric circles and their diameters. If X is not one of the poles these lines will project into two mutually perpendicular sets of coaxal circles. Any circle of latitude cuts all circles of longitude at right angles, and this remains true of the projected figure.

References

1. H. S. M. Coxeter. *Introduction to Geometry* (Wiley, 1961).
2. H. G. Forder. *Geometry* (Hutchinson, 1960).
3. H. M. Cundy and A. P. Rollett. *Mathematical Models*, 2nd edition (Oxford, 1961).
4. C. S. Ogilvy. *Excursions in Geometry* (Oxford, 1969).

6

Some ruled surfaces

1. *The doubly ruled quadric surface*

We first note that a unique line can be drawn through a point P in 3-space to cut a pair of non-intersecting or skew lines m and n. In Figure 1 (a) join P to all points of m, thus defining a plane. Similarly, P and n define a second plane, and the two planes intersect in the required common transversal through P. In the second diagram the dotted line is to be ignored for the moment. Starting with each point of a line l we can draw through it a common transversal of the 3 skew lines l, m

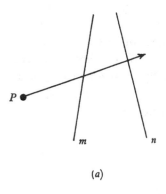

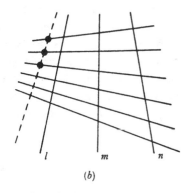

(a) (b)

Fig. 1

and n This infinite set of lines generates a ruled surface. We now try to find the degree of this surface, i.e. the number of points in which it is cut by a general straight line a.

In Figure 2 we select an origin O on line a, and measure distances along the line from this origin. Starting with P at distance x from O we construct a point Q at distance y as follows. Through P draw the line cutting l and m. Through the point B in which it cuts l draw the line cutting n and a. Its intersection with a is Q, the point at distance y from O. Our construction of Q from P is unique, and if used in reverse yields a unique P for a given Q. It therefore seems reasonable to assume a relationship of the form

$$axy + bx + cy + d = 0,$$

since this is the most general algebraic relationship in which x determines y uniquely and vice versa.

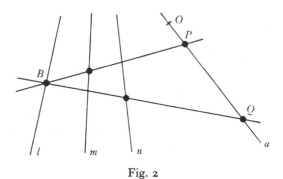

Fig. 2

The quadratic equation

$$at^2 + (b+c)t + d = 0,$$

has 2 roots. If x equals either of these, the construction starting from P leads back to this point, for $y = x$ and $Q \equiv P$. This means that PB cuts n as well as l, m and a, i.e. each root of the quadratic leads to a common transversal of the 4 lines. Thus a intersects just 2 lines defining the ruled surface, and it is of degree 2; it is called a quadric surface.

We return to the second diagram of Figure 1 and insert the dotted line cutting 3 of the transversals of l, m and n. This can be done in a single infinity of ways. This dotted line cuts the quadric in 3 points, so if it had been treated as the line a above there would be 3 values of t satisfying the quadratic equation. The only way in which this can happen is for all coefficients to vanish, and then it is satisfied by all values of t. In other words the dotted line lies in the surface, cutting not just 3, but all the transversals of l, m and n. So we have 2 sets of lines, any line of one set cutting each line of the other. The section of the quadric surface by a plane is a conic. The special plane defined by one line of each system cuts the surface in just this pair of lines; this is a degenerate conic, indicating that the plane touches the quadric at the point of intersection of the lines.

We shall not prove the following result, but it is interesting to note that, if universal joints are inserted at all the intersections of lines in Figure 1 (b), the surface can be deformed continuously without tearing

53

of the joints or bending of the lines. The surface is illustrated in Figures 3(*b*) and 4. By squashing in the vertical direction it flattens out to a disc-like ellipse, touched by each generator. Deformations along the other axes produce plane hyperbolas; Hilbert and Cohn-Vossen[1] illustrate these changes.

2. *Quadrics in general*

The best known quadric is the ellipsoid of Figure 3(*a*). Its equation is

$$\frac{x^2}{a^2}+\frac{y^2}{b^2}+\frac{z^2}{c^2} = 1, \tag{1}$$

sections by planes perpendicular to any of the 3 axes being ellipses.

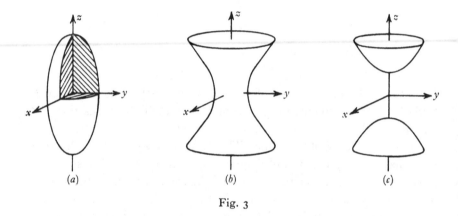

(*a*)　　　　　(*b*)　　　　　(*c*)

Fig. 3

Figure 3(*b*) is the ruled quadric we met above, a hyperboloid of one sheet with equation

$$\frac{x^2}{a^2}+\frac{y^2}{b^2}-\frac{z^2}{c^2} = 1. \tag{2}$$

Figure 4 shows the rulings upon this surface. We consider the pair of planes

$$\frac{x}{a}-\frac{z}{c} = \mu\left(1-\frac{y}{b}\right), \quad \frac{x}{a}+\frac{z}{c} = \frac{1}{\mu}\left(1+\frac{y}{b}\right). \tag{3}$$

For given μ they define a straight line, and as μ varies a family of such lines results. The 'product' of the two equations (3) leads to (2), so that a point satisfying both of them lies on the surface. Thus (3) gives one family of generating lines. By interchanging the signs in the 2 right-hand

54

side terms the second family of lines results. If we try to obtain lines on (1) by the same method we get

$$\frac{x}{a} - \frac{iz}{c} = \mu \left(1 - \frac{y}{b} \right), \quad \frac{x}{a} + \frac{iz}{c} = \frac{1}{\mu} \left(1 + \frac{y}{b} \right). \qquad (4)$$

Thus it is a ruled surface, but the lines are not real ones because of the complex coefficients in (4).

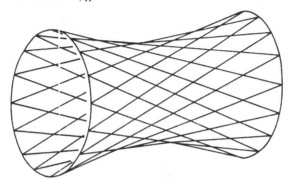

Fig. 4

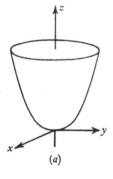

(a)

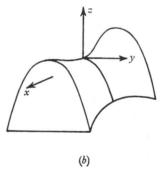

(b)

Fig. 5

The third general type of quadric with equation

$$\frac{x^2}{a^2} + \frac{y^2}{b^2} - \frac{z^2}{c^2} = -1,$$

a hyperboloid of two sheets, is also without real generators; it is illustrated in Figure 3(c).

Two special types of quadric are shown in Figure 5. The first has an equation of the form

$$z = \frac{x^2}{a^2} + \frac{y^2}{b^2}, \qquad (5)$$

55

and is an elliptic paraboloid without real generators, it is shown in Figure 5(a). Figure 5(b) shows the hyperbolic paraboloid

$$z = \frac{x^2}{a^2} - \frac{y^2}{b^2},$$

a saddle-shaped surface with two sets of real generators; these are shown in Figure 6.

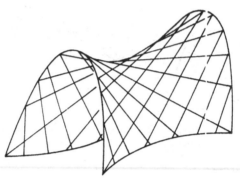

Fig. 6

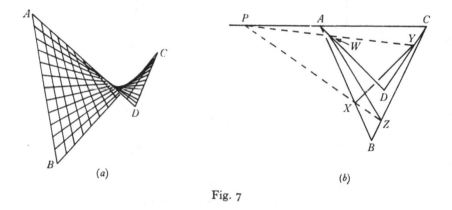

(a)

(b)

Fig. 7

Figure 7(b) illustrates the construction of a hyperbolic paraboloid from a skew quadrilateral $ABCD$ with sides of equal length. One set of generators is typified by XY, where $AX = DY$. Similarly, if $CZ = DW$, then WZ gives the other set. We prove that ZW and XY intersect as follows. If ZX cuts CA at P then, by the theorem of Menelaus

$$\frac{CZ}{ZB} \cdot \frac{BX}{XA} \cdot \frac{AP}{PC} = -1.$$

Replacing lengths by equal lengths gives

$$\frac{CY}{YD} \cdot \frac{DW}{WA} \cdot \frac{AP}{PC} = -1,$$

and the converse theorem shows that P, Y and W are collinear. Thus, as ZX, YW intersect at P, they are coplanar so that XY, WZ, lying in their plane, also meet.

Hilbert and Cohn-Vossen[1] and Sommerville[2] discuss many other aspects of quadric surfaces.

3. The theorems of Pascal and Brianchon

Figure 8(a) shows a hexagon with vertices 1 to 6 inscribed in a conic. Pascal's celebrated theorem states that, if opposite sides

12 and 45 meet in X,

23 and 56 meet in Y,

34 and 61 meet in Z,

then these points of intersection are collinear.

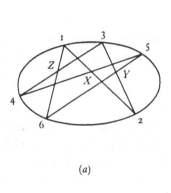

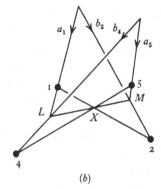

(a)   (b)

Fig. 8

Pascal's theorem can be deduced from the existence of generators on a quadric surface. We regard the conic as the intersection of a plane π and the surface. Through points 1, 3, 5 we select generators a_1, a_3, a_5 from one system; and through 2, 4, 6 pass b_2, b_4, b_6 from the other. Let

a_1, b_4 intersect in L,

b_2, a_5 intersect in M,

a_3, b_6 intersect in N.

57

Then in Figure 8(b) the lines a_1, b_4, a_5, b_2 form a skew quadrilateral with one diagonal LM. It follows at once that 12 and 45 intersect on LM, i.e. X lies on LM, similarly Y, Z lie on MN, NL. As X, Y and Z also lie in plane π, they lie on its intersection with plane LMN, i.e. they are collinear.

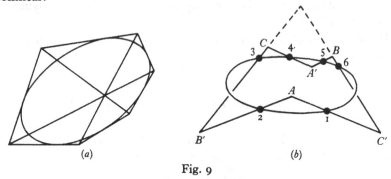

Fig. 9

Brianchon's theorem states that the diagonals of a hexagon circum-scribing a conic are concurrent, it is illustrated in Figure 9(a). We again consider a quadric surface and its generators. However, the relation of this quadric to the plane Brianchon configuration does not appear until the end of the argument.

The second diagram of Figure 9 shows three generators of one type through points 1, 3, 5 and 3 of the other through 2, 4, 6. The 6 points 1 to 6 again lie in a plane section π of the surface. As BC' and $B'C$ are generators of different types they intersect, and are therefore coplanar. Hence BB' and CC' meet; two similar proofs show that all three of AA', BB', CC' meet in a point.

We now assume that the tangent planes to the surface at all points of the conic in which it is cut by π meet in a point P (the pole of π). Projecting from P on to a plane, the conic yields a conic. The generator AC' lies in the tangent plane at 1, and this goes through P. Thus the projection of AC' touches the projected curve, and the six generators project into a circumscribed hexagon. The lines AA', etc., project into its diagonals which therefore concur, thus the plane projection is just Brianchon's configuration.

4. The astigmatic surface

Figure 10(a) illustrates the process of image formation by a lens. All rays leaving O at a given moment reach points of a spherical wave-front s

58

at some later time. The lens deforms this wave-front into surface t with individual rays still perpendicular to it. If t is part of a sphere of radius R, its equation will be

$$z(2R - z) = x^2 + y^2,$$

and this can be expanded in the form

$$z = \frac{x^2 + y^2}{2R} + A(x^2 + y^2)^2 + \ldots,$$

where $A = 1/8R^3$. The normals to t then pass through a true point image on the axis of the lens. In practice A has a different value, and normals from different circular zones of the non-spherical surface of revolution t meet at different points on the axis. They all touch a caustic surface with a cusp at the so-called paraxial image, formed by rays very close to the axis. This is the phenomenon of spherical aberration.

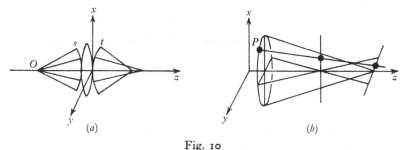

Fig. 10

In an astigmatic lens there is no longer axial symmetry, but we do assume that A is negligible. This situation is common in the human eye, and is also encountered if the object O lies off the axis of a symmetrical lens. The equation of t is

$$z = \frac{x^2}{2R_1} + \frac{y^2}{2R_2}. \tag{6}$$

This is the elliptic paraboloid (5), and we assume that R_1 exceeds R_2. We consider the elliptic zone lying in a plane at distance d from the origin. As θ varies the ellipse is described by the point P

$$\{\sqrt{(2R_1 d)} \cos \theta, \quad \sqrt{(2R_2 d)} \sin \theta, \quad d\}. \tag{7}$$

The equation of the normal to $z = f(x, y)$ at (x_1, y_1) is given by

$$(x - x_1)\Big/\frac{\partial f}{\partial x_1} = (y - y_1)\Big/\frac{\partial f}{\partial y_1} = (z - z_1)\Big/-1,$$

and applied to (7) this gives

$$\frac{x-\sqrt{(2R_1 d)}\cos\theta}{\sqrt{(2d/R_1)}\cos\theta}=\frac{y-\sqrt{(2R_2 d)}\sin\theta}{\sqrt{(2d/R_2)}\sin\theta}=\frac{z-d}{-1}. \tag{8}$$

If we put $z=d+R_1$ in (8) we find that $x=0$ for all values of θ. The normals to our elliptic section of (6) thus cut a horizontal straight line that itself intersects the z axis at distance R_1 from the centre of the ellipse. Putting $z=d+R_2$ gives $y=0$, so the normals also cut a vertical line at distance R_2. These normals form the astigmatic surface shown in Figure 11.

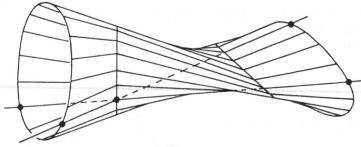

Fig. 11

This surface is of degree 4, and through each point of its 2 focal lines pass 2 generators. One such pair have been dotted in over regions where they are concealed by the surface. Optically we get 2 perpendicular line images of a point object at distances $(d+R_2)$ and $(d+R_1)$ from the lens. These lines move as d, and therefore the zone of t being considered, vary. However, the maximum values taken by d is small, so this broadening of the 2 line images is negligible.

Such a lens gives a sharp image of a vertical object at the first line and an equally sharp image of a horizontal object at the second. At the intermediate distance

$$z=d+\sqrt{(R_1 R_2)},$$

it is easily proved that normals to the zone at distance d from the origin cut a circle of radius

$$\sqrt{(2d)}(\sqrt{R_1}-\sqrt{R_2}),$$

with its centre on the axis, and in a plane perpendicular to the axis. The largest d, corresponding to rays from the circumference of the lens, defines the so-called circle of least confusion.

60

5. *The pitch of a ruled surface*

We first consider the lines touching a twisted curve in 3-space. They form a surface consisting of two distinct leaves intersecting in the curve itself. In Figure 12 one of the tangent lines has been emphasised; the portions on either side of the point of contact lie one in each leaf of the surface. A plane cuts the surface in a curve with a cusp lying on the original space curve, which is therefore called a cuspidal edge.

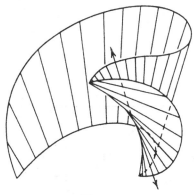

Fig. 12

Such a surface is said to be developable. Its characteristic property is that, as we move along any of its generating lines, the tangent plane remains fixed. The surface can be flattened out on to a plane without distortion.

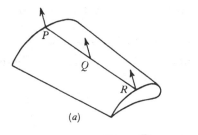

(a)   (b)

Fig. 13

In Figure 13(a) an aeroplane wing is sketched. Until recently such wings were portions of developable surfaces. Thus, besides being ruled by lines like *PQR*, the normals at points *P*, *Q*, *R* are of fixed direction.

61

This makes it easy to machine model wings for wind tunnel tests. The roughly cut shape is set up with the tangent plane at points P, Q, R horizontal. A tool rotating about a vertical axis produces a flat along this line as the shape moves perpendicular to the picture plane. After producing a series of such flats, the ridges between them are removed by hand scraping.

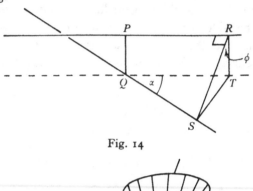

Fig. 14

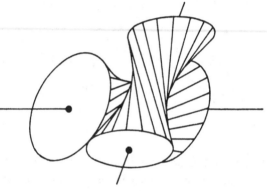

Fig. 15

In the general ruled surface the tangent plane rotates as we move along a generator. In Figure 14 PR and QS are neighbouring generators, their closest distance PQ being of length d. The tangent plane at Q contains QS and PQ, while that at S contains QS and RS, R being the point of PR nearest to S. Since the angle α is small we have

$$\tan \phi \simeq \frac{\alpha . QS}{d} = QS \times \text{pitch}, \quad \text{where} \quad \text{pitch} = \frac{\alpha}{d}.$$

Thus pitch relates rotation of the tangent plane to distance moved along a generator. Measured from the tangent plane at the point of striction, where neighbouring generators are closest, ϕ goes from $-\frac{1}{2}\pi$ to $\frac{1}{2}\pi$ as we move from $-\infty$ to ∞ along the generator. For a developable surface the pitch is everywhere zero, i.e. the quantity α is of second order in d.

The locus of points like P is called the line of striction; for the hyperboloid of Figure 3(b) it is the ellipse lying in the plane $z = 0$. If the hyperboloid is a solid of revolution, i.e. if $a = b$, it can be proved that the pitch equals c.

In Figure 15 the hyperboloids of revolution touch along a common generator and the points of striction coincide. For this to be possible the tangent planes must rotate at the same rate along the generator, i.e. the values of 'c' are equal. If the values of 'a' are also equal the surfaces are congruent, and the common generator makes equal angles with the lines of striction. Rotation about their axes then produces a pure rolling motion of one on the other.

If the values of 'a' differ pure rolling would cause points of striction to separate. Rotation about the axes now produces rolling combined with sliding of one generator along the other to maintain strictional contact. In spite of this sliding, pairs of 'hypoid' gears with teeth cut along generators are often used to couple non-intersecting shafts. The use of different values of 'a' enables a speed change to be effected as well. Hilbert and Cohn-Vossen[1] discuss these kinematic problems in greater detail.

References

1. D. Hilbert and S. Cohn-Vossen. *Geometry and the Imagination* (Chelsea, 1952).
2. D. M. Y. Sommerville. *Analytical Geometry of Three Dimensions* (Cambridge, 1934).

7

Random walks

1. *Coin tossing and random walks*

We assume that the penny being tossed has a probability p of giving heads and $(1-p)$ or q of giving tails. This implies that, in a large number of throws N, there will be about Np heads. We shall be interested in the difference between the observed number of heads h and the expected number np, where n is not necessarily large.

In Figure 1 we illustrate a one-dimensional random walk. Starting at the origin steps are of unit length to left or right; at each stage the toss of a coin determines the next step, heads indicating moves to the right.

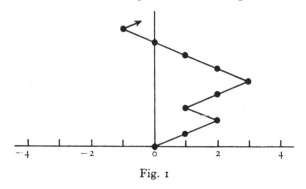

Fig. 1

Although along a line, it is easier to illustrate the walk if time is introduced as a second coordinate. Thus the walk above results from the sequence of moves $$R\,R\,L\,R\,R\,L\,L\,L\,L\,R\dots.$$

The final distance to the right after n moves will be $(2h-n)$ if h heads have appeared.

We shall be concerned with the probability of returning to the zero line, and in the proportion of the time spent on one side. For a symmetrical penny with $p = 0.5$, we might suppose that in most walks of n steps about half the time is spent on the right. However, this is not so; in the majority of walks much more time is spent on one side than on the other. In order to eliminate positions on the zero line when doing such calculations we adopt the following convention: a point on the zero line is regarded as being on the right if the previous position was on that side and vice versa.

The symmetrical two-dimensional walk involves a step to north, south, east or west at each stage, the probability of each being $\frac{1}{4}$. An analogous three-dimensional walk can also be envisaged. The probability of a return to the starting point is again of interest.

A modified two-dimensional walk inside the region bounded by a closed curve can also be considered. The walk lasts until the curve is reached, and it is therefore called an absorbing barrier. This problem is related to the solution of the Laplace equation in potential theory.

Historically the study of random walks originates from the discovery of Brownian motion. Small particles suspended in a liquid exhibit a curious zig-zag motion, even when every precaution is taken to eliminate convection currents. The explanation in terms of the bombardment of the particle by molecules of the liquid is as follows. At any instant there are a large number of impacts occurring in random directions. On average these balance out, but occasionally a preponderance of impacts in a given direction does occur. There is then a resultant acceleration in that direction.

Einstein published a theory of the Brownian motion in 1905, the year in which he announced his discovery of Special Relativity. In that year too he published a classical paper on the photoelectric effect; this, like the one on Brownian motion, introduced probability considerations into mathematical physics. Since then such considerations have become of increasing importance, and the study of random walks arises in numerous physical problems.

2. *The binomial distribution*

If a penny is thrown n times, what is the probability that just h heads appear? Such an outcome implies a sequence such as

$$H\,H\,T\,H\,T\,T \ldots H\,T,$$

containing heads h times. The probability of getting exactly this sequence is given by the product

$$p \cdot p \cdot q \cdot p \cdot q \cdot q \cdot \ldots \cdot p \cdot q,$$

65

where each H leads to a factor p and each T to a q. Since there are h p's and $(n-h)$ q's this probability is

$$p^h q^{n-h}.$$

However, other patterns also give h heads, in fact the total number of such patterns will be

$$\binom{n}{h} = \frac{n!}{h!\,(n-h)!},$$

the number of ways of choosing h out of n places for the heads to occur. As all these patterns have the same probability of occurring, the probability of h heads is

$$\binom{n}{h} p^h q^{n-h}. \tag{1}$$

The following table is for $p = 0.5$ and $n = 10$:

Table 1

No. of heads	Probability
0 or 10	0·0010
1 or 9	0·0098
2 or 8	0·0440
3 or 7	0·1172
4 or 6	0·2050
5	0·2460

From this table we can find the probability of getting less than 3 heads; it is 0·0548, obtained by adding the first 3 terms.

If we multiply (1) by t^h and sum from 0 to n, we obtain the binomial expansion of

$$(q + pt)^n, \tag{2}$$

and call (2) the probability generating function. It enables us to find the average or mean number of heads in n throws. This mean value is obtained by multiplying (1) by h and summing for h from 0 to n, giving

$$0 . q^n + 1 \binom{n}{1} pq^{n-1} + 2 \binom{n}{2} p^2 q^{n-2} + \ldots + np^n. \tag{3}$$

This is a weighted sum of the numbers from 0 to n, the weights, which add up to 1, being the probabilities of occurrence of the numbers they multiply. If we differentiate (2) with regard to t and put $t = 1$ we get np. On the other hand if (2) is expanded before differentiating and putting $t = 1$ we get (3). Thus the mean number of heads is np.

66

Much the same technique enables us to find the mean value of $(h-np)^2$, two differentiations are needed and the result is npq. It is useful to note that observed values seldom differ from the expected number np by more than twice the square root of npq. In other words most trials give between

$$np - 2\sqrt{(npq)} \quad \text{and} \quad np + 2\sqrt{(npq)} \tag{4}$$

heads. Interesting too is the fact that the spread of values implied by (4) varies as $\sqrt{n}$. If we use h/n to estimate an unknown p, the uncertainty in the answer varies inversely as $\sqrt{n}$.

3. *Approximating to the binomial distribution*

In principle, (1) enables us to evaluate probabilities, or sums of probabilities, in simple coin tossing. For large n these calculations are not convenient, and we examine two well-known approximations.

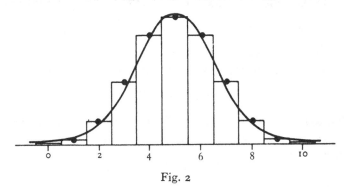

Fig. 2

In Figure 2 the values of Table 1 have been plotted as ordinates. Corresponding to the ordinate at 2, a column of the same height, and with base from 1·5 to 2·5, has been drawn and so on for all ordinates. The probability of obtaining 2 heads or less is the sum of the heights of the first 3 ordinates, or the area of the histogram taken from $-0·5$ to $+2·5$.

We see that the histogram is approximately bell-shaped, and this led de Moivre to relate it to the bell-shaped curve $\exp(-x^2)$. More precisely, he showed that, if we define a variable x by

$$x = \frac{h-np}{\sqrt{(npq)}}, \tag{5}$$

67

the curve

$$y = \frac{1}{\sqrt{(2\pi)}} e^{-\frac{1}{2}x^2}, \tag{6}$$

is close to the histogram. To approximate to the area under the histogram up to 2·5 we put $h = 2·5$ in (5) giving $-1·58$. Then the integral of (6) from $-\infty$ to $-1·58$ gives 0·057, a value quite close to 0·0548 found from Table 1.

The expression (6) is called the normal probability function, and for large n with p not too far from 0·5 it provides a good approximation to the binomial distribution (1). Since many applications of importance involve a large n with a small p, it is necessary to have a separate approximation for such cases. Let us consider an unusual event such as the issue of an unperforated sheet of postage stamps. The number of sheets n issued per year is large, and p, the probability that a given sheet is faulty, is small. In practice we do not usually have access to either number, but instead may well know $m = np$, the average number of faulty sheets issued per year.

We write the generating function (2) in the form

$$(q+pt)^n = (1-p+pt)^n = [1+(m/n)(t-1)]^n.$$

When n tends to infinity the limit of this expression is known to be $\exp(mt-m)$. This, when expanded, gives

$$e^{-m}e^{mt} = e^{-m} + me^{-m}t + (m^2/2!)e^{-m}t^2 + \dots.$$

The coefficient of t^r in the generating function is the probability of r occurrences, just as it was in expression (2). So from the series this probability must be

$$\frac{m^r e^{-m}}{r!}. \tag{7}$$

This result is known as the Poisson distribution of probabilities.

To illustrate its use let us suppose that $m = 5$. What is the probability that, in a given year, less than 3 faulty sheets will be issued? The answer is

$$e^{-5} + \frac{5}{1!}e^{-5} + \frac{25}{2!}e^{-5} = 0·125,$$

so that for 1 year in 8 there will be 2 or fewer faulty issues.

While the Poisson distribution might be expected to apply to many rare events, it is often found that results observed do not agree well with its predictions. The reason is that it assumes a constant p in all trials. In practice a faulty perforating machine might go undetected for a period. During this time the value of p would be much higher than usual. Such fluctuations of p usually lead to a greater scatter of observed values about the mean than that predicted by the Poisson formula.

It is interesting to note an analogue of (4) for the Poisson distribution. The limits

$$m - 2\sqrt{m} \quad \text{and} \quad m + 2\sqrt{m},$$

include the majority of observed numbers of an event in a given period, where m denotes the average number in that time. Thus, if the average rate of occurrence of a particular type of industrial accident is 100 per year, we should expect fluctuations over the range 80 to 120. As indicated above, observed values may show still wider swings because the simple Poisson model does not describe the situation adequately.

4. Returns to zero in the random walk

We consider a random walk of $2n$ steps, and define 2 probabilities.

u_{2n} = probability of reaching the zero line at $2n$th step.

v_{2n} = probability of reaching the zero line for the first time at the $2n$th step.

The first of these involves n left and n right moves and hence

$$u_{2n} = \binom{2n}{n} p^n q^n = \binom{-\frac{1}{2}}{n} (-4pq)^n.$$

The second form for u_{2n}, obtained by algebraic manipulation, leads to

$$U(x) = 1 + u_2 x^2 + u_4 x^4 + \ldots = (1 - 4pqx^2)^{-\frac{1}{2}}, \tag{8}$$

the generating function for the u's. We define also

$$V(x) = v_2 x^2 + v_4 x^4 + \ldots.$$

The generating function $U(x)$ was constructed from known probabilities. With its aid we shall determine $V(x)$, then its series expansion will give us the values of the v's. A return to the zero line at the $2n$th step can arise in the following different ways.

(1) No previous return to zero.

(2) First return to zero at $(2n-2)$, and a return 2 after.

(3) First return to zero at $(2n-4)$, and a return 4 after.

$\ldots\ldots\ldots\ldots\ldots\ldots\ldots\ldots\ldots\ldots\ldots\ldots\ldots\ldots\ldots\ldots\ldots\ldots\ldots\ldots\ldots\ldots\ldots\ldots$

$\ldots\ldots\ldots\ldots\ldots\ldots\ldots\ldots\ldots\ldots\ldots\ldots\ldots\ldots\ldots\ldots\ldots\ldots\ldots\ldots\ldots\ldots$

(n) First return to zero at 2, and a return $(2n-2)$ after.

Figure 3 illustrates event number (3) in this list. The dotted path includes no returns to zero. There are numerous ways in which, after the return at $(2n-4)$, we return again at $2n$. Two of these are shown, one involving also a return at $(2n-2)$.

The probability of event (3) is the product of the probability of a first return at $(2n-4)$, and of a further return in 4 steps, i.e. it is

$$v_{2n-4}u_4.$$

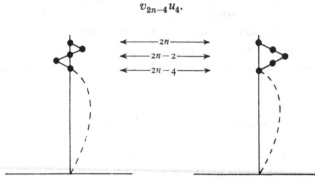

Fig. 3

We note that both the types of path shown in the figure are covered by the term u_4. We compute the probabilities of each of the events (1) to (n) and add them to get

$$u_{2n} = v_{2n} + v_{2n-2}u_2 + v_{2n-4}u_4 + \ldots + v_2 u_{2n-2}. \tag{9}$$

Consider the coefficient of x^{2n} in the product $U(x)V(x)$; for $n > 0$ this is given by the right-hand side of (9). It follows that

$$U(x) = 1 + U(x)V(x),$$

and we get at once

$$V(x) = 1 - \frac{1}{U(x)} = 1 - (1 - 4pqx^2)^{\frac{1}{2}}.$$

The binomial expansion now gives

$$v_{2n} = -\binom{\frac{1}{2}}{n}(-4pq)^n,$$

and we can evaluate the v's.

The probability of at least 1 return in $2n$ steps is

$$v_2 + v_4 + v_6 + \ldots + v_{2n},$$

since there must be a first return, and this can occur at steps 2, 4, ..., $2n$.

70

If n is large this sum is virtually equal to $V(1)$. The probability of an eventual return to zero in a long walk is thus

$$V(1) = 1 - (1 - 4pq)^{\frac{1}{2}} = 1 - [(p+q)^2 - 4pq]^{\frac{1}{2}} = 1 - |p - q|.$$

For a symmetrical walk $p = q$, and the probability of eventual return is unity, i.e. return to zero is certain.

In the two-dimensional symmetrical walk, return to the starting point is also certain. In three-dimensions however, the probability of eventual return is only 0·35. Dynkin and Uspenskii[1] give a very interesting account of two-dimensional walks, particularly for the case of a bounded region. The results just mentioned are discussed by Feller[2].

5. The time spent on one side

Feller[2] proves the results of this section; they are discussed here because of their rather unexpected nature. We consider a walk of $2n$ steps, and, with the allocation of points on the zero line adopted earlier, there will be $2r$ positions on the right and $(2n - 2r)$ on the left. The probability of this event can be shown to be

$$\binom{2r}{r} \binom{2n - 2r}{n - r} \frac{1}{2^{2n}}. \tag{10}$$

The method of proof is very similar to that used in the last section, and the formula for v_{2n} found there is a central requirement.

The expression (10) gives the values of Table 2 if $n = 10$ (20 steps).

Table 2

$2r$	Probability
0 or 20	0·1762
2 or 18	0·0927
4 or 16	0·0736
6 or 14	0·0655
8 or 12	0·0617
10	0·0606

Thus the least likely outcome is that half the time be spent on each side. It is nearly 3 times as likely that all the time will be on the right. Denoting the proportion of time on the right by $x = r/n$, we can prepare the frequency diagram of Figure 4. It is U-shaped, in sharp contrast to the normal curve.

The probability of being on the right for 4 or less steps, obtained by adding the first 3 terms of Table 2 is 0·3425. We seek an analytic function which, like the normal curve of Section 3, makes such computations easier.

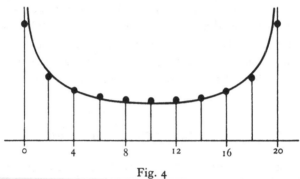

Fig. 4

Stirling's asymptotic formula is

$$n! \sim n^n \sqrt{(2\pi n)}\, e^{-n},$$

and it can be used to replace each of the factorials in (10). Simple manipulation gives

$$\frac{1}{n\pi \sqrt{[x(1-x)]}},$$

whose graph is a U-shaped curve, infinite at $x = 0$ or 1. Writing $dx = 1/n$, we find that the chance of spending a fraction X or less of the time on the right is about

$$\int_0^X \frac{dx}{\pi \sqrt{[x(1-x)]}} = \frac{2}{\pi}\sin^{-1}\sqrt{X},$$

the Arc Sine Law of Chung and Feller.

We set $X = \frac{5}{20}$ for the probability of spending 0, 2 or 4 steps on the right, a continuity correction of half an interval of size 2 being added to 4 to give the top line of X. This correction should be compared with integration up to 2·5 in Section 3, when a probability of 2 or less was being estimated. The result is 0·3333, reasonably close to our exact value of 0·3425. What is the probability of spending less than 1% of the time on the right in a long walk? The formula gives 0·064, so this apparently unlikely situation will arise oftener than once in 20 walks.

6. A Monte Carlo solution of the Laplace problem

We wish to find a function $\psi(x, y)$ at points inside a closed curve, with its value specified at each point of the boundary, and with

$$\nabla^2 \psi \equiv \frac{\partial^2 \psi}{\partial x^2} + \frac{\partial^2 \psi}{\partial y^2} = 0, \qquad (11)$$

satisfied at all internal points.

Figure 5(a) shows a square mesh of points superimposed on the curve, points on the stepped boundary of this mesh lie close to the curve. The continuous problem is replaced by that of finding $V(P)$ at each internal mesh point like P. At boundary points V takes the value that ψ has at the nearest point of the true curved boundary. It remains to find a relationship between neighbouring V's analogous to (11). Let the mesh interval be h, and suppose that K, L, M and N are the nearest points to a typical mesh point J.

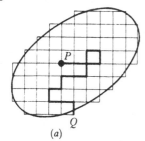
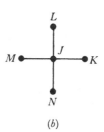

(a) $\qquad$ (b)

Fig. 5

We approximate as follows

$$\left(\frac{\partial V}{\partial x}\right)_{u-\frac{1}{2}h} \simeq \frac{V(J) - V(M)}{h}, \qquad \left(\frac{\partial V}{\partial x}\right)_{u+\frac{1}{2}h} \simeq \frac{V(K) - V(J)}{h},$$

where (u, v) are the coordinates of J. These in turn give

$$h^2 \left(\frac{\partial^2 V}{\partial x^2}\right)_u \simeq h\left(\frac{\partial V}{\partial x}\right)_{u+\frac{1}{2}h} - h\left(\frac{\partial V}{\partial x}\right)_{u-\frac{1}{2}h} \simeq V(M) + V(K) - 2V(J).$$

A similar result in the y direction leads us to replace (11) by

$$4V(J) = V(K) + V(L) + V(M) + V(N). \qquad (12)$$

There is a value of V to be found at each internal point, and each such point also supplies an equation like (12). While a direct solution of the

73

set of equations is feasible, a number of iterative ways of approximating to the answer have been developed.

We adopt another approach; to find $V(P)$ we conduct a large number of symmetrical walks from P. These terminate as soon as a boundary point Q is encountered. If n_i walks out of a total of N terminate at the ith boundary point, where V takes the known value V_i, then

$$V(P) \simeq \sum_{(i)} \frac{n_i V_i}{N},$$

the sum being over all boundary points.

The justification is as follows. We define $U_i(P)$ as the probability that a walk starting at P ends on the ith boundary point. Then the function U_i takes the value 1 at the ith boundary point and zero at all others. Moreover, the probability of moving from J to each of K, L, M and N is $\frac{1}{4}$, for our walk is symmetric. Thus the probability of ending at the ith boundary point after starting from J is given by

$$U_i(J) = \tfrac{1}{4}U_i(K) + \tfrac{1}{4}U_i(L) + \tfrac{1}{4}U_i(M) + \tfrac{1}{4}U_i(N).$$

Comparison with (12) shows that the function U_i is also an approximate solution of the Laplace problem with the boundary values defined above. Our random walks enable us to estimate $U_i(P)$ by

$$U_i(P) \simeq \frac{n_i}{N}.$$

We next consider the function whose value at P is defined by

$$W(P) = V_1 U_1(P) + V_2 U_2(P) + \dots + V_i U_i(P) + \dots, \qquad (13)$$

the sum being over all boundary points. Its value at the ith boundary point is V_i, for there all U's except U_i vanish, and this one is unity. So $W(P)$ has the same boundary values as $V(P)$. Moreover, as

$$\nabla^2 W(P) = \sum_{(i)} V_i \nabla^2 U_i(P) \simeq 0,$$

this function satisfies the Laplace equation in finite difference form. It is known that the problem has a unique solution, so that $W(P)$ coincides with $V(P)$. We estimate it by

$$V(P) = \sum_{(i)} V_i U_i(P) \simeq \sum_{(i)} \frac{n_i V_i}{N}.$$

While such a method would certainly not be used in this case, it does illustrate an approach of great value. In complicated physical situations

Monte Carlo procedures of this kind are often the only possible way of obtaining solutions. With the advent of the electronic computer they have been widely used in such fields as nuclear physics.

References

1. E. B. Dynkin and V. A. Uspenskii. *Random Walks* (D. C. Heath and Co., 1963).
2. W. Feller. *An Introduction to Probability Theory and its Applications*, vol. 1, 2nd edition (Wiley, 1957).

8

The four-colour problem

1. *Introduction*

Cayley proposed the problem of determining the minimum number of colours needed for a map, no two adjacent regions being of the same colour. The sea surrounding the land area is also to be coloured; Figure 1 (*a*) is of a simple map needing 4 colours. The colours marked 2 do not count as touching at the multiple vertex of Figure 1 (*b*). Such vertices can always be removed as in (*c*), and the resulting map, with only triple vertices, is called regular. A solution for the transformed regular map evidently gives a solution for the original one, but the converse does not hold. All areas are assumed to be simply connected, i.e. there are no ring-shaped regions.

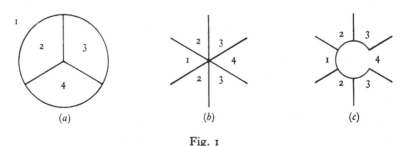

(*a*) (*b*) (*c*)

Fig. 1

We saw above that 4 colours may be necessary, and we prove later that 5 suffice. We shall also show that, if there are less than 12 regions, 4 colours will do. Appel[7] has proved, almost a century after Cayley's proposal, and by using 1000 hours of computer time, that 4 always suffice. A number of equivalent statements of the problem can be formulated; we shall discuss one later. The 2- and 3-colour problems are easier and we shall derive the conditions under which these smaller numbers are sufficient.

We can think of our map as lying on a sphere; in fact the problem is a topological one. This means that any continuous distortion of the spherical surface without tearing does not affect its solution. However, the problem of a map lying on an anchor ring is different. Oddly enough it is easily solved; we shall prove that 7 colours always suffice. It is possible to draw 7 hexagons on the anchor ring, each of which touches the other 6. Thus 7 colours are necessary for this map; we next examine its derivation.

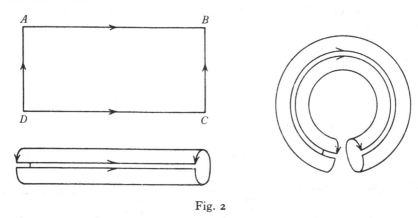

Fig. 2

Figure 2 shows how an anchor ring can be formed from a rectangular sheet of rubber. This is first rolled into a cylinder so that edges AB, CD come into contact. Then the cylinder is bent into a ring and the circular

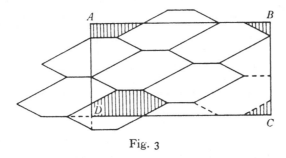

Fig. 3

ends joined. We note that the arrows on these end circles run in the same direction. The possibility of a closing up of the rubber sheet with a reversal of directions is discussed by Hilbert and Cohn-Vossen[4].

In Figure 3 we have a hexagon surrounded by 6 others. Various parts of the cluster lie outside the rectangle $ABCD$, but they have counterparts within this area indicated by dotted lines. As a consequence, folding of $ABCD$ as in Figure 2 transfers the cluster of 7 hexagons to an anchor ring. We note, for instance, that the region containing D is built up of

77

four separate shaded portions, one at each vertex of the rectangle. We see too that, because of the way it is formed on the ring, it touches the other 6 hexagons; and this is true for all 7 hexagons.

We consider next a solid figure-of-eight, or alternatively a sphere with two non-intersecting holes bored through it. This problem, like that of the sphere with one hole, i.e. the anchor ring, is also solved. Eight colours always suffice, and some maps cannot be coloured in less. Similar results have been found for many other surfaces obtained by this hole-boring process.

2. *Euler's theorem*

We shall need Euler's formula

$$F + V - E = 2, \tag{1}$$

for a map of F regions with V vertices and E edges. This was given as a result for convex polyhedra in Chapter 1. Figure 4 illustrates how a cube can be opened out into a plane map. The face $EFGH$ is removed and the 4 faces attached to $ABCD$ stretched till they fold out flat. The

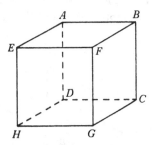 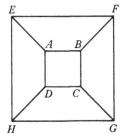

Fig. 4

resulting map has the same edge and vertex numbers as the cube. The missing face $EFGH$ can be taken as corresponding to the outside of the map so that both map and cube have the same F number. A mapping of this kind is possible for any convex polyhedron, so that result (1) applies equally to the 2 structures. We shall prove it for plane maps, and start by adding lines to triangulate all regions with more than 3 sides. For an n-sided region we can join any one vertex to $(n-3)$ others, thus creating $(n-3)$ new edges and $(n-3)$ new regions. As a result the left-hand side of (1) has not been altered.

78

We now work round the outer edge of the map removing one triangle at a time. Figure 5 illustrates the three possible ways in which a triangle is eliminated. In Figure 5(a) we remove edge AB losing also one region, so that $(F + V - E)$ is unaltered. In Figure 5(b) we delete edges AB and BC, also losing a region and a vertex, so again $(F + V - E)$ is unchanged. In Figure 5(c) we lose 3 sides, 2 vertices and 1 region, so that once more the left-hand side of (1) is unchanged. We carry out removals until a single triangle remains, then $E = 3$, $V = 3$, $F = 2$ (the outside area being counted as a region). So the left-hand side of (1) must be 2, and our proof is complete.

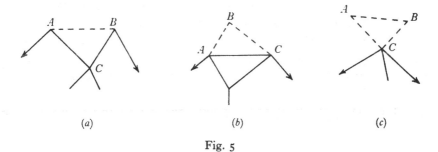

(a) (b) (c)

Fig. 5

For a map on a sphere with holes (1) is replaced by

$$F + V - E = K, \qquad (2)$$

and K is called the characteristic of the surface. It is 2 for the sphere, o for the anchor ring, -2 for the figure-of-eight and $(2 - 2n)$ for a sphere with n holes. It is of interest to note that there is a series of one-sided surfaces for which K takes odd values; these are discussed by Arnold[1] and Hilbert and Cohn-Vossen[4]. An ingenious proof of (2) for the anchor ring is given by Goodstein[3].

In the special case of a regular map we have 3 edges at each vertex. A count of edges gives

$$3V = 2E, \qquad (3)$$

the 2 allowing for the fact that each edge is counted twice in $3V$. Suppose there are n_3 triangles, n_4 quadrilaterals, and so on, then

$$F = n_3 + n_4 + n_5 + \dots . \qquad (4)$$

Counting edges by face-polygons gives

$$2E = 3n_3 + 4n_4 + 5n_5 + \dots . \qquad (5)$$

79

From (2), (3), (4) and (5) we deduce the useful result

$$-3n_3 - 2n_4 - n_5 + n_7 + 2n_8 + \ldots = -6K, \qquad (6)$$

valid for regular maps.

3. The five-colour theorem

Results in this section will be proved for regular maps; as remarked earlier they then hold for any map. Our first argument shows that 6 colours suffice for the plane map. The same method applied to the anchor ring gives 7 colours. We then apply a subtler argument to show that, for the plane, 5 colours will always be adequate.

From equation (6) with $K = 2$ (the plane) we see that at least one of the numbers n_3, n_4 or n_5 is non-zero, as otherwise the left-hand side would not be negative. Suppose that n_5 is not zero, so there is at least one pentagon P. Remove the dotted side, as in Figure 6(a). The resultant map has one region less and we assume that 6 colours suffice for it.

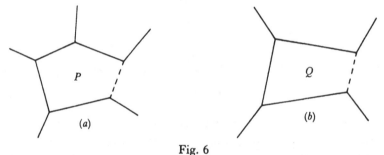

Fig. 6

At most, 4 of these will touch P, with one of them inside it. We can restore the dotted side, leave all colours outside P unchanged, and paint P with the remaining colour. This supplies the basis of an inductive proof, for we have reduced the problem of colouring a map of n regions in 6 colours to the same problem for a map of $(n-1)$ regions.

Had there been no pentagons, but at least one quadrilateral Q, the second diagram indicates how the same procedure could be used. Similarly, if there are no P's or Q's, a triangle must be present, and can be eliminated.

If $K = 0$ (the anchor ring) the left-hand side of (6) is zero and two cases arise. When a region of 7 or more sides occurs there is at least one of 5 or fewer, and vice versa. Otherwise the map must consist only of hexagons. A hexagon can be eliminated in this case, and we assume the remaining map can be coloured in 7 colours. Only 6 touch the eliminated hexagon, the side can be restored and the seventh colour inserted in the hexagon. Again, an inductive proof follows, and if there are no hexagons, there must be regions of 5 or fewer sides. These can be eliminated in the same way, so that, at any stage, the problem can be reduced to one for a map with one region less.

Returning to plane maps we can show that, in Figure 7(a), at least one of the pairs of regions A, D and B, C are not in contact. For if A touches D then region B is surrounded by the chain A–P–D and cannot touch C or vice versa. This result does not hold for the anchor ring.

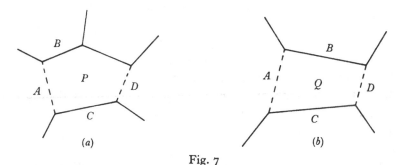

Fig. 7

Remove the boundaries separating A and D (assumed not to be in contact) from P. Assume that 5 colours suffice for the new map; only 4 will be involved around P, and one of these will belong to A, P and D. We restore the boundaries and leave all colours unaltered outside P. Since A and D, now separate regions again, do not touch, this restoration does not upset the basic colouring condition. We can use the fifth colour for region P. The second diagram applies when there are no pentagons, but at least one quadrilateral; a triangular region is removed by deleting one side. Again an inductive proof, also for 5 colours, follows.

Finally, consider a plane map with less than 12 regions. From (6) there must be at least one quadrilateral or one triangle, since n_5 is less than 12. Taking the case of a Q, as in the second diagram of Figure 7, then one of the pairs A, D or B, C are not in contact. Deleting sides on the assumption that A, D do not touch, the reduced map is assumed to be painted in 4 colours, of which only 3 touch Q. The basis of the induction follows as before.

4. Heawood's formula

Heawood found a remarkable formula for all surfaces with $K \leqslant 0$. He showed that

$$[7/2 + \sqrt{(49 - 24K)/2}], \tag{7}$$

colours suffice for such surfaces, the square brackets denoting that the integer part of the expression inside is to be taken. If $K = 0$ the formula gives 7, and we have proved that some maps require this number. Ringel and Youngs[6] have shown that, for all K, there are maps needing this number of colours.

To derive (7) we start with (6) written in the form

$$\sum_{j=3}^{\infty} (j-6)n_j = -6K, \tag{8}$$

and assume the existence of a map whose r-coloration cannot be reduced to that of a map with one region less in r colours. We deduce an upper bound for r. If this upper bound is exceeded, our assumption is false, and all maps can be reduced. Thus an inductive colouring is possible.

If the map has a region of $(r-1)$ or fewer sides, the method of the previous section shows that its colouring can be reduced to that of a map with one region less. Hence for all j with non-vanishing n_j

$$j > r-1 \quad \text{or} \quad j-6 \geqslant r-6. \tag{9}$$

Moreover, the map must have more than r regions or there is an immediate colouring in r colours, so

$$\Sigma n_j \geqslant r+1. \tag{10}$$

Expressions (8), (9) and (10) lead to

$$-6K = \Sigma(j-6)n_j \geqslant (r-6)\Sigma n_j \geqslant (r-6)(r+1),$$

the second result depending on the assumption that r exceeds 5. This in turn implies that $K \leqslant 0$; the inequality gives

$$r^2 - 5r + 6(K-1) \leqslant 0. \tag{11}$$

Provided that $K \leqslant 0$ (11) is satisfied for $r = 0$, and remains valid until

$$r > \tfrac{1}{2}\{5 + \sqrt{(49 - 24K)}\}. \tag{12}$$

Adding 1 to this quantity, and taking the integer part we get (7). This number of colours causes (11) to fail, and by the above reasoning an inductive r-coloration is possible. We made two assumptions in deriving this result; that $r > 5$ and that $K \leqslant 0$. The smallest value taken by (7) is 7, provided the second condition holds. Thus the first assumption is not a restriction. The second is vital, as it excludes the all-important case of plane maps.

5. The two-colour theorem

In the case of two colours the necessary and sufficient condition on the map is that, at each vertex, there are an even number of edges.

Figure 8(a) shows a vertex with colours 1 and 2 alternating round it. This alternation is only possible for even vertices, so the condition is necessary.

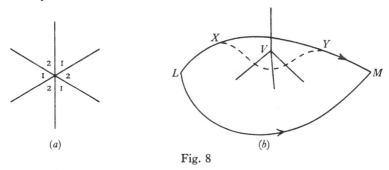

(a) (b)

Fig. 8

Allotting colour 1 to any region we can reach any other by many paths. Selecting any one path we change colour each time we cross an edge, and eventually all regions will be coloured. For the process to succeed we must be sure that, starting from L in one region, any two paths leading to M in another, give this latter region the same colour.

Figure 8(b) shows two such paths. We can deform the upper continuously into the lower one. In the deformation we need only examine what happens each time a vertex like V is crossed. In the diagram points X and Y are separated by one edge before V is crossed and three after. They are thus of opposite colours in either case. The argument is general; if k edges are crossed before a vertex is encountered and l edges after, then $k + l = 2n$. Hence, k and l are both odd or both even. In the first case there is a colour change on both new and old routes, in the second no change occurs on either.

Thus our colouring process determines all regions uniquely once a single one has been allotted. The alternating process guarantees that adjacent regions always differ, and the problem is solved.

83

6. The three-colour theorem

We restrict ourselves to regular maps, and prove that the necessary and sufficient condition for 3 colours to suffice is that all regions have an even number of sides. The proof depends upon constructing a dual map. We select a point in each region to form the vertices of this new map. Edges of the new map cross edges of the old, and the triple vertices of the old map imply that all regions of the dual map are triangles.

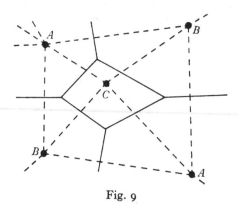

Fig. 9

A colouring of regions of the first map in 3 colours leads to a colouring of vertices of the dual map. This has the property that, for each of its triangles, the vertices carry different colours. In Figure 9 the vertex inside the quadrilateral is coloured C, and the 4 surrounding vertices alternate from A to B. We deduce a face colouring of the dual map in 2 colours from its vertex colouring by A, B and C. If A, B, C occur clockwise round a triangle it is white, otherwise black. The rotational rule means that 2 black or 2 white regions never touch.

Thus, starting with a 3-colour map, we derive a 2-colour solution for the dual map. For this to be possible all its vertices must be even by Section 5. Hence it is necessary that all regions of the 3-colour map have an even number of sides.

To prove the converse, we note that, if all regions of the original map have an even number of sides, then the dual map, having only even vertices, can be coloured in 2 colours. We allot a clockwise direction to white and anticlockwise to black. Starting with a white triangle we place colours A, B and C clockwise at its vertices. The 3 adjacent triangles are

84

black and the remaining vertex of each is coloured to produce anti-clockwise ordering. This process is continued until all the vertices are coloured. There can be no contradictions because of the fact that adjacent triangles always carry opposite colours, and therefore opposite directions of rotation. Finally, we observe that the vertex colours of the dual map can be regarded as face colours of the original. At any of its triple vertices each of the colours A, B and C occur, so adjacent colours always differ.

7. The Tait–Wolinskii theorem

Tait showed that, for regular maps, the 4-colour problem is equivalent to an edge-colouring problem using 3 colours, so that, at each vertex, the 3 edges differ. Wolinskii, a young Russian killed in World War 2, rediscovered this result, and gave the following elegant proof.

We first show that a 4-colour solution for faces leads to a 3-colour solution for edges. Let the face colours be denoted by the 4 number pairs

$$(0, 0), \quad (0, 1), \quad (1, 0), \quad (1, 1). \tag{13}$$

We define an addition operation for pairs of these symbols using addition modulo 2, for instance

$$(0, 1) + (1, 1) = [1, 2] = [1, 0].$$

We readily verify that the 6 pairs give 3 different sums

$$A = [0, 1], \quad B = [1, 0], \quad C = [1, 1]. \tag{14}$$

Edges are coloured A, B or C, the colour being determined by adding face symbols on either side of an edge.

If we go round any vertex adding the edge symbols and using the same rule adopted for face symbols we obtain the number pair $(0, 0)$. For, by the definition of the edge symbols, this is equivalent to adding together the 3 face symbols twice each, and this must give $(0, 0)$ because of modulo 2 addition, whichever faces colours are involved. Now, if 2 of the edge symbols at this vertex coincide, their sum is $(0, 0)$, again because of modulo 2 addition. Thus the sum of all 3 could not be $(0, 0)$, and coincidence of edge symbols at a vertex is ruled out, i.e. we have obtained a proper edge colouring.

We now have to deduce a face colouring in 4 colours from an edge colouring in 3. Starting at any face we allot colour $(0, 0)$ and take any route we like from face to face. On leaving face coloured (p, q) we cross edge coloured $[E, F]$. The colour of the new face entered is given by

$$(r, s) = (p, q) + [E, F] = (p+E, q+F),$$

with our usual addition rule.

We have to ensure that this process allots colours uniquely, and consider the path deformation shown in Figure 10. The upper path from L to M on its way into the lower will cross a number of vertices like V. On the old path the addition to face colour symbol on going from X to Y is one of the pairs A, B or C. On the new path it is the sum of the other 2. The sum of the number pairs A, B and C is $(0, 0)$, and hence, remembering our modulo 2 addition, the sum of any 2 equals the third. Thus the colour change on going from X to Y is the same by either path, and our proof is complete.

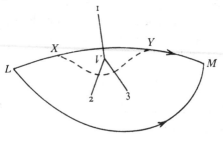

Fig. 10

An interesting consequence of this theorem is that 4 colours suffice for any regular map all of whose regions have their numbers of sides divisible by 3. We can colour the edges of such a map in 3 colours as follows. Allot any colour to the first edge and then follow various paths along edges. If the colours are denoted by 0, 1, 2, then when we encounter a new edge on the path being pursued its colour is decided as follows. At its junction with the previously coloured edge a third edge will lie either to right or left of the path. If to the right add 1 to the old colour number, if the left subtract 1. The resulting number denotes the colour for the new edge. It must be interpreted modulo 3, thus if it is -1 we add 3 to get colour 2, if it is 3 we subtract 3 to give colour 0. It can be proved (see Dynkin and Uspenskii[2]) that any 2 paths give a unique colour to a given edge, so that no contradictions will arise. As there is a 3-colour solution for edges, there is a 4-colour solution for faces.

We close by mentioning a remarkable colour theorem proved by

Dynkin and Uspenskii[2]. If a sphere has on it a 3-colour map, then there is at least one pair of points at the opposite ends of a diameter lying in regions of the same colour. Another excellent account of colouring problems is given by Stein.[5]

References

1. B. H. Arnold. *Intuitive Concepts in Elementary Topology* (Prentice-Hall, 1962).
2. E. B. Dynkin and V. A. Uspenskii. *Multicolor Problems* (D. C. Heath and Co., 1963).
3. R. L. Goodstein. *Fundamental Concepts of Mathematics* (Pergamon, 1962).
4. D. Hilbert and S. Cohn-Vossen. *Geometry and the Imagination* (Chelsea, 1952).
5. S. K. Stein. *Mathematics the Man-made Universe* (W. H. Freeman and Co., 1963).
6. G. Ringel and J. W. T. Youngs. 'Solution of the Heawood Map Colouring Problem', *Proc. Nat. Acad. Sci. U.S.A.* **60**, p. 438 (1968).
7. K. Appel. 'The proof of the 4-colour theorem', *New Scientist*, October 1976.

9

Dissection problems in two and three dimensions

1. *Introduction*

The word dissection is applied to a wide range of mathematical procedures and problems. The classical dissection puzzle is typified by Figure 1, where a square has been cut into five pieces and reassembled to form a regular octagon. In such problems the emphasis is on the smallest number of pieces needed for a solution, and Lindgren[6] discusses them in detail. Later we give a proof of Bolyai's theorem, that any plane polygon can be dissected into any other of equal area.

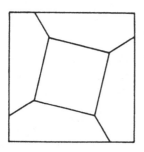 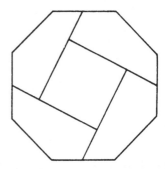

Fig. 1

Figure 2 illustrates another type of problem, that of dissecting a rectangle into unequal squares; the 33 × 32 rectangle provides 9 squares with the sides indicated. Similarly, one could seek a dissection of a square into unequal squares; Gardner[4], Stein[8] and Meschkowski[11] give excellent accounts of these tiling problems.

Another application is shown in Figure 3, based on the dissection of a square into either 4 right isosceles triangles or into 4 of these and a regular octagon. The second dissection is applied to all members of a tessellation of squares, and the triangular pieces reassembled to form the smaller squares. Other tessellations of the plane can be generated in the same way; an interesting account is given by Kraitchik[5]. We shall apply this process to solid tessellations.

In three dimensions there is no simple counterpart of Bolyai's theorem, and Dehn proved that it is not possible to dissect a regular tetrahedron into a cube of the same volume. A complete theory has appeared quite recently, clearly outlined by Boltyanskii[1]. In spite of the greater difficulty of the general theory, many ingenious puzzles in 3-space have been developed. The Soma puzzle given by Gardner[4], and the similar but more difficult cube puzzle discussed by Steinhaus[9], are worth trying. Lindgren[6] gives a 7-piece dissection of a $2 \times 1 \times 1$ block into a cube.

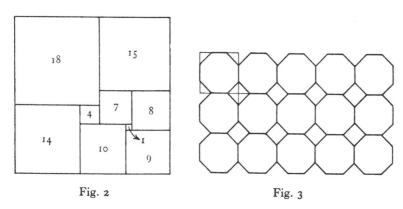

Fig. 2   Fig. 3

Figure 4 shows 2 attractive dissections of the cube into 4 congruent parts. Ehrenfeucht[3] discusses the first, for which one of the 4 parts is drawn below; we return to the second later. Here each part consists of a triangular dipyramid; for the one drawn, ABC is the common base, and P, Q the 2 vertices.

The arithmetic results

$$3^3 + 4^3 + 5^3 = 6^3,$$

$$1^3 + 6^3 + 8^3 = 9^3,$$

$$1^3 + 12^3 = 9^3 + 10^3,$$

can be illustrated by dissection. Lindgren[6] describes an 8-piece dissection of a $6 \times 6 \times 6$ cube that can be reassembled to form 3 cubes of sides 3, 4 and 5. The other two are to be found in Cadwell[2].

We conclude this introduction with a dissection proof of the fact that the sum of the cubes of the first n integers is the square of their sum. Cubes of sides 1, 2, 3, ..., n are first cut into square tiles of unit thickness.

89

From one of the 2×2 tiles a 1×1 square is removed, one of the 4×4 tiles has a 2×2 square removed and so on. Figure 5 (for $n = 4$) shows how the pieces are arranged to produce a square of unit thickness, and with side $(1 + 2 + \ldots + n)$. The use of dissection methods in proof will be illustrated further by a celebrated result due to Besicovitch.

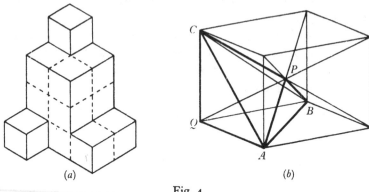

Fig. 4

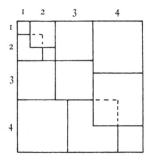

Fig. 5

2. *The two-dimensional dissection theorem*

This theorem states that the necessary and sufficient condition for dissecting polygon P into polygon Q is the equality of their areas. Hadwiger and Glur recently showed that a solution can be found so that the edges of each piece in P are parallel to its edges in Q. This implies that moving the piece from P to Q requires a parallel translation alone, or such a move together with a half-turn.

We observe from Figure 6(*a*) that a triangle can be dissected into three pieces that form a rectangle; both parallel translations and half-

90

turns arise in the reassembly. The second diagram shows that parallelograms on the same base and between the same parallels can be dissected into each other; no half-turns are involved here.

In Figure 7 the rectangles with diagonals AC, EG are of the same area. This will also be true of rectangles BE, CF; and we show that these can be dissected into each other From the area equalities we see that

$$\frac{DE}{DC} = \frac{DA}{DG} \quad \text{and} \quad \frac{BH}{HF} = \frac{HC}{HE}.$$

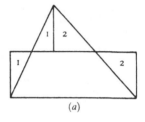

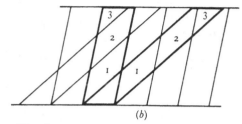

(a)

(b)

Fig. 6

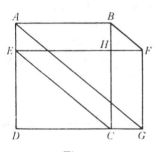

Fig. 7

As a result AG is parallel to both EC and BF. Using the parallelogram dissection above we transform

rectangle BE to parallelogram AF,
parallelogram AF to parallelogram BG,
parallelogram BG to rectangle CF.

We now have dissected rectangle AC into rectangle EG using only parallel translations.

We now proceed to the main theorem, and start by dividing P into triangles. Each triangle is then dissected and reassembled to form part of a column of unit width, and with height equal to the area of P. The construction for a typical triangle (No. 3) is shown in Figure 8. It is

91

first transformed to a rectangle as above, then the rectangle is transformed to a parellelogram with its base parallel to that of the column being constructed. In the second illustration the parallelogram goes into a rectangle, and the rectangle into a new one of base unity. This rectangle is then moved into position in the column, triangle No. 4 is dealt with, and so on.

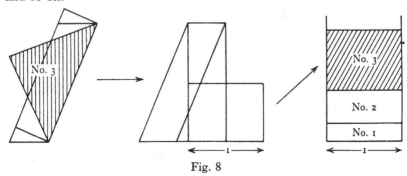

Fig. 8

A similar process applied to Q gives a column of the same dimensions. This is superimposed on the P column and all its cuts added to those already existing in P. The more numerous and smaller pieces that result can now be reassembled into either of the original polygons. Moreover, in moving these pieces only parallel translations and half-turns arise. Hence all cuts in P are parallel to their positions in Q.

3. Three-dimensional tessellations

One of our earliest geometrical experiences is the filling of space with equal cubes. Many other packings exist, and we restrict the field by allowing only one or more kinds of Archimedean solid. In addition we require each edge to be surrounded by the same set of solids throughout the tessellation. Thus for our packing of cubes each edge is surrounded by 4 cubes. Andreini proved that there are just 4 more tessellations of this type. We do not give this proof, but show how they may be derived from the cubic filling by dissection and reassembly. The 5 Andreini tessellations consist of:

(1) cubes,
(2) octahedra and cuboctahedra,
(3) truncated octahedra,
(4) octahedra and tetrahedra,
(5) tetrahedra and truncated tetrahedra.

92

In Figure 9(a) the dotted equilateral triangle indicates the start of the process that turns a cube into the cuboctahedron shown in Figure 11(a) of Chapter 1. This dotted section cuts off a tetrahedron, and 8 of them will form a regular octahedron. By applying the same dissection to all the cubes in a cubic space packing we can therefore derive a packing involving cuboctahedra and octahedra. The second diagram of Figure 9 shows 4 cuboctahedra stacked together. It is easy to visualise the octahedron, half of which fills the central gap in this layer.

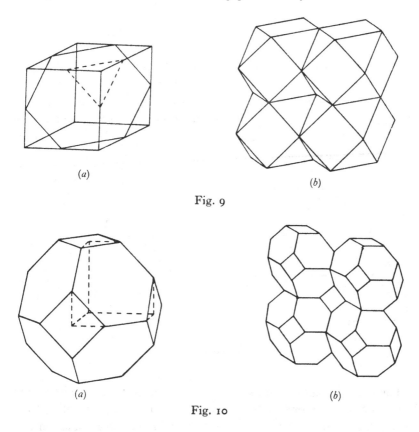

(a)

(b)

Fig. 9

(a)

(b)

Fig. 10

Returning to Figure 9(a) we see that a plane parallel to the dotted triangle can be chosen to cut the cube into two equal pieces, each with a regular hexagonal face. This section is obtained by joining the mid-points of sides of the cube's Petrie polygon, mentioned in Chapter 1.

The truncated octahedron of Figure 10(a) has 6 square faces and 8 regular hexagonal faces. The squares arise by cutting off the 6 corners of a regular octahedron in such a way that the 8 faces, originally equi-

93

lateral triangles, become regular hexagons. The dotted lines indicate that this solid consists of 8 identical shapes each of the form obtained by bisecting a cube as above. This dissection of all cubes in a packing can be carried out and the resulting pieces reassembled to form truncated octahedra. A group of 4 such solids in the second diagram shows clearly that another one will fit into the central gap. A number of biological applications of this packing are discussed by D'Arcy Thompson.[10]

4. The remaining Andreini tessellations

We first note that appropriate parallel shifts will distort a cubic pack into one of parallelopipeds with rhombic faces. These shifts can be chosen to make the acute angles of all faces equal to $\frac{1}{3}\pi$. Figure 11(a) shows such a parallelopiped divided into a regular octahedron and two regular tetrahedra. This dissection leads to a method of filling space with these regular solids.

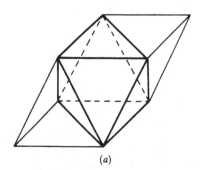
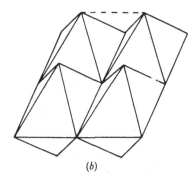

(a)   (b)

Fig. 11

The second diagram shows 4 octahedra in edge contact. The dotted line indicates how one of the gaps between them can be filled by a regular tetrahedron. If all 4 gaps on the upper surface are filled in this way a central pit remains; this will accommodate just one-half of a regular octahedron. A series of these new octahedra form the next layer. Rouse Ball[7] gives a fine illustration of this packing. We note that an octahedron of this packing, together with the 8 tetrahedra touching its faces, constitute the stella octangula shown in Figure 6 of Chapter 1. Figure 12 indicates a regular tetrahedron cut from a cube. The 4 remaining portions are each one-eighth of a regular octahedron. So this

94

mixed tessellation can also be derived from a cubic pattern without first distorting to rhombic parallelopiped form.

Figure 13(*a*) shows a rhombic parallelopiped cut into 2 regular tetrahedra and 2 truncated tetrahedra. The latter solid is obtained from a tetrahedron by cutting off corners suitably, it has 4 triangular and 4 hexagonal faces, all regular. The associated space filling is illustrated by a block of 5 truncated tetrahedra in contact. One of them is surrounded by the rest and is just visible in the diagram. The gaps between them are filled by the regular tetrahedra, giving another mixed tessellation.

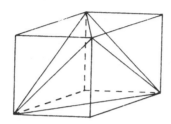

Fig. 12

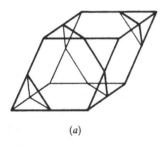

(*a*)

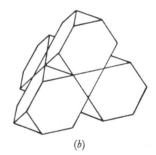

(*b*)

Fig. 13

5. *The rhombic dodecahedral pack*

The rhombic dodecahedron, illustrated in Figure 11(*b*) of Chapter 1, has 12 rhombic faces, and is the dual of the cuboctahedron. While not an Archimedean solid it provides another important tessellation of space. This is related to close packing of spheres and to the bee's cell, topics discussed by Rouse Ball[7] and D'Arcy Thompson.[10]

A cube is divided into 6 congruent square pyramids by joining its centre to each vertex. If these are stuck on the 6 faces of another cube, as in Figure 14(*a*), a rhombic dodecahedron results. Thus, if alternate

95

cubes in a pack are so dissected, and the bits reassembled with uncut cubes, a rhombic dodecahedral pack results. A group of 4 dodecahedra in contact is shown in the diagram.

This packing can be generated in another way, depending on the dissection of the cube shown in Figure 4(b). Eight of the resulting pieces fit together to form a rhombic dodecahedron, so another splitting of the basic cubic tessellation, this time involving all cubes, produces the same result.

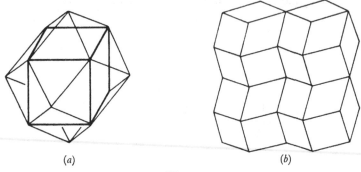

(a)   (b)

Fig. 14

6. Rotating a rod in minimum area

In moving a rod through angle π about its centre it sweeps out a circular area. Can the same reversal of direction be achieved, but with a smaller area? Figure 15(a) represents a three-cusped hypocycloid. It has the property of cutting off a constant length PQ along its tangents. Starting with the point of contact P at cusp 2, and rotating the tangent anticlockwise, it moves until Q touches the curve at cusp 3, with P lying at the mid-point of arc 1–2. Then P moves on to touch the curve at 1, Q being at the mid-point of arc 2–3. Finally, Q moves up to 2 and the tangent PQ has reversed its direction, sweeping out the area of the curve. For some time it was thought that this was the minimum area solution, but Besicovitch proved the surprising result that, by a suitable combination of motions, the area swept out could be made as small as required.

We first show that a rod can be moved to a parallel position sweeping out a small area. In Figure 15(b) a rotation through angle ϕ carries P_1 to P_2. The rod then moves along its length until its centre lies on the line of the new position required. This brings P_2 to P_3, and a reverse

96

rotation of ϕ takes P_3 to P_4. Finally, a shift along the rod's length takes P_4 to P_5. The total area swept out is $2l^2\phi$, where $2l$ is the length of the rod. This can be made small if ϕ is small, but at the cost of moving a considerable distance from P_2 to P_3.

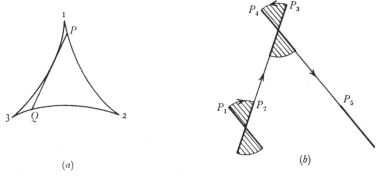

(a)　　　　　　　　　　　　　　(b)

Fig. 15

Next we prove that a triangle can be dissected, and the pieces reassembled by parallel translation so as to overlap. The new total area is $2/(m+2)$ times the area of the triangle, where m is an arbitrary integer.

We illustrate the case $m = 3$, and Figure 16 shows triangle ABC divided into 2^m smaller triangles. In addition we insert $(m+1)$ guide lines at equal intervals, and parallel to its base.

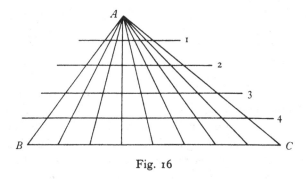

Fig. 16

The first step reassembles the 4 pairs of adjacent triangles as in Figure 17. The pairs overlap along the second guide line. We consider the total lengths they cut off on each of the 4 guide lines and the base of length b. Before overlapping these lengths were

$$\tfrac{1}{5}b, \quad \tfrac{2}{5}b, \quad \tfrac{3}{5}b, \quad \tfrac{4}{5}b, \quad b.$$

97

Afterwards 4 overlaps, each of $b/20$, occur at all levels but the first, and the totals of intercepts become

$$\tfrac{1}{5}b, \quad \tfrac{1}{5}b, \quad \tfrac{2}{5}b, \quad \tfrac{3}{5}b, \quad \tfrac{4}{5}b.$$

The next stage overlaps 2 sets of pairs at the third level to produce Figure 18. Total lengths along guide lines and base are now

$$\tfrac{1}{5}b, \quad \tfrac{1}{5}b, \quad \tfrac{1}{5}b, \quad \tfrac{2}{5}b, \quad \tfrac{3}{5}b,$$

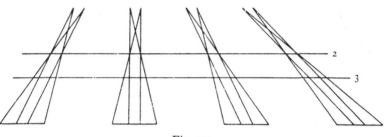

Fig. 17

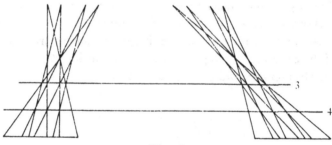

Fig. 18

as $\tfrac{1}{5}b$ is subtracted from the last 3 entries. Finally we overlap the 2 sets of 4 triangles on the fourth guide line obtaining Figure 19. Total lengths intercepted are now $\tfrac{1}{5}b$ on all but the base, where $\tfrac{2}{5}b$ remains. From these totals the area of the figure is readily found to be $\tfrac{1}{5}bh$, where h is the height of the triangle. So the original area has been multiplied by $\tfrac{2}{5}$. For general m this factor becomes $2/(m+2)$, and can be made as small as we wish. Note that in the overlap process only parallel translations of triangles have occurred.

We could rotate a rod of length less than h from AB to AC, sweeping out an area less than that of the triangle, but still of appreciable size. Instead we achieve the same final result by successive rotations through the subtriangles alternating with the translations they suffer in the overlap process. The total area swept out in rotation will then be less than $2/(m+2)$ times that of the original triangle, for it is less than the total overlapped area. There are $(2^m - 1)$ parallel shifts, and the method outlined earlier enables these to be done so that the total area covered is less than

$$bh/(m+2)+(2^m-1)2l^2\phi.$$

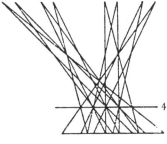

Fig. 19

Choose m large enough to make the first term as small as required, and then choose ϕ to make the second equal to it. We have thus turned the rod through angle BAC in as small an area as we care to specify. Motions of the rod along its length for enormous distances are involved. Cunningham's method[12] avoids such large displacements.

References

1. V. G. Boltyanskii. *Equivalent and Equidecomposable Figures* (D. C. Heath and Co., 1963).
2. J. H. Cadwell. 'Some Dissection Problems involving Sums of Cubes.' *Math. Gaz.* 48, no. 366, 391 (1964).
3. A. Ehrenfeucht. *The Cube made Interesting.* (Pergamon Press, 1964).
4. M. Gardner. *More Mathematical Puzzles and Diversions* (Bell, 1963).
5. M. Kraitchik. *Mathematical Recreations* (George Allen and Unwin, 1943).
6. H. Lindgren. *Geometric Dissections* (van Nostrand, 1964).
7. W. W. Rouse Ball. *Mathematical Recreations and Essays*, 11th edition (Macmillan, 1940).
8. S. K. Stein. *Mathematics the Man-made Universe* (W. H. Freeman and Co., 1963).
9. H. Steinhaus. *Mathematical Snapshots* (Oxford, 1960).
10. D'Arcy W. Thompson. *On Growth and Form*, abridged edition (Bonner) (Cambridge, 1961).
11. H. Meschkowski. *Unsolved and Unsolvable Problems in Geometry* (Oliver and Boyd, 1966).
12. F. Cunningham. 'The Kakeya problem for simply connected and for star-shaped sets', *Amer. Maths. Monthly*, 78, p. 114 (1971).

10

Newton's polygon and plane algebraic curves

1. *The Folium of Descartes*

Descartes, the discoverer of coordinate geometry, studied the curve of Figure 1 (*a*). Its equation is

$$x^3 + y^3 = 3axy, \qquad (1)$$

and we see that neither x nor y can be expressed explicitly in terms of the other. Newton carried out a classification of cubics, and devised a powerful method for plotting curves. Before applying it to (1) we shall make a few general comments on curve tracing.

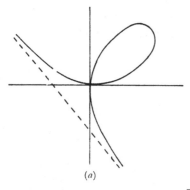

(*a*)
(*b*)

Fig. 1

The general equation of a straight line is

$$Ax + By + C = 0,$$

and if we substitute for y in (1) we get a cubic equation in x indicating that there are 3 points of intersection. However, if the line touches the curve, or goes through its double point, 2 of the roots will coincide. Still more special is an inflectional tangent which both touches and crosses a curve, in this case 3 roots coincide at the point of contact. We see later that the Folium has such a tangent. We may lose 2 roots too if our line, besides going through a double point, touches the curve there. It is a general result that, for a curve through the origin, the tangent(s) there come from the homogenous group of terms of lowest degree in its equation. Here this group consists of the right-hand side of (1), so the axes $x = 0$, $y = 0$ touch the curve at the origin.

Some lines lose points of intersection with a curve in quite a different way. Thus the line

$$x + y + C = 0,$$

when combined with (1), leads to a quadratic. The vanishing of the cubic term indicates that, as C varies, these lines all go through a point at infinity on the curve. One of them, with $C = a$, when combined with (1) leads to the result $a^3 = 0$, indicating that all 3 roots are at infinity.

This line, shown dotted in Figure 1 (a), is an asymptote; it touches the curve at infinity. It is in fact a rather special asymptote, for the curve has an inflection at the infinite point of contact. This accounts for the loss of 3 roots rather than 2, and for the fact that the curve comes in from infinity on the same side of the line at either end.

Newton's polygon ABC is shown in Figure 1 (b). The points A, B and C arise from the 3 terms of (1) and are placed on a square mesh as follows. The term in x^3 gives A with coordinates (3, 0) on this mesh, y^3 gives C or (0, 3), while xy gives B or (1, 1). Thus it is the indices of x and y that determine points on the mesh. Each side of the polygon indicates a branch of the curve. We now show how these are used, leaving proofs until the next section.

Side AB has no marked points below it, and gives a branch at the origin. The terms corresponding to A, B are retained from (1) giving

$$x^3 - 3axy = 0 \quad \text{or} \quad x^2 = 3ay,$$

the equation of a parabola touching $y = 0$. Similarly, side BC leads to the branch

$$3ax = y^2,$$

a parabola touching the y axis. Figure 2 indicates these branches in sketches (a) and (b). The Folium resembles these curves along their fully marked portions; we assume that $a > 0$.

The side CA has only points below it and slopes at $135°$, indicating linear asymptotes. Their directions are given by

$$x^3 + y^3 \equiv (x + y)(x + \omega y)(x + \omega^2 y) = 0,$$

101

and only one indicated a real line. To find the corresponding asymptote we include all 3 terms, and write (1) as

$$x+y = \frac{3axy}{x^2-xy+y^2} \simeq \frac{-3ax^2}{3x^2} = -a. \qquad (2)$$

The approximate result follows by substituting $y = -x$, and the asymptote is $x+y+a = 0$.

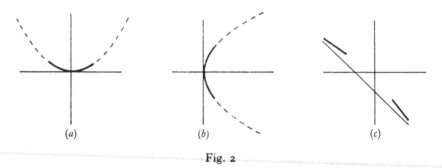

(a)   (b)   (c)

Fig. 2

It is useful to take this approximation one stage further. This time y is replaced by $-(x+a)$ giving

$$x+y = \frac{3axy}{x^2-xy+y^2} \simeq \frac{-3ax(x+a)}{3x^2+3ax+a^2} \simeq -a+\frac{a^3}{3x^2}.$$

The final result is obtained by simple division. The term in $1/x^2$ indicates that the curve lies above the asymptote for both large positive and negative x, as in Figure 2(c).

There are some general questions we ask before proceeding to join up the various branches. Does the curve cut either axis (apart from doing so at the origin)? Does the asymptote have any finite intersections with the curve? The negative answers to these questions help by reducing possible modes of joining branches, leaving only the one shown.

Had there been oblique tangents at the origin, their points of intersection with the curve would also supply useful information. As interchange of x and y leaves (1) unaltered there is symmetry about the line $y = x$. It is useful to determine the point for which $y = x$, apart from the origin.

Whenever a curve has a multiple point at the origin it is worth seeing if a parametric representation can be found. The line $y = mx$ will have several of its intersections with the curve at $(0, 0)$; if, as here, only one other remains, the curve can be expressed parametrically. The Folium can be defined by

$$x = \frac{3am}{1 + m^3}, \quad y = \frac{3am^2}{1 + m^3},$$

a result that makes accurate plotting easy.

2. *Newton's polygon*

Having plotted the indices of all terms of a curve's equation on a square mesh there will be a unique convex polygon joining some of the marked points, and including the rest within it. Let PQ be one of its sides such that all other marked points lie on the opposite side of PQ to the origin. The point R is typical of these other points, the corresponding terms in the equation being

$$P \equiv ax^\alpha y^\beta, \quad Q \equiv bx^\gamma y^\delta, \quad R \equiv cx^\xi y^\eta.$$

The line PQ will have negative slope, so that, if $\alpha > \gamma$ then $\delta > \beta$. After dividing out a common factor, the 3 terms P, Q and R give

$$ax^{\alpha-\gamma} + by^{\delta-\beta} + cx^{\xi-\gamma} y^{\eta-\beta}.$$

The approximation (4) obtained by setting the sum of the first 2 terms equal to zero can be used to eliminate y in the third. The index of x after this step will be

$$\xi - \gamma + (\eta - \beta)\frac{\alpha - \gamma}{\delta - \beta}.$$

This exceeds the index of x in the first term by

$$\xi - \alpha + (\eta - \beta)\frac{\alpha - \gamma}{\delta - \beta}, \tag{3}$$

and if (3) is positive, the third term, involving a higher power of x than the first, can be neglected for small values of x.

Introducing coordinates (X, Y) in the polygon diagram, the equation of PQ is

$$(\delta - \beta)(X - \alpha) + (\alpha - \gamma)(Y - \beta) = 0.$$

The perpendicular distance from R to this line is

$$\{(\delta - \beta)(\xi - \alpha) + (\alpha - \gamma)(\eta - \beta)\}/\sqrt{\{(\delta - \beta)^2 + (\alpha - \gamma)^2\}},$$

a positive quantity, as R and the origin lie on opposite sides of the line. Thus (3) is positive, and we can approximate the curve by

$$ax^{\alpha-\gamma} + by^{\delta-\beta} = 0, \tag{4}$$

103

for points near the origin. By moving PQ parallel to itself until it meets a marked point we determine which other term should be included to approximate one stage further. Note that, if the curve does not pass through the origin, no branches of this sort occur in the polygon.

A precisely similar argument shows that a side of the polygon with all marked points lying below it gives an approximation valid for large x and y. If this line slopes at $135°$, the corresponding terms will be of the same total degree in x and y, and, in general, linear asymptotes result. Sides at other slopes lead to curvilinear asymptotes. The way in which asymptotes parallel to the axes are indicated will appear in one of the examples.

3. *A curvilinear asymptote*

We consider the curve
$$4x^2 - 2xy^2 + y^3 = 0,$$

the Newton polygon, again a triangle, is shown in Figure $3(b)$. The side AB corresponds to
$$4x^2 + y^3 = 0,$$

an approximation near the origin leading to Figure $4(a)$.

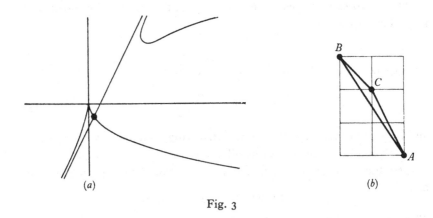

(a) $\qquad\qquad\qquad\qquad\qquad (b)$

Fig. 3

The side BC at $135°$ indicates a linear asymptote; writing the dominant terms B and C first, we have

$$y^2(y - 2x) + 4x^2 = 0.$$

Putting $y = 2x$ everywhere except in the bracket leads to the asymptote

104

$y = 2x - 1$. Replacing y by $(2x - 1)$ and using the binomial expansion or simple division gives the better approximation

$$y = 2x - 1 - \frac{1}{x}.$$

This implies that the curve lies below the line for positive x, and vice versa; Figure 4 (b) illustrates this result. We note that, in the general case, before substituting a better value for y another term besides A would be included with B and C. This is selected by further parallel displacement of BC. Here there are no other terms left to consider.

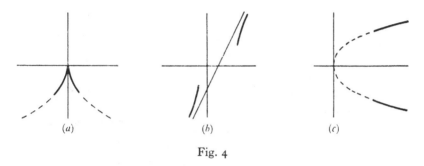

(a)   (b)   (c)

Fig. 4

The side CA indicates another infinite branch,

$$4x^2 - 2xy^2 = 0 \quad \text{or} \quad y^2 = 2x,$$

Figure 4 (c) shows this parabolic asymptote.

The curve does not cut the axes other than at $(0, 0)$. To find its intersection with the linear asymptote we put $y = 2x - 1$, to get

$$0 . x^3 + 0 . x^2 + 4x - 1 = 0,$$

indicating, as it should, two roots at infinity, together with the finite point $(\frac{1}{4}, -\frac{1}{2})$. It is now easy to sketch the curve.

4. Asymptotes parallel to the axes

Our next example $\quad 2x - x^2y - xy^2 - y^4 + xy^4 = 0,$

gives the polygon of Figure 5 (b).

The branch AB is $\qquad 2x = y^4,$

105

and supplies Figure 6(a). The horizontal side BC gives a vertical asymptote

$$xy^4 - y^4 \equiv y^4(x-1) = 0,$$

or $x = 1$. Moving BC parallel to itself we first encounter the point corresponding to xy^2, so that a better approximation is

$$y^4(x-1) - xy^2 = 0 \quad \text{or} \quad x = 1 + (1/y^2),$$

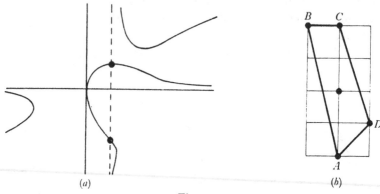

(a) (b)

Fig. 5

obtained by putting $x = 1$ except in the bracket itself. Thus, as shown in the sketch in Figure 6(b), the curve lies to the right of $x = 1$ for both positive and negative y.

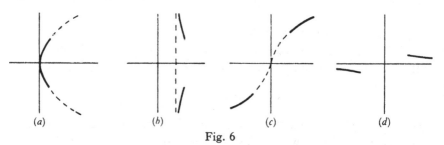

(a) (b) (c) (d)

Fig. 6

Side CD gives the curvilinear asymptote

$$x = y^3$$

of Figure 6(c). Branch DA has a positive slope, and the horizontal axis itself is an asymptote, the terms D and A give

$$2x - x^2 y = 0 \quad \text{or} \quad y = 2/x$$

so that the curve lies above the axis for $x > 0$, and vice versa, as in Figure 6(d).

106

The line $x = 1$ cuts the curve at the finite points with $y = -2, 1$. There is still some uncertainty left as to how the various branches should be joined. We note that the line $y = 1$ cuts the curve twice at $x = 1$, and at no other point. We can now produce a sketch of the curve.

5. Oblique tangents at the origin

We consider
$$(y-x)^2(y+x) = x^4 + 4y^4, \qquad (5)$$

sketched in Figure 7(a). Side AB indicates an intersection with the y axis at distance $\frac{1}{4}$; CD leads to the point $(1, 0)$ on the curve. The only infinite branch is
$$x^4 + 4y^4 = 0.$$

This has no real factor, so the curve is of finite extent.

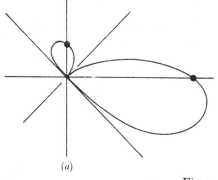
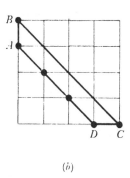

(a)   (b)

Fig. 7

The side DA includes four terms, which together give
$$(y-x)^2(y+x) = 0,$$

indicating oblique tangents $y = \pm x$ at $(0, 0)$. We note that a segment sloping at $135°$ always leads to oblique tangents. Parallel translation of DA brings in all terms of (5). Putting $y = -x$ everywhere except in the critical bracket we get
$$y + x = \tfrac{5}{4}x^2,$$

so the curve lies above this tangent on either side of zero.

Putting $y = x$ except in its own bracket leads to
$$(y-x)^2 = \tfrac{5}{2}x^3 \quad \text{or} \quad y = x \pm \sqrt{\tfrac{5}{2}}\, x^{\frac{3}{2}}.$$

107

This indicates a cusp, for there are no points near the origin for negative x. For $x > 0$ there are pairs of points one on each side of the tangent.

We note that neither tangent cuts the curve at points other than the origin, and a sketch is readily prepared. The intersection of the curve and the line $y = mx$ again leads to a parametric form.

6. A rhamphoid cusp

We consider the curve
$$(y-x^2)^2 = xy^2 - y^4, \tag{6}$$

sketched in Figure 8(a). Side BC of the polygon gives zero as the only real intersection with $x = 0$. The side CA leads to

$$x^4 + y^4 = 0,$$

without real factors, and therefore indicating a finite curve.

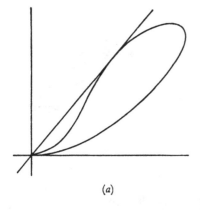

(a)   (b)

Fig. 8

The side AB contains three terms giving

$$(y-x^2)^2 = 0.$$

Had this been the product of different factors it would lead to two parabolic branches touching the x axis. As it is we must go to the next stage
$$(y-x^2)^2 = xy^2,$$

and replace the y on the right-hand side by x^2, getting

$$y = x^2 \pm x^{\frac{5}{2}}.$$

108

For $x < 0$ there are no real points, for $x > 0$ they occur in pairs one on each side of $y = x^2$, leading to a rhamphoid cusp. The simpler cusp of the last example, with the two branches of the curve on opposite sides of the tangent, is called ceratoid.

The line $y = mx$ cuts the curve in two points given by

$$x^2(1+m^4) - x(2m+m^2) + m^2 = 0, \qquad (7)$$

a quadratic with real roots for values of m lying between 0 and the real root of the cubic

$$4m^3 - m - 4 = 0,$$

obtained by equating the discriminant of (7) to zero. This root is close to $1 \cdot 08$, and indicates the tangent from the origin to the curve shown in the figure. As a further guide to sketching, the quadratic can be solved for $m = 0 \cdot 5$. Here, although the curve is not rational, i.e. does not admit of a parametric representation, its intersections with lines through a multiple point still provide valuable information for plotting purposes.

7. A singularity at infinity

The curve

$$y - 2x + x(y-x)^2 = 0, \qquad (8)$$

has a single tangent at the origin. The improved approximation to the curve there is

$$y = 2x - x^3,$$

so that the inflectional tangent is crossed by the curve at zero.

The side BC of the polygon in Figure $9(b)$ is of positive slope and indicates that the vertical axis is an asymptote, the two terms give

$$y + xy^2 = 0 \quad \text{or} \quad x = -1/y,$$

the curve is to the left of the axis for $y > 0$, and vice versa.

Side CD indicates oblique asymptotes, and as the direction is squared we expect them to be parallel. The next approximation is (8) itself, putting $y = x$ except in the critical bracket gives

$$x(y-x)^2 - x = 0 \quad \text{or} \quad y = x \pm 1.$$

We improve on this by putting $y = x+1$, noting that there are no further terms to include at this stage. The result is

$$y = x + \sqrt{\{1 - (1/x)\}}$$

109

so that the curve lies above the line for $x < 0$ and vice versa. For the other asymptote the situation is similar.

The side AD indicates points $(\pm\sqrt{2}, 0)$ on the curve. There are no non-zero intersections with $y = 2x$, the tangent at the origin; or finite intersections with any of the asymptotes. Changing the sign of both x and y in (8) leaves it unaltered, so the curve has central symmetry about the origin.

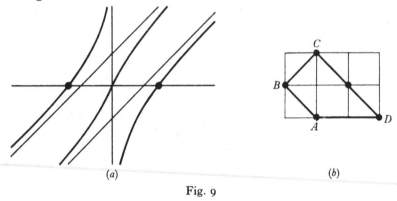

(a)   (b)

Fig. 9

The parallel asymptotes arise from a double point of the curve at infinity. In our last example we shall meet a squared asymptotic direction factor with a quite different result.

8. Contact with the line at infinity

We consider
$$x^2 - y^3 + (x^2 - y^2)^2 = 0.$$

Behaviour at the origin corresponds to BC, so there is a simple cusp there. Sides AB and CD lead to one real axial point $(0, 1)$.

Side AD has three points on it, leading to

$$(y - x)^2(y + x)^2 = 0.$$

Parallel translation shows that C must next be included to give

$$(y - x)^2(y + x)^2 - y^3 = 0.$$

Putting $y = x$ except in the critical bracket gives

$$4(y - x)^2 = x.$$

110

This time the squared term leads, not to parallel asymptotes, but to an asymptotic parabola, with its axis parallel to $y = x$. Similarly the other squared factor gives

$$4(y+x)^2+x = o.$$

These parabolas have been sketched as guide curves in Figure 10(a). Whereas in Section 7 the repeated factor indicates a double point at infinity, here it arises because the curve touches the line at infinity. In fact it does so twice, and at each point can be replaced by an approximating parabola. The curvilinear asymptotes we encountered previously arose because the curves involved touched the line at infinity at either $x = o$ or $y = o$.

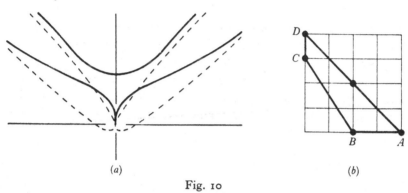

(a)   (b)

Fig. 10

Newton's polygon seems to be little known, although Frost[2] gives a large number of examples of its use. In addition, Walker[4], besides discussing algebraic plane curves systematically, includes a section on this topic. We mention too the interesting introduction to algebraic curves by Forder[1]. Although not directly related to the present topic, fascinating discussions of plane curves both algebraic and transcendental, are to be found in Lockwood,[3] Yates[5] and Zwikker[6].

References

1. H. G. Forder. *Geometry* (Hutchinson's Home University Library, 1960).
2. P. Frost. *Curve Tracing* (Chelsea Publishing Co., 1960).
3. E. H. Lockwood. *A Book of Curves* (Cambridge, 1961).
4. R. J. Walker. *Algebraic Curves* (Dover, 1962).
5. R. C. Yates. *Curves and their Properties* (Michigan, 1947).
6. C. Zwikker. *The Advanced Geometry of Plane Curves and their Applications* (Dover, 1963).

11

The plane symmetry groups

1. *Symmetry and isometric motion*

The patterns shown in Figures 11 to 16 of this chapter (pp. 123–128) are all possessors of obvious regularity or symmetry. At the same time it is evident that the type of symmetry varies from pattern to pattern. We shall prove that the 7 one-dimensional patterns of Figure 11 include all possibilities of this kind. There are 2 possible types of point symmetry, these are shown in Figure 8; each type has a single infinity of variants. Of two-dimensional patterns there are just 17 types. We shall not enumerate these in detail; however, they are all illustrated in Figures 12 to 16.

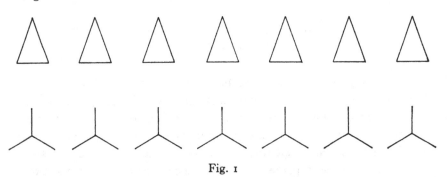

Fig. 1

We start by defining symmetry in terms of rigid or isometric motions of the plane. It is evident that both one- and two-dimensional schemes can be moved in various ways until the pattern in its new position lies over the old. Besides translations, this can be achieved in some cases by suitably rotating the plane. Further, some patterns are invariant when reflected in certain lines. We say that a pattern is characterised by the totality of isometric motions that bring it into self-coincidence.

In connection with this definition of symmetry we note that the two patterns of Figure 1 belong to the same class. While rotations through $\frac{2}{3}\pi$ leave the pattern elements of the second unaltered, they do displace

112

the pattern as a whole, and from our point of view they add nothing to the symmetry properties of the first diagram.

An isometric transformation or rigid motion of the plane preserves distances and angles. There are two distinct types, illustrated in Figure 2. Diagrams (*a*) and (*b*) show the same sense of rotation in their lettering, and (*a*) can be laid exactly over (*b*) by a suitably chosen isometry, as we shall see in the next section. However, (*c*) has an opposite rotation, and no rigid motion of (*a*) in its own plane can secure coincidence with it. It is possible to rotate (*a*) into the dotted triangle. Then a further change, a reflection in $A''B''$, turns this into (*c*). We can regard this last change as a rigid motion of the plane obtained by rotating it about the line $A''B''$ through angle π. Isometries obtained by motions lying entirely in the plane are called proper. Those requiring a rotation of the plane out of itself are called improper.

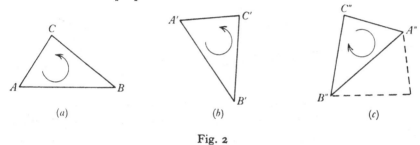

Fig. 2

A fundamental property of isometries is that any two carried out in succession define a third. Thus if isometry **U** takes figure F into figure F', while **V** takes F' into F'', their combined effect, taken in this order, is an isometry carrying F into F''. We denote this 'product' isometry by the symbol **VU**, our convention being that operation **U** is applied before **V**. As a proper isometry **P** does not affect rotational sense, while an improper isometry **I** alters it, we see that

$$P_1 P_2 \text{ is proper,}$$
$$PI \text{ or } IP \text{ is improper,}$$
$$I_1 I_2 \text{ is proper.}$$

2. *Rotations and glide reflections*

We first prove that any proper isometry is equivalent to a rotation. Let the isometry move triangle ABC into $A'B'C'$. In Figure 3(*a*) we join AA' and BB' and construct their perpendicular bisectors meeting in O.

113

Then, as triangles AOB, $A'OB'$ have equal sides, they are congruent. By adding angle BOA' to their angles at O we see that angles AOA' and BOB' are equal. A rotation through this angle α about O thus carries AB into $A'B'$. By adding angle OAB to BAC and $OA'B'$ to $B'A'C'$, and noting that $OA = OA'$, $AC = A'C'$, we prove that triangles OAC, $OA'C'$ are congruent. It follows that $OC = OC'$ and that angle COC' is also equal to α. Hence the same rotation carries point C into C'.

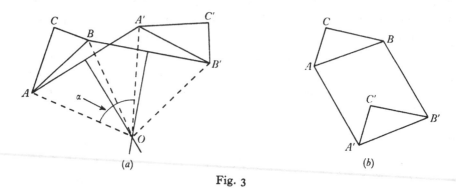

(a) (b)

Fig. 3

The above construction fails if the perpendicular bisectors are parallel, as in Figure 3 (b). We agree to call this translatory motion a rotation through zero angle about a suitable point at infinity. This then justifies our claim that any proper isometry is a rotation.

It is easy to construct an example to show that the products S_1S_2 and S_2S_1 of 2 rotations or spins S_1 and S_2 differ, or we can prove this as follows. Let O_1, O_2 be the centres of rotation, and consider the effects of the 2 products on O_1. We note that S_1 leaves O_1 unchanged, or

$$S_1(O_1) = O_1.$$

Hence
$$S_2S_1(O_1) = S_2(O_1),$$

$$S_1S_2(O_1) = S_1(S_2(O_1)),$$

and the two mappings of O_1 differ, for S_1 displaces all points except O_1, in particular it moves $S_2(O_1)$. It is useful to note that, for either product, the resultant angle of rotation is the sum of the two constituent angles.

We define a basic improper isometry called a glide reflection. In Figure 4, triangle ABC is reflected in line l, and then moved parallel to it through distance d. The improper motion taking ABC into $A'B'C'$ is a glide reflection defined by l (with a direction along it) and d.

114

Any improper isometry is obtainable by a glide reflection. The direction of l bisects the angle between AB and $A'B'$, and it passes through the mid-point of AA'. The point in which a line through A', parallel to l, cuts the circle on AA' as diameter determines d. So the glide reflection needed to take ABC into $A'B'C'$ is completely specified.

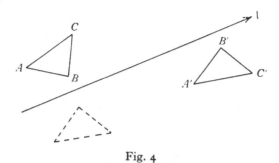

Fig. 4

3. *Groups of isometries*

A group of operations is a set possessing the following properties:

(i) The application of two operations in sequence produces the same result as some other operation of the group. In symbols, if the result of U followed by V is another element W, we write $W = VU$.

(ii) There is an identity operation I with the properties

$$UI = IU = U$$

for all members U of the group.

(iii) Each member U has an inverse called U^{-1} in the group with the properties $UU^{-1} = U^{-1}U = I$.

(iv) Operations combine associatively, so that $U(VW) = (UV)W$.

We first consider all plane translations. Any two translations can be combined by the parallelogram law to give a resultant translation, so that (i) is satisfied. It is also true here that $T_1T_2 = T_2T_1$, so we say the operators are commutative. There is a translation through zero distance that we call I, and if T is a translation through distance d, then T^{-1} is a parallel translation through $-d$. The result (iv) holds, so that we may speak of the group of plane translations.

In an exactly similar way all rotations form a group; this includes translations as special cases. We say that the translation group is a subgroup of the group of plane rotations. We note that the latter group

115

is not commutative. Since the product of a pair of improper isometries is proper, the improper motions do not form a group. Adding them to the rotation group we arrive at all possible isometric operations, the group of plane isometries.

All the above groups are continuous. This means that we can find operations that will map a point P as near as we care to specify to any point of the plane. The symmetry groups are discrete, and although a point P will have an infinite number of mappings, no two of these are closer together than some finite distance.

4. Examples of discrete groups

We consider the first diagram of Figure 11. Here there is a basic translation T through distance d, taking the pattern along through one of its elements. The infinite set of operations forming the group consist of all powers of T, i.e. of all shifts of a positive or negative multiple of d. They are

$$..., \ T^{-2}, \ T^{-1}, \ I, \ T, \ T^2, \ T^3, \ .... \tag{1}$$

Next we consider the third diagram. Besides the operations of (1) there are reflections in a series of axes at spacing $\frac{1}{2}d$. If R denotes reflection in any one of these axes, Figure 5 shows that $T^k R$ is a reflection

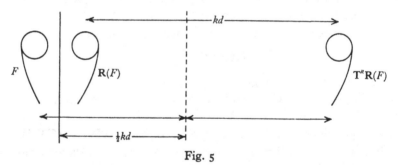

Fig. 5

in an axis at distance $\frac{1}{2}kd$ from that of R. This is another axis of the system. Thus by premultiplying R by the sequence (1) we generate all reflections of the group. We say that (1) is a subgroup of this new group; by combining (1) with any member outside itself all others are generated. These others, being improper, do not form a subgroup. We refer to them as the coset corresponding to subgroup (1).

We note that any member of (1) when applied to a reflection axis produces another such axis. Similarly, if a discrete group possesses a

116

centre of rotation; translations, rotations and reflections of the group will generate further centres when applied to it. We shall frequently use this property in the process of enumeration.

A pattern is characterised by the set of isometric motions that bring it into self-coincidence, and this set forms a discrete group. Thus the enumeration of patterns is equivalent to the enumeration of discrete groups of isometric motions in the plane.

5. *Similarity transformations*

The theory of plane isometries is developed by Yaglom[4], who describes some fascinating applications to geometrical problems. We content ourselves with examining the idea of a similarity transformation applied to an isometry. This concept proves very useful as a tool in enumerative proofs. Let $\mathbf{T}$ be a translation through distance d in direction l, and $\mathbf{S}$ be a rotation. We consider the sequence of three operations denoted by $\mathbf{STS^{-1}}$, and prove that it is also a translation through d, but parallel to a line $\mathbf{S}(l)$ obtained by rotating l by $\mathbf{S}$.

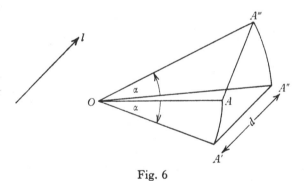

Fig. 6

The proof follows from Figure 6, where $\mathbf{S^{-1}}(A)$ gives A', by a negative rotation of magnitude α about O. Translation $\mathbf{T}$ moves A' through d parallel to l giving A''. Then $\mathbf{S}(A'')$ gives A''' by the rotation through α taken positively. The result, A to A''', is a translation through d, and its direction makes angle α with l, i.e. it is l's direction after modification by $\mathbf{S}$.

It is left to the reader to verify that, for rotations $\mathbf{S_1}$ and $\mathbf{S_2}$, the product $\mathbf{S_2 S_1 S_2^{-1}}$ is a rotation whose magnitude is that of $\mathbf{S_1}$, and whose

117

centre is $S_2(O_1)$, the image of O_1 by rotation S_2. The product RTR^{-1} is a translation whose direction is obtained by applying reflection R to that of T.

6. The point groups

The plane symmetry groups are discrete, and we base our enumeration upon their translational structure. There are three cases to consider:
  (i) No translations present, the point groups.
  (ii) Only one direction of translation, the frieze groups.
  (iii) More than one direction of translation, the wallpaper groups.
The first of these is the subject of this section. Such a group can contain no glide reflections, for the square of a glide reflection is readily seen to be a translation. If there is more than one reflection, the group contains their product, a rotation. The case of a single reflection is shown in Figure 7(a). The group contains two operators, I and R, with $R^2 = I$. We know that all others have rotations, and our first step will be to prove that there can be only one centre of rotation.

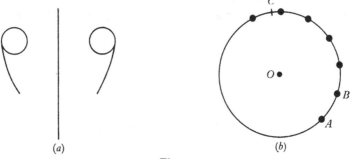

(a)   (b)

Fig. 7

Suppose there are two rotations or spins S_1 and S_2 with different centres O_1 and O_2, and of magnitudes α_1 and α_2. We have already said that the operation
$$S = S_2 S_1 S_2^{-1}$$
is a rotation through α_1 about the point $S_2(O_1)$. Now we consider $S_1^{-1}S$; this is a rotation of magnitude $(-\alpha_1 + \alpha_1)$ or zero, i.e. it is a translation. By (i) it must be the zero translation so that

$$S_1^{-1}S = I \quad \text{or} \quad S = S_1.$$

The last result, obtained by multiplying both sides of the first expression by S_1, implies that S and S_1 have the same centre or that $S_2(O_1) \equiv O_1$,

118

a contradiction. Thus we are forced to drop the assumption of more than one centre of rotation.

Our next step is to prove that all rotations about the one centre are by multiples of an angle $2\pi/n$, where n is an integer greater than 1. The point A of Figure $7(b)$ has among its mappings by the group one of minimum separation from A itself, for our group is discrete. Call it B, and let C be any other mapping of A. Let angle $AOB = 2\pi/n$, and the corresponding isometry be $\mathbf{S}$. We wish to prove that AOC is an integer multiple of this angle. If not, then we can find an integer k so that

$$\frac{2\pi k}{n} < \widehat{AOC} < \frac{2\pi(k+1)}{n}.$$

The operation $\mathbf{S}^{-k}$ brings C to a point inside the smaller arc AB, contradicting our definition of B. Thus all images of A under the group consist of an equally spaced set of points. One of the multiples of $2\pi/n$ must be

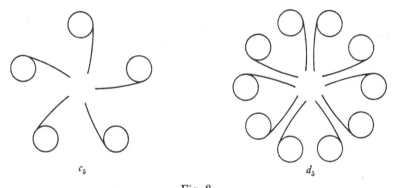

c_5 d_5

Fig. 8

2π, for A comes back to its old position after one revolution, and n must be an integer. The first diagram of Figure 8 illustrates the point group c_n for $n = 5$.

We must now see how reflections can be added. If an axis of reflection did not go through O, then the reflection of O in this axis would furnish a second centre of rotation. This is inadmissible, so any axes of reflection pass through O. Any one axis will provide $(n-1)$ others under the rotational operations of the group. A reflection $\mathbf{R}$, followed by a rotation $2\pi/n$ about a point on the reflecting axis, is easily seen to be equivalent to a reflection in an axis at π/n to the first one. This gives a second set of axes bisecting the first.

Next we observe that the product of any two reflections is a rotation through twice the angle between their axes (see Figure $9(a)$). So any two

119

axes of reflection of the group must be separated by half a multiple of $2\pi/n$. This means that all such reflections have been accounted for. We have the point group d_n with $2n$ axes of reflection, illustrated above for $n = 5$. If we consider c_1 as consisting of the identity operation only, and d_1 to correspond to Figure 7(a), then c_n and d_n include all point groups. Leonardo da Vinci is said to have enumerated these groups.

7. The frieze groups

These correspond to (ii) at the beginning of Section 6, involving a single direction of translation. It is readily found that the translations are all integer multiples of a basic distance d; this follows closely the similar investigation for rotations in the point groups. The fundamental pattern is thus the first of Figure 11 and it is called F_1. We now examine what rotations can be added. If there is a rotation S, the translation $\mathbf{STS}^{-1}$ obtained by transforming T by this operation must be parallel to

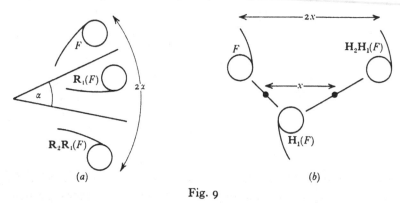

(a)

(b)

Fig. 9

T, for only one direction of translation is allowed. Thus S must be of magnitude o or π, i.e. only half-turns can be admitted. We derived d_n from c_n by introducing reflections. The introduction of half-turns into F_1 gives F_2 in a very similar manner. Figure 9(b) shows that the product of two half-turns is a translation through twice the distance between their centres. As $2x = kd$ then $x = k.\frac{1}{2}d$, and the half-turn centres at a spacing of $\frac{1}{2}d$ are directly comparable with the set of reflection axes in d_n.

We now admit reflections, and the similarity transformation $\mathbf{RTR}^{-1}$ shows that the direction of T must be unaltered by any R, just as it was by any S. Therefore reflection axes must be parallel to T, or perpendicular

120

to it. A pair of parallel reflecting axes generate a translation through twice the distance between them. From this fact we make two deductions. First, there can only be one reflection axis parallel to $\mathbf{T}$, for two would generate a translation perpendicular to it. Secondly, if there is a reflection axis perpendicular to $\mathbf{T}$, there will be an array at spacing $\frac{1}{2}d$. Again, the arguments that show the existence and completeness of this array are similar to those used in moving from c_n to d_n, or in introducing half-turns into F_1. A reflection, followed by a perpendicular translation d, gives a reflection in an axis moved through $\frac{1}{2}d$. Thus there are a pair of axes $\frac{1}{2}d$ apart. Then $\mathbf{T}$ and its powers operating on these generate the whole array. A pair of parallel reflections is equivalent to a translation through twice the distance between their axes. So any two reflections must be separated by half a multiple of d, the group's translation distance. Therefore all reflections are included in the above array.

An axis of reflection introduced parallel to $\mathbf{T}$ in F_1 gives F_1^1, and a set of axes perpendicular to it results in F_1^2. It is easy to see that, if both types of axes occur, their intersections become half-turn centres, so we really get a derivative of F_2 called F_2^1. We must consider the introduction of a parallel axis alone in F_2. Such an axis would have to pass through the half-turn centres, or another line of these centres would result by reflection in it. It is then readily seen that the product of a reflection and a half-turn would be equivalent to a reflection in an axis perpendicular to $\mathbf{T}$ so we are back to F_2^1. Finally, we consider an array of perpendicular axes only. This array must either pass through the array of half-turn centres or bisect this array. Otherwise reflections of these centres would produce a new set. The first case again implies reflection in the line of centres, i.e. it leads to F_2^1. The second gives a new scheme, that of F_2^2.

To round off the enumeration the possibility of glide reflections $\mathbf{G}$ must be considered. It is not difficult to show that only one axis, and that one parallel to $\mathbf{T}$, is possible. Further, half-turn centres, if present, must lie upon it. Now $\mathbf{G}^2$ is a translation, and therefore

$$\mathbf{G}^2 = \mathbf{T}^{2k} \quad \text{or} \quad \mathbf{G}^2 = \mathbf{T}^{2k+1}.$$

In the first case $\mathbf{T}^{-k}\mathbf{G}$ is a glide reflection whose square is the identity Thus it is a simple reflection, and $\mathbf{G}$ can add nothing to the patterns already found, for introducing this reflection is equivalent to including $\mathbf{G}$. In the second case $\mathbf{T}^{-k}\mathbf{G}$ is a glide through $\frac{1}{2}d$. We omit details, but the possibility of combining this operation with the six existing patterns leads to a single new one called F_1^3. In its first pattern unit the result of the reflection part of the glide has been dotted in.

8. The wallpaper groups

We shall not attempt the enumeration of the cases under (iii) of Section 6. The first step is to prove that there are two directions of translation in terms of which all others may be expressed. The method is an extension of that used to show that only one basic rotation entered with the point groups. The resultant simplest wallpaper pattern is W_1 of Figure 12.

We then enquire what rotations can be allowed; we shall prove that only the angles $2\pi/n$ need be considered for $n = 2, 3, 4$ and 6. The corresponding basic patterns are the first ones in Figures 13 to 16.

In Figure 10 let P be a centre of rotation, angle $2\pi/n$, and let Q be the nearest centre to it of the same type. Then rotating the pattern through $2\pi/n$ about Q must produce another centre of this type at R. Rotation about R will put yet another at S. If n is greater than 6, PR will be less than PQ, a contradiction of our assumption that Q is closer to P than any other '$2\pi/n$' centre. With $n = 6$, S and P coincide; $n = 5$ makes PS less than PQ so it is not a possible value. The values $n = 4, 3, 2$ do not produce contradictions.

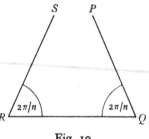

Fig. 10

With W_1 (no rotations) and W_2 (half-turns) the pattern mesh can be defined by the contents of a single parallelogram. The pattern parallelogram of W_2 has half-turn centres at its vertices, the mid-points of its edges, and the point of intersection of its diagonals. The introduction of reflections requires rhombic or rectangular regions. In two cases, W_1^3 and W_2^4, glide reflections produce new patterns. In each pattern one of the basic parallelograms has been inserted. If rotations occur, they cover this parallelogram with copies of the smaller area indicated by dotted line(s). Reflection and glide axes are shown as full lines, they bisect these smaller areas, or the basic parallelograms in Figure 12.

For W_3 the pattern can be considered as built up of repetitions of a 60° rhombus with triple centres at its vertices. Reflection in short and long diagonals produces two variants. With W_4 a quarter-square of the basic pattern can be considered, it has half-turn centres at two vertices and quarter-turns about the other two. Reflections in the two types of

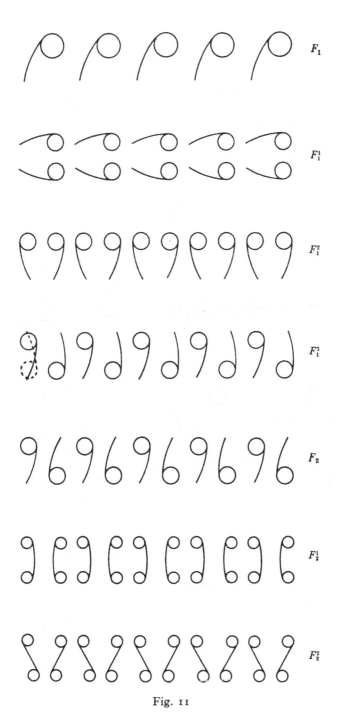

Fig. 11

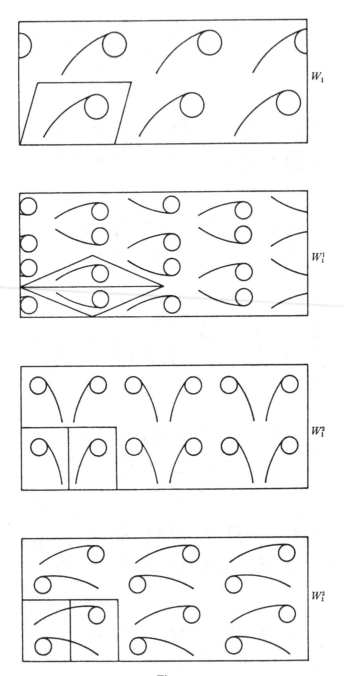

W_1

W_1^1

W_1^2

W_1^3

Fig. 12

124

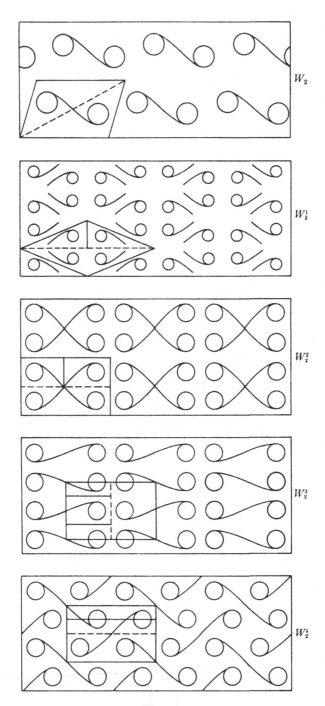

Fig. 13

125

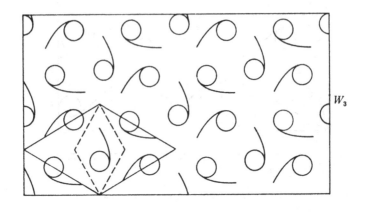

W_3

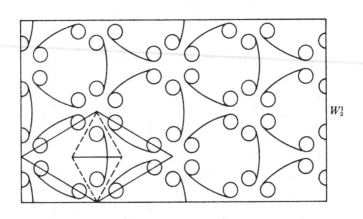

W_3^1

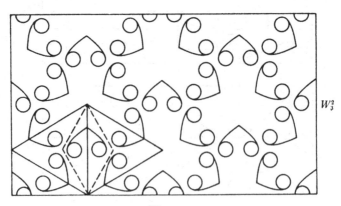

W_3^2

Fig. 14

126

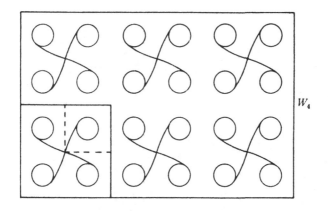

W_4

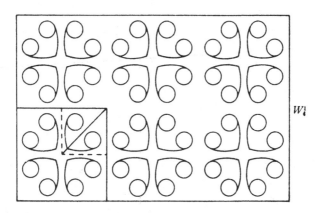

W_4^1

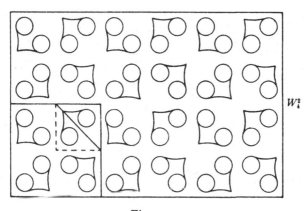

W_4^2

Fig. 15

127

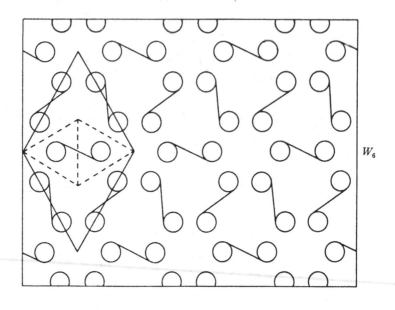

W_6

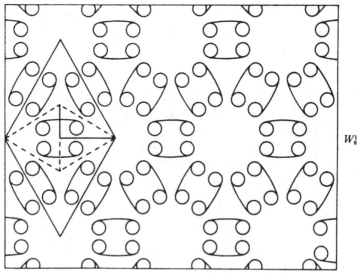

W_6^1

Fig. 16

diagonal lead to W_4^1 and W_4^2. The pattern W_6 can be built up by repetitions of an equilateral triangle. There are one-third-turn centres at two vertices, a one-sixth-turn centre at the third, and a half-turn centre at the mid-point of one side. Reflection in the height perpendicular to this side produces W_6^1, the richest of all the patterns in symmetry elements.

Féjes Tóth[1] gives a fairly complete account of the wallpaper patterns, while Hilbert and Cohn-Vossen[2] and Weyl[3] discuss them in outline. Weyl also derives the space analogues of the point groups found here. These are 3 in number, and are associated with the Platonic solids, being called the tetrahedral, octahedral and icosahedral groups respectively. The analogue in 3-space of the 17 wallpaper groups are the 230 space groups of crystallographic theory. It is interesting to note that the Moors used all 17 groups in ornamental decorations in the Alhambra at Granada.

References

1. L. Féjes Tóth. *Regular Figures* (Pergamon Press, 1964).
2. D. Hilbert and S. Cohn-Vossen. *Geometry and the Imagination* (Chelsea, 1952).
3. H. Weyl. *Symmetry* (Princeton, 1952).
4. I. M. Yaglom. *Geometric Transformations* (Random House, New Mathematical Library, 1962).

12

The real number system

1. Rationals and irrationals

We are not concerned with an axiomatic approach to the real numbers, but rather with their classification. Starting with the positive integers, the negative integers and the rationals result from the attempt to solve the equations
$$x+n = 0, \quad nx-m = 0.$$

The first defines $-n$, and the second the rational number m/n. The need for a further extension of the number system was evident to the Greeks. Geometrically the quantity $\sqrt{2}$ is capable of simple construction, but the following proof, attributed to Pythagoras, shows that there is no rational number whose square is 2.

Suppose $\sqrt{2} = p/q$, where p and q have no common factor. Then we have
$$2q^2 = p^2$$
and reason as follows. Since 2 divides the left side, it also divides the right. This is a perfect square, and so must be divisible by 4. The left is therefore also divisible by 4, and q^2 must have 2 as a factor. We have arrived at the conclusion that 2 divides both p and q, contradicting our initial assumption. The only way out of this impasse is to agree that $\sqrt{2}$ is not a rational number.

It is possible to choose rationals whose squares are as close as we please to 2. It therefore seems reasonable to introduce a new class of numbers called irrational. We can regard the irrational number $\sqrt{2}$ as the limit of a sequence of rationals, chosen so that their squares are closer and closer to 2. We looked at a continued fraction for $\sqrt{3}$ in Chapter 2. That for $\sqrt{2}$ is

$$1 + \cfrac{1}{2 + \cfrac{1}{2 + \cfrac{1}{2 + \dots}}}$$

By using more and more terms of this fraction we get the sequence of rationals
$$\frac{1}{1}, \quad \frac{3}{2}, \quad \frac{7}{5}, \quad \frac{17}{12}, \quad \frac{41}{29}, \quad \dots,$$

whose squares are

$$2-1, \quad 2+\tfrac{1}{4}, \quad 2-\tfrac{1}{25}, \quad 2+\tfrac{1}{144}, \quad 2-\tfrac{1}{841}, \quad \ldots.$$

The formation of successive values of the numerators is illustrated by the results

$$7 = 2\times3+1, \quad 17 = 2\times7+3, \quad \ldots,$$

and so on. The same recurrence relation is used for denominators.

Other limiting processes, such as the summation of more and more terms of the series for e given below, lead to a sequence of rationals which converge to an irrational limiting value. The Wallis infinite product for π given in Chapter 14 is another example. We shall prove that π is irrational in Chapter 14; we now show that e is not a rational number.

We define e by the infinite series

$$e = 1 + \frac{1}{1!} + \frac{1}{2!} + \frac{1}{3!} + \cdots,$$

and assume it is the rational number p/q. We have the result

$$q!\left(e - 1 - \frac{1}{1!} - \frac{1}{2!} - \cdots - \frac{1}{q!}\right) = \frac{1}{q+1} + \frac{1}{(q+1)(q+2)} + \cdots. \quad (1)$$

Now the expression on the left is an integer, since $e = p/q$, while the right is positive and less than.

$$\frac{1}{q+1} + \frac{1}{(q+1)^2} + \cdots = \frac{1}{q} < 1.$$

The contradiction causes us to dismiss the assumption of rationality.

In general it is difficult to carry through proofs of this kind. Thus we do not know if Euler's number

$$\gamma = \lim_{n\to\infty}\left\{1 + \frac{1}{2} + \cdots + \frac{1}{n} - \log n\right\}$$

or $\pi^{\sqrt{2}}$ is rational or irrational.

We end this section by outlining an approach to irrational or incommensurable ratios in similiarity. The Greek geometers saw the need for a logical treatment, and their genius evolved one of astonishing modernity in outlook. We first prove that between any two real numbers α and β we can find a rational m/n. The axiom of Archimedes asserts that, given two real numbers x and y, there is an integer n such that $nx > y$. This means that we can define m, n by

$$n(\beta-\alpha) > 1, \quad m < n\beta \leqslant m+1,$$

131

for if the n chosen to satisfy the first makes $n\beta$ an integer the sign $\leqslant$ still allows the second to be satisfied. The inequalities

$$\alpha < \beta - \frac{1}{n} \leqslant \frac{m+1}{n} - \frac{1}{n} = \frac{m}{n}, \quad \frac{m}{n} < \beta,$$

follow at once. This property of the rational numbers is summarised by saying that they are everywhere dense in the real number system. We now apply this result to prove the fundamental theorem concerning intercepts on a transversal by a set of parallel lines.

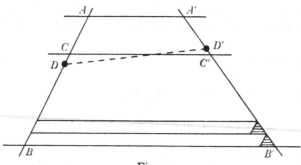

Fig. 1

Our problem is to prove that $AC/CB = A'C'/C'B'$, given that AA', BB' and CC' are parallel. If the first ratio is the rational p/q, we divide AB into $(p+q)$ equal parts, and AC will occupy just p of these. Draw lines through the points of division parallel to AA'. It is easy to prove by way of a set of congruent triangles that they cut off equal intervals on $A'B'$. Two such triangles are shaded in Figure 1. There will be p equal intervals in $A'C'$, and $B'C'$ occupies q more; the proof follows at once.

In the case where AC/AB is irrational, let us assume that

$$AC/AB < A'C'/A'B'.$$

We can find a rational r so that

$$\frac{AC}{AB} < r < \frac{A'C'}{A'B'}. \tag{2}$$

Let $AD = rAB$, and draw DD' parallel to AA'. By the result proved for the rational case $A'D' = rA'B'$. Moreover, (2) tells us that

$$AD > AC, \quad A'D' < A'C'.$$

This implies that DD' cuts CC', in direct contradiction to their parallelism. In a similar way we derive a contradiction from the assumption that $AC/AB > A'C'/A'B'$, and conclude that these ratios must be equal.

2. *The approximation of irrationals by rationals*

To convert p/q to decimal form we use the division process. There are q possible remainders, so after at most q steps one must recur, and the process is then cyclic. Thus rationals always give terminating or recurring decimals. The representation is unique except for the possibility exemplified by

$$0\cdot3999\ldots = 0\cdot4000\ldots.$$

The converse is also true; a recurring decimal necessarily represents a rational number. For example

$$0\cdot1\dot{2}\dot{3} = \frac{123}{1000} + \frac{123}{1000^2} + \ldots = \frac{123}{1000} \cdot \frac{1}{1 - \frac{1}{1000}} = \frac{123}{999}.$$

While a decimal representation is most convenient for practical purposes, the well-known ratio $22/7$ shows that rational approximations have their uses. They are of considerable importance from a number-theoretic standpoint.

Dirichlet's theorem states that an irrational number ξ has an infinite number of rational approximations m/n, in lowest terms, and such that

$$\left|\xi - \frac{m}{n}\right| < \frac{1}{n^2}. \tag{3}$$

This is not possible for a rational number p/q in lowest terms, as

$$\left|\frac{p}{q} - \frac{m}{n}\right| = \left|\frac{pn - qm}{qn}\right| \geqslant \frac{1}{qn},$$

since the top line of the second expression is a non-zero integer unless $q = n$ and $p = m$. For property (3) to hold, n must be less than q and there are only a finite set of possibilities. We say that irrationals can be approximated to order 2 but rationals only to order 1. To prove Dirichlet's theorem we first show that, for a given integer k, there is an approximation m/n with $n \leqslant k$ such that

$$\left|\xi - \frac{m}{n}\right| < \frac{1}{nk}. \tag{4}$$

133

Consider the set of $(k+1)$ numbers

$$0, \ \xi - [\xi], \ 2\xi - [2\xi], \ \ldots, \ k\xi - [k\xi], \tag{5}$$

all less than 1, together with the set of k regions

$$(0, \ 1/k), \ (1/k, \ 2/k), \ \ldots, \ (1 - 1/k, \ 1). \tag{6}$$

We invoke Dirichlet's pigeon-hole principle, used in Chapter 3. As each of (5) lies in one of the regions (6), there must be one region with at least 2 numbers of the set in it. Let them correspond to integers r and s, so that we have

$$(r-s)\xi - [r\xi] + [s\xi] < 1/k. \tag{7}$$

Putting $n = (r-s)$, $m = [r\xi] - [s\xi]$ and dividing (7) by n, we get (4). Should m and n have a common factor, so that $m = fM$ and $n = fN$, with M and N co-prime, (4) becomes

$$\left| \xi - \frac{M}{N} \right| < \frac{1}{fNk} < \frac{1}{Nk},$$

and so is unaltered in form.

To prove Dirichlet's theorem we suppose that there are only a finite set of approximations satisfying (3). Each one gives rise to a non-zero error, for the number ξ is not rational. Let e be the error of smallest modulus and choose $k > 1/|e|$. By the result just proved we can find an approximation satisfying (3), such that

$$\left| \xi - \frac{m}{n} \right| < \frac{1}{nk} < \frac{1}{k} < |e|.$$

This approximation, having an error less than $|e|$, cannot be one of the original set. Thus the assumption of a finite number of approximations satisfying (3) leads to a contradiction.

The most powerful approach to questions of this kind is by way of continued fractions. They can be used to show that (3) can be replaced by the stronger result

$$\left| \xi - \frac{m}{n} \right| < \frac{1}{\sqrt{5}\,n^2}.$$

It can be proved that, for some irrationals, $\sqrt{5}$ cannot be replaced by a larger number, so this result is the best possible. Hardy and Wright[2], Drobot[6] and Niven[4] deal with the continued fraction approach, while Niven[5] gives a detailed account of Dirichlet's theorem.

3. *Algebraic numbers*

We can regard the irrational $\sqrt{2}$ as defined by

$$x^2 - 2 = 0.$$

More generally we say that the algebraic equation

$$a_0 + a_1 x + a_2 x^2 + \ldots + a_m x^m = 0 \tag{8}$$

defines a set of algebraic numbers, one for each real root. The coefficients are integers or rational numbers; the numbers so defined are said to be of order m. The rationals are algebraic numbers of order unity.

Suppose we add together algebraic numbers α and β of orders m and n respectively. Then we shall prove that the result is another such number, in general of order mn. We have

$$a_m \alpha^m = -a_0 - a_1 \alpha - \ldots - a_{m-1} \alpha^{m-1}, \tag{9}$$

$$a_m \alpha^{m+1} = -a_0 \alpha - a_1 \alpha^2 - \ldots - a_{m-1} \alpha^m, \tag{10}$$

$$a_m \alpha^{m+2} = -a_0 \alpha^2 - a_1 \alpha^3 - \ldots - a_{m-1} \alpha^{m+1}. \tag{11}$$

The equation (9) can be used to eliminate α^m from (10), while (9) and (10) will remove α^m and α^{m+1} from (11). Thus we can express α^m and higher powers of α as sums of rational multiples of $1, \alpha, \alpha^2, \ldots, \alpha^{m-1}$. A similar result holds for β, and we now turn to consider powers of $(\alpha + \beta)$.

We can find rationals c_{pqr} so that

$$(\alpha + \beta)^j = c_{j,0,0} + c_{j,1,0}\alpha + c_{j,1,1}\alpha\beta + \ldots + c_{j,m-1,n-1}\alpha^{m-1}\beta^{n-1}, \tag{12}$$

for $j = 0, 1, 2, \ldots, mn$. They are obtained by expanding by the binominal theorem, and eliminating powers of α greater than $(m-1)$ and β greater than $(n-1)$, by the method outlined above. Thus the powers of $(\alpha + \beta)$ have been expressed in terms of $\alpha^s \beta^t$ for s from 0 to $(m-1)$, and t from 0 to $(n-1)$. These mn quantities can be eliminated from the $(mn+1)$ equations (12) to give the determinant

$$\begin{vmatrix} (\alpha+\beta)^{mn} & c_{mn,0,0} & \cdots & c_{mn,m-1,n-1} \\ (\alpha+\beta)^{mn-1} & c_{mn-1,0,0} & \cdots & \cdots \\ \vdots & & & \\ 1 & c_{0,0,0} & \cdots & \cdots \end{vmatrix} = 0.$$

135

Expansion in terms of the first column yields an algebraic equation in $(\alpha + \beta)$ of degree mn with rational coefficients.

Thus the sum of two algebraic numbers is an algebraic number, and the same holds for the other operations of arithmetic. The algebraic number system is closed, and its members form a field. There is a still more general form of closure. We might ask if the numbers x defined by the quadratic

$$x^2 + xz - y = 0, \qquad (13)$$

with its coefficients positive algebraic numbers specified by

$$y^2 - 2 = 0, \quad z^2 - z - 1 = 0,$$

are still algebraic numbers. The answer can be found by eliminating y and z to get

$$x^8 + 2x^7 - x^6 - 2x^5 - 3x^4 - 4x^3 - 6x^2 + 4 = 0.$$

Two of the roots of this equation are the numbers defined by (13). This result is general, and nothing new emerges when we allow the coefficients of (8) to be algebraic numbers themselves.

However, there is another category of real numbers to which π and e belong. This class is called transcendental, and its members do not satisfy any algebraic equation like (8). We shall prove the existence of such numbers in two quite different ways. The problem of proving that a particular number is transcendental is a difficult one. Hermite proved e transcendental in 1873; Lindemann extended his method, and used

$$e^{i\pi} + 1 = 0$$

to prove π transcendental in 1882. The one general result known was discovered independently by Gelfond and Schneider in 1934, after Hilbert had propounded the problem in 1900. If α and β are algebraic, neither 0 or 1, and β is irrational, then α^β is transcendental. For instance, we can be sure that $2^{\sqrt{2}}$ is transcendental, but 2^e is still in doubt, as e, although irrational, is not algebraic.

An interesting result, due to Mahler, concerns polynomials $f(x)$ that take integer values when $x = 1, 2, \ldots$. As an example consider $x(x+1)/2$, taking the values $1, 3, 6, 10, \ldots$. Mahler proved that the number

$$0 \cdot f(1) f(2) f(3) \ldots,$$

is transcendental. In our example the construction leads to the number $0 \cdot 1361015 \ldots$.

Those algebraic numbers that can be constructed using ruler and compass only are called Euclidean. We see heuristically that such

136

numbers are combinations of quadratic surds, for intersections of lines and circles with lines and circles lead to linear or quadratic equations. The Greeks considered three problems of geometrical construction that defied all attempts at solution for 2,000 years. They were the duplication of the cube, the trisection of the angle, and the squaring of the circle. The equations to be solved are

$$x^3 - 2 = 0, \quad 4x^3 - 3x + c = 0, \quad x^2 - \pi = 0.$$

It can be shown that the first two do not reduce to quadratic form, i.e. they do not define Euclidean numbers. These problems are thus insoluble. The third demands the construction of a transcendental number, again an impossible task. Courant and Robbins[1] and Klein[3] discuss these three problems in more detail; Klein[3] proves that π is transcendental.

4. Liouville's theorem

In 1851 Liouville found the first demonstrably transcendental number. He proved the following theorem: an algebraic number defined by an equation of degree r is not approximable to an order higher than r. The number ξ satisfies

$$f(\xi) = a_0 + a_1 \xi + \ldots + a_r \xi^r = 0.$$

We can find M so that

$$|f'(x)| < M \quad \text{if} \quad |x - \xi| < 1.$$

Consider the difference

$$f\left(\frac{p}{q}\right) - f(\xi) = f\left(\frac{p}{q}\right) = \left(\frac{p}{q} - \xi\right) f'(x), \qquad (14)$$

for some value x between ξ and p/q by the first Mean Value theorem. Now

$$\left| f\left(\frac{p}{q}\right) \right| = \left| a_0 + a_1 \frac{p}{q} + \ldots + a_r \frac{p^r}{q^r} \right| = \frac{\text{Integer}}{q^r} \geqslant \frac{1}{q^r},$$

since the integer cannot vanish or ξ would be rational. From this result and (14) it follows that

$$\left| \frac{p}{q} - \xi \right| > \frac{1}{Mq^r}. \qquad (15)$$

If ξ is approximable to order $(r+1)$, then, for some constant c, and for infinitely many fractions p/q in lowest terms

$$\left| \frac{p}{q} - \xi \right| < \frac{c}{q^{r+1}}. \qquad (16)$$

137

This requires $q < Mc$ in view of (15), so that (16) has only a finite number of possibilities.

Liouville found a class of numbers that could be approximated to an arbitrarily chosen order, and could not therefore be algebraic by the above theorem. The simplest member of this class is the sum of the series

$$\frac{1}{10^{1!}} + \frac{1}{10^{2!}} + \frac{1}{10^{3!}} + \cdots.$$

Suppose we wish to approximate to order k. Choose any j greater than k, and sum the first j terms. The resulting rational p/q will have $q = 10^{j!}$. The error will be

$$\frac{1}{10^{(j+1)!}} + \frac{1}{10^{(j+2)!}} + \cdots = \frac{1}{q^{(j+1)}} + \frac{1}{q^{(j+1)(j+2)}} + \cdots < \frac{2}{q^{(j+1)}} < \frac{2}{q^k},$$

as we chose j in excess of k. There are an infinity of possible values of j, so the number is approximable to order k for any k whatever.

As recently as 1955 Roth proved the much more powerful result that no algebraic number can be approximated to an order higher than 2. We have shown that all irrationals can be approximated to this order; the rationals are only approximable to order 1.

5. Cantor's construction of transcendental numbers

Cantor called a set of objects countable if they can be placed in one-to-one correspondence with the positive integers. He proved that the rational numbers are countable as follows. We associated with p/q the index $(p+q)$ and arrange the rationals in groups with the same index. For instance, with index 4 we have the rationals 3/1, 2/2, 1/3. The numbering or counting of the rationals commences with the single member 1/1 of index 2. The second and third are 2/1, 1/2 of index 3, then integers 4, 5 are allotted to the class of index 4, and so on. Fractions like 2/2 not in their lowest terms are omitted.

A similar procedure enables us to count the system of algebraic numbers. We define as the index of the algebraic number or numbers specified by (8) the quantity

$$m + |a_0| + |a_1| + \ldots + |a_m|.$$

There is no loss of generality in assuming that the a's are integers. For a given value of this index, which is necessarily an integer, there are a finite number of possible equations; these define a finite number of

138

algebraic numbers. Some will have been defined by equations belonging to smaller indices. These are rejected, and the remainder arranged in increasing numerical order. Then, starting at the lowest index and working upwards, the process used for rationals can be repeated.

Cantor then assumed that all the real numbers in the range (o, 1) can be counted. On this assumption they can be listed as follows

$$x_1 = \text{o·}d_{11}d_{12}d_{13}...$$
$$x_2 = \text{o·}d_{21}d_{22}d_{23}...$$
$$x_3 = \text{o·}d_{31}d_{32}d_{33}...$$
$$x_4 = \text{o·}d_{41}...$$
$$.........................$$

From this set we construct a new number, not in the set itself. It commences o· and its first decimal digit is any one differing from d_{11}. Its second can be any digit differing from d_{22} and so on. This new number differs in at least one digit from each of the x's. Thus we have constructed a new number, so that our assumption that all numbers could be counted must be abandoned. Since the algebraic numbers are countable, other types (i.e. transcendental numbers) must exist.

An approach based on measure theory makes Cantor's last step unnecessary. The measure of a class of numbers is defined as the limit of the sum of a set of intervals including every number of the class. We prove that the measure of a countable class of numbers is zero. Enclose a_n, the nth member of the class, in the interval $(a_n - \epsilon/2^n, a_n + \epsilon/2^n)$. The total length of the intervals used will be

$$\epsilon + \tfrac{1}{2}\epsilon + (\tfrac{1}{2})^2\epsilon + ... = 2\epsilon,$$

a quantity that can be made as small as we choose.

Now heuristically we see that the measure of the class of all the numbers in (o, 1) must exceed zero. As the measure of the algebraic numbers in this interval, or indeed in any interval, is zero, the existence of other numbers follows. We say that almost all numbers are transcendental, since the exceptions are of measure zero.

6. Normal numbers

Borel introduced quite a different way of looking at the real number system. He defined as simply normal a number whose decimal expansion contains each of the digits from o to 9 in just $\tfrac{1}{10}$ of all possible positions.

Most rational numbers do not have this property but the number 0·i234567890 is simply normal. However, like all other rationals, it fails to pass his criterion for (full) normality. This is that any 2-digit group, for example, 37 shall occur in just $\frac{1}{100}$ of all possible 2-digit selections. A 3-digit group occurs in $\frac{1}{1000}$ of all such groups and so on. We note that similar definitions could be made for a radix other than 10.

Borel proved that almost all numbers are normal to any radix, i.e. that the measure of those numbers not normal to radix r is zero. In spite of this it is not easy to prove that a given number is normal, thus π and e have defied investigation so far. Nor is it easy to write down a demonstrably normal number. Such a one is

$$0 \cdot 1234567891011121 3 \ldots, \qquad (17)$$

where the decimal digits arise by placing the ascending integers in juxtaposition.

In order to prove this we concentrate our attention on the triplet 327, and count its occurrences in the integers from $10^{(k-1)}$ to $10^k - 1$. These occurrences are of four types:

(i) $|327 \leftarrow (k-3) \text{ digits} \rightarrow|$,

(ii) $| \leftarrow l \text{ digits} \rightarrow 327 \leftarrow (k-l-3) \text{ digits} \rightarrow|$,

(iii) $|\leftarrow (k-1) \text{ digits} \rightarrow 3||27 \leftarrow (k-2) \text{ digits} \rightarrow|$,

(iv) $|\leftarrow (k-2) \text{ digits} \rightarrow 32||7 \leftarrow (k-1) \text{ digits} \rightarrow|$,

and type (ii) occurs for values of l from 1 to $(k-3)$. In case (i) there are 10 possible digits in each of the $(k-3)$ arbitrary positions, so we have $10^{(k-3)}$ triplets of this type. In case (ii) the first digit must not be zero, but takes all other values; allowing for the various values of l we get $9.(k-3).10^{(k-4)}$ possibilities. In case (iii) a fraction of $\frac{1}{10}$ of the numbers preceding a number starting with 27 will end in 3. Hence we get $\frac{1}{10}$ times $10^{(k-2)}$ or $10^{(k-3)}$ of this type. Similarly case (iv) gives $\frac{1}{100}$ times $10^{(k-1)}$ triplets 327. The grand total is found to be $(9k+3).10^{(k-4)}$. We must now count the possible number of triplets. Each digit except for the last two is the first one of a triplet. The number of digits is k times the number of integers, this is $k(10^k - 10^{(k-1)})$ or $9k.10^{(k-1)}$. Thus the proportion of 327's among the triplets will be

$$\frac{9k+3}{9000k - 2.10^{-k+4}},$$

a quantity that tends to $\frac{1}{1000}$ for large k. Thus, over all values of k, 327 occurs in the correct proportion of cases. A similar proof applies for

140

quadruples and so on. A few triples like 991 will occur at the junctions of different 'k' blocks, but these instances are such a small part of the whole as to be negligible. Niven[4] gives an excellent account of normality. We note that the number (17) is irrational, for its digits cannot recur. Mahler's result, quoted in Section 3, states that it is transcendental. So this number has two of the properties possessed by almost all real numbers; it is both transcendental and normal.

References

1. R. Courant and H. Robbins. *What is Mathematics?* (Oxford, 1943).
2. G. H. Hardy and E. M. Wright. *An Introduction to the Theory of Numbers*, 4th edition (Oxford, 1960).
3. F. Klein. *Famous Problems of Elementary Geometry* (Dover, 1956).
4. I. Niven. *Irrational Numbers*, Carus Monograph no. 11 (Wiley, 1956).
5. I. Niven. *Numbers: Rational and Irrational* (Random House New Mathematical Library, 1961).
6. S. Drobot. *Real Numbers* (Prentice-Hall, 1964).

13

A theorem of combinatorial geometry

1. Helly's theorem in two dimensions

Helly's theorem belongs to the somewhat ill-defined sphere of combinatorial geometry. It also has applications in analysis, particularly in the
theory of approximations. We shall prove the result in a simple form,
and mention an extension. The n-dimensional version is stated later.

We have a finite collection of convex regions such that any 3 of the
regions have a point in common. The theorem states that they all have
a point in common. Figure 1 (a) shows that the number 3 cannot be
replaced by 2. Each pair of the circles has a common point, but no point

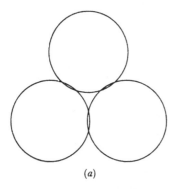

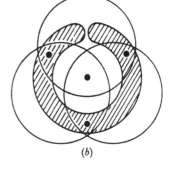

(a)   (b)

Fig. 1

belongs to all 3. Figure 1 (b) shows that convexity of the regions is
essential. Each pair of circles have a point in common with the shaded
zone, and there is a point common to all 3 circles. However, there is no
point lying in all 4 regions.

We start by proving the result for 4 regions R_1 to R_4. Let A_1 lie in
R_2, R_3 and R_4, A_2 lie in R_1, R_3 and R_4, and so on. Then every part of the
triangle A_1, A_2, A_3 lies in R_4, for all its vertices are in this region, and the
region is convex. Similar results hold for the other 3 triangles. There are
two essentially different dispositions of the points A_1 to A_4. Either they

142

form a convex quadrilateral, or one of them lies inside the triangle formed by the other three. In the first case the point of intersection of the diagonals belongs to all 4 triangles, and hence lies in all the regions. In the second case, if A_4 lies inside $A_1 A_2 A_3$ then A_4 belongs to all 4 triangles and is again within R_1 to R_4.

We now assume that the theorem holds for all sets of n regions. Consider $(n+1)$ regions R_1 to R_{n+1} satisfying the conditions of the theorem. Since R_n, R_{n+1} and R_1 have a common point, the intersection R of the first two is not empty. Moreover, R, whose points lie in both convex regions, is itself convex; this is proved as follows. A necessary and sufficient condition for convexity is that the joint of any 2 points in the region lies entirely within it. If we select points 1 and 2 of R they lie in convex region R_n, and therefore all points of line (1, 2) lie in R_n. Similarly line (1, 2) lies entirely in R_{n+1} and hence also in R.

Consider the set of n convex regions R_1 to R_{n-1} and R. By the result just proved for 4 regions we see that R_i, R_j, R_n and R_{n+1} have a common point, for any 3 of them have such a point. Hence too R_i, R_j and R have a point in common because of the way R is defined. This is true for all i, j from 1 to $(n-1)$, therefore this set of n regions satisfies the conditions of the theorem. The theorem is assumed to be true for any set of n regions, and they thus have a point in common. This lies in R, and hence belongs to both R_n and R_{n+1}, i.e. all $(n+1)$ regions have a common point, and our inductive proof is complete.

The theorem is also true for an infinite set of regions provided they are bounded. This assumption was not made above, and our proof applies to any finite set of regions, bounded or unbounded.

2. A centring theorem

We consider chords PQ of a closed convex curve through a fixed point O. In the special case where O is the mid-point of each chord we have a centrally symmetric curve. We shall prove that, in all cases, O can be chosen so that the centre of every chord is within one sixth of its length of O. Put otherwise, we can always find O so that both PO and OQ are greater than or equal to $\frac{1}{3}PQ$.

In Figure 2(a) the point A is joined to each point X of the curve and $AY = \frac{2}{3}AX$, so that Y generates a similar and similarly situated convex curve. In Figure 2(b) we have applied this construction at any 3 points A, B and C. We next show that G, the centroid of triangle ABC lies

143

inside or on all 3 derived convex curves. The mid-point L of BC lies inside or on the outer curve, since it is convex. It is a fact of elementary geometry that $AG = \frac{2}{3}AL$. Hence AG is less than or equal to AY, i.e. G lies inside or on the derived curve at A. By symmetry, it does so for the curves at B and C.

Now we apply Helly's theorem to the infinite set of bounded convex curves obtained from all points of the original convex curve by the '$\frac{2}{3}$' construction. Any 3 have a point in common by the proof just given.

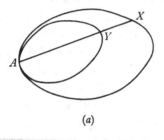 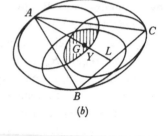

(a)   (b)

Fig. 2

Hence all have a point O in common. Draw a chord PQ through O, and consider the '$\frac{2}{3}$' curve for P. It contains O, and therefore OQ equals or exceeds $\frac{1}{3}PQ$. In exactly the same way we show that PO does so too.

For some convex regions the fraction $\frac{1}{3}$ can be increased, thus $\frac{1}{2}$ is possible for the ellipse. For the equilateral triangle no such improvement in choosing a 'centre' is possible. It has 3 chords trisected at its in-centre, any other point has chords through it divided in a ratio less than $\frac{1}{3}$.

3. A theorem of Blaschke

We first define the support line at any point P of a closed convex curve. It is any line through P with all points of the convex region lying on one side of it. At points where the slope is continuous it is the tangent at P. If there is a slope change at P, any of a pencil of lines lying between the two tangents will satisfy the condition.

The breadth of the curve in a given direction is the distance between the unique pair of parallel support lines perpendicular to this direction. Figure 3 illustrates three cases of this definition of breadth. As the chosen direction varies the breadth will change. We call the minimum value taken the width of the curve. Thus a curve of width unity can just be covered by a parallel-sided strip one unit wide.

144

Blaschke proved that a convex curve of width 1 or more always contains a circle of radius ⅓. That this figure cannot be increased can be seen by considering the equilateral triangle. If its height is unity, the largest possible circle contained in it is its in-circle of radius ⅓.

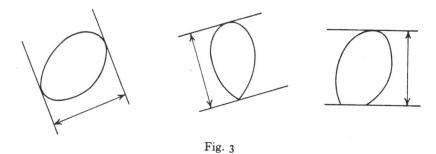

Fig. 3

We shall show that the point O of the last section is at least ⅓ of a unit from every point A of the curve. It then follows that we have a circle centre O with the Blaschke property. Taking a support line l' in Figure 4 parallel to the support line l at A, let it meet the curve at Q. Draw the chord QOP, cutting l at P'. The line XOY is perpendicular to l and l'.

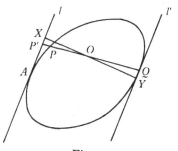

Fig. 4

We have $OP/PQ \geqslant$ ⅓, and adding equal quantities to top and bottom lines, we see that $OP'/P'Q \geqslant$ ⅓. Then by similar triangles, $OX/XY \geqslant$ ⅓, and, as XY is at least 1, this gives $OX \geqslant$ ⅓. Finally, AO equals or exceeds OX, and the theorem is proved.

4. Krasnosellskiï's theorem

We are concerned with non-convex polygons with the property of being star-shaped with regard to an interior point P. This means that we can see the whole of each side from P; or that, if we join P to any point Q of the boundary in Figure 5 (a), the half-line PQ does not cut the polygon again.

The second diagram illustrates a region possessing no point with respect to which it is star-shaped. The triangles formed by the dotted

145

extensions of the sides would each have to contain any point from which all sides can be seen, and they do not intersect.

Krasnosellskii showed that, if each set of three sides is visible from some point, then all the sides can be seen from a suitably chosen point.

In Figure 6(a) we allot a clockwise direction of rotation to the sides, so that the interior lies to the right of each side. To each side corresponds the half-plane lying on its right; one has been shaded in the diagram. These half-planes are unbounded convex regions. A point

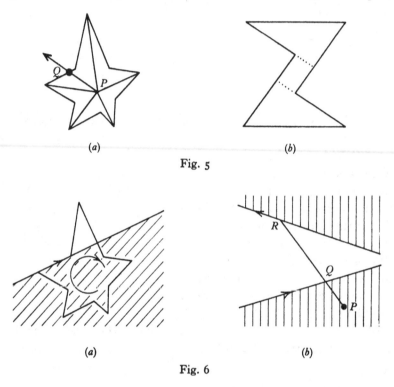

(a)   (b)

Fig. 5

(a)   (b)

Fig. 6

from which a side can be seen must lie in the half-plane corresponding to that side. Hence all sets of three half-planes have a point in common. It follows that they all have a common point P. It is easy to prove that this lies inside the polygon. We show that a ray PQ cannot have another intersection besides Q with the boundary, i.e. that all sides can be seen from P.

The proof is by *reductio ad absurdum*; we assume that PQ cuts the boundary again at R. In Figure 6 (b) there is no point of the boundary between Q and R. Hence QR is outside the polygon, and the half-plane

146

defined by the side containing R lies on the opposite side of R to Q. But this means that it cannot contain P, a contradiction. Suppose that there are other points of the boundary between Q and R; let S be the one nearest Q. In exactly the same way we see that P lies outside the half-plane with S on its boundary.

The general non-convex polygon may have no point like P. If it has two such points, called P and Q, it is easy to see that a side visible from both P and Q remains visible from all points of the interval PQ. In fact the set of points with regard to which a region is star-shaped is either empty or convex.

5. *Jung's theorem and covering problems*

We call the maximum breadth of a convex figure its diameter. In fact we can define diameter for a quite general set of points as the maximum distance separating any pair of these points. Jung's theorem states that a plane point set of diameter unity can be covered by a circle of radius $1/\sqrt{3}$.

We consider a set of 3 points of diameter 1 or less. If they form an obtuse triangle they are included in a semi-circle on the longest side as diameter, its radius is not greater than $\frac{1}{2}$ and is therefore less than $1/\sqrt{3}$. Consequently the 3 circles of radius $1/\sqrt{3}$ with centres at the vertices all contain the mid-point of the longest side. If the triangle is acute it has one angle A lying between $60°$ and $120°$, so that its sine equals or exceeds $\frac{1}{2}\sqrt{3}$. The radius of its circumcircle is given by

$$2R = \frac{a}{\sin A}$$

and $a \leqslant 1$. Thus $R \leqslant 1/\sqrt{3}$, and this time the 3 circles introduced above all contain the circumcentre.

Returning to our set of points we draw a circle of radius $1/\sqrt{3}$ centred on each; by the results just proved and Helly's theorem they possess a common point. This point is no further than $1/\sqrt{3}$ from any of the set, so it is the centre of a covering circle of this radius. We call this circle a universal cover for sets of diameter unity. An equilateral triangle of side unity has a circumradius of $1/\sqrt{3}$, so this is the smallest possible such circle. There are other smaller universal covers, we shall prove that a regular hexagon inscribed in this circle is one. The distance between a pair of parallel sides of this hexagon is unity. The problem of the universal cover of smallest area is not yet solved. The second diagram of Figure 7, a regular hexagon with two corners removed, is another cover, and still smaller ones have been found.

To prove the result for the regular hexagon we start with three pairs of distant parallel lines with the set of points inside the large hexagon they define. The lines are chosen so that angles of this hexagon are all $\frac{2}{3}\pi$. Sides are moved in parallel to themselves until they encounter a point of the set. We arrive at an equiangular hexagon whose parallel sides are not more than unit distance apart.

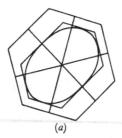 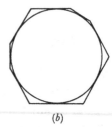

(a)   (b)

Fig. 7

There was one degree of freedom in the choice of the original lines, we define the angle that one of the three directions makes with the horizontal as θ. Lengths of sides of the hexagon will be continuous functions of θ. We consider a pair of opposite sides $a(\theta)$ and $b(\theta)$ and define the continuous function

$$d(\theta) = a(\theta) - b(\theta).$$

For a given value θ_0 we assume that a exceeds b, so that $d(\theta_0) > 0$. We have

$$d(\theta_0+\pi) = a(\theta_0+\pi) - b(\theta_0+\pi) = b(\theta_0) - a(\theta_0) = -d(\theta_0),$$

since an increase of θ by π interchanges opposite sides. Thus $d(\theta)$ is of opposite signs at θ_0 and at $(\theta_0+\pi)$. By a well-known result for continuous functions it must take the value zero at some intermediate point. So we can find an equiangular hexagon with an opposite pair of sides equal. It is a matter of simple geometry to prove that all three pairs of opposite sides are necessarily equal, i.e. that it has central symmetry.

Through the centre we draw the three perpendiculars to pairs of opposite sides. These are perpendicular to the sides of a regular hexagon with the same centre and of circumradius $1/\sqrt{3}$ illustrated in Figure 7 (a). The centre is at a distance of $\frac{1}{2}$ from the sides of the regular hexagon.

148

Its distance from a side of the centrally symmetric hexagon circumscribing the set cannot exceed $\frac{1}{2}$. Otherwise this hexagon would possess a pair of parallel sides more than unit distance apart. So the regular hexagon covers the circumscribing hexagon, and hence the set of diameter unity.

6. Some n-dimensional theorems

We give without proofs extensions of some of the results discussed above. Helly's theorem states that, if a set of bounded convex regions in n-space is such that any $(n+1)$ have a point in common, then a point can be found lying in all of them.

Blaschke's theorem for a convex body of width exceeding 1 in n-space states that it contains a sphere of diameter

$$\frac{\sqrt{(n+2)}}{n+1} \, (n \text{ even}) \quad \text{or} \quad \frac{1}{\sqrt{n}} \, (n \text{ odd}).$$

Jung's theorem states that a hypersphere of diameter

$$\sqrt{\left(\frac{2n}{n+1}\right)},$$

is a universal cover for a set of points of unit diameter. The regular n-simplex of side

$$\sqrt{\{\tfrac{1}{2}n(n+1)\}},$$

also provides a universal cover. In 2-space an equilateral triangle of side $\sqrt{3}$ will cover the regular hexagon found in Section 5, so we have a proof of the simplex result in the plane.

7. Approximating a function by polynomials

We have n points (x_i, y_i) lying on the curve $y = f(x)$, and seek the 'best' approximation to these points by a straight line. Any line $y = mx + c$ will give an error

$$e_i = |y_i - mx_i - c|,$$

at the ith point. We choose as best the line that minimises the maximum error, i.e. the maximum value of the set e_1 to e_n.

For three points the optimum solution is shown in Figure 8(a). The absolute values of errors are equal and they alternate in sign. Assuming that x_1, x_2, x_3 are in ascending order of magnitude we solve the three equations

$$y_1 = mx_1 + c + \epsilon,$$

$$y_2 = mx_2 + c - \epsilon,$$

$$y_3 = mx_3 + c + \epsilon,$$

for m, c and ϵ. It is not possible to decrease the error at point 1 without increasing the error at either 2 or 3. A decrease at 1 means that the line moves up there; if it stays fixed or moves down at 2 it moves down at 3 so

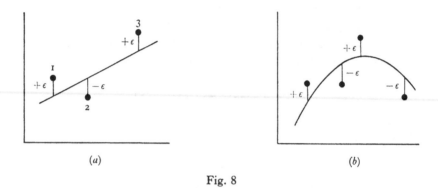

(a)   (b)

Fig. 8

increasing the error there. Otherwise the line must go up at 2, with an increase in 2's error. Attempts to decrease the maximum absolute error of the set are therefore unavailing.

To determine the optimum line for n points we first carry out this process for all possible sets of 3. Each one gives an ϵ, and one set will have a value ϵ_0 of maximum modulus. Then this line is optimum, giving ϵ_0 as the smallest possible worst error for this set of data.

The proof runs as follows. We associate with each of the n points a line joining $(x_i, y_i - \epsilon_0)$ and $(x_i, y_i + \epsilon_0)$. A typical point and the associated vertical line are shown in Figure 9(a). The solution

$$y = mx + c$$

must cut this vertical line if no error exceeds ϵ_0. This leads to the inequalities

$$y_i - mx_i - \epsilon_0 \leqslant c \leqslant y_i - mx_i + \epsilon_0.$$

The second diagram is one in which each line of the (x, y) plane has its 'c' plotted against its 'm'. The above inequalities imply that all possible

150

lines with error less than ϵ_0 lie in the shaded region between a pair of parallel lines in the (m, c) plane.

Now there are n vertical intervals, and a corresponding set of n shaded strips in the (m, c) diagram. There is a line cutting any set of 3 vertical intervals, for each set is fitted by a line with an error not exceeding ϵ_0. So any set of 3 of these convex strips have a common point. In most cases a set of 3 will have a common area, but in the set that gave ϵ_0 there will be only a single point in all 3 regions. Helly's theorem tells us that all n regions have a common point; from the observation just

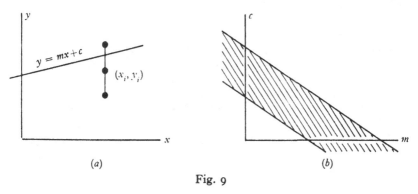

(a)   (b)

Fig. 9

made it must correspond to the unique (m, c) determined by the worst case. Since the point lies in all other regions in the (m, c) diagram, the corresponding line cuts all vertical lines in the (x, y) plane. Thus this line gives a maximum error of ϵ_0, and as one set of 3 points actually realises this maximum, there is no better fit.

In practice there may be far too many sets of 3 to make this plan possible, and iterative procedures have been worked out to avoid evaluating sets of 3 unnecessarily. Our proof above is an existence theorem, not a procedure we would necessarily employ on extensive data.

Figure 8(b) shows 4 points fitted with equal errors of alternating sign by the curve
$$y = A + Bx + Cx^2,$$
a parabola. Four points are now needed to determine A, B, C and ϵ. The (m, c) plane is replaced by an (A, B, C) 3-space. The three-dimensional form of Helly's theorem is now available to assure us that the best parabolic fit is given by trying all sets of 4 and selecting the parabola for that set giving the greatest numerical value of ϵ. Here, even more than in the previous case, some shortcutting procedure is desirable. The method

151

is quite general and indicates the best fit of degree k. This is obtained by fitting all possible sets of $(k+1)$ points with equal errors of alternating signs. The worst such set gives the curve of degree k appropriate to all data points.

Topics akin to those we have discussed in earlier sections are treated admirably by Hadwiger, Debrunner and Klee[2], Lyusternik[3] and Yaglom and Boltyanskii[4]. Eggleston's[1] book is more difficult, but it is a standard work on convexity.

References

1. H. G. Eggleston. *Convexity*. Cambridge Tract no. 47 (Cambridge, 1958).
2. H. Hadwiger, H. Debrunner and V. Klee. *Combinatorial Geometry in the Plane* (Holt, Rinehart and Winston, 1964).
3. L. A. Lyusternik. *Convex Figures and Polyhedra* (Dover, 1963).
4. I. M. Yaglom and V. G. Boltyanskii. *Convex Figures* (Holt, Rinehart and Winston, 1961).

14

The number π

1. Historical background

The number π has been the subject of a series of investigations spanning a period of more than 2,000 years. Its early importance lay in its relationship to the circle. With the development of the calculus its fundamental nature became apparent, and the connection with the circle is today seen to be of secondary importance. In Biblical times the value 3 was in use as an approximation; it is implied by verse 23 of I Kings 7 and verse 2 of II Chronicles 4. Since then a host of better approximations have been devised. The ratio 22/7 is the best known and, apart from decimal values, no other is in use today. Among many contenders in the past we mention $\sqrt{10}$. Although not a particularly good approximation, it was used in India, China and throughout the ancient world. We shall mention later the value 355/113, a very good approximation, discovered by Tsu Ch'ung-chih in about 470 A.D. The Indian mathematician Ramanujan, discovered by Hardy while working in obscurity in his own country, found the approximation $\sqrt[4]{(2143/22)}$, a quantity in error by 10^{-9}.

Serious attempts to compute π go back at least to Archimedes. By inscribing and circumscribing regular polygons of 96 sides in and around a circle he proved that
$$3\tfrac{10}{71} < \pi < 3\tfrac{1}{7}.$$

In this, as in so many other respects, the Greek approach is characteristically modern.

A long list of successively more accurate computations could be given, we shall mention only a few. Vieta, who found the beautiful formula discussed in Section 2, determined its value to 10 decimal places. Like most of the early computers he used a method based on the length of side of a regular n-gon inscribed in a circle. In Figure 1, AB, the side of an n-gon, is of length s_n, while BC, corresponding to $2n$ sides is s_{2n}. We have the results

$$OD^2 = 1 - \tfrac{1}{4}s_n^2, \quad s_{2n}^2 = \tfrac{1}{4}s_n^2 + (1 - OD)^2,$$

153

leading to the recurrence relation

$$s_{2n} = \sqrt{\{2 - \sqrt{(4 - s_n^2)}\}}.$$

Each application doubles the number of sides of the inscribed polygon. After enough steps we have N sides, and assume that $2\pi = Ns_N$.

Machin, in about 1700, used formula (3) of the next section to derive π to 100 decimals. From this time formulae of similar type were used, and Shanks determined 607 decimals in 1853. The digital computer has now taken over; its value has recently been determined to no less than 100,000 decimal digits. Such results enable counts of digit patterns to be made in connection with tests of normality (see Section 6 of Chapter 12). It is not yet known if π is a normal number.

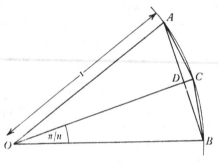

Fig. 1

We turn to the status of π in the real number system. Greek attempts to draw a square equal in area to a circle by ruler and compass methods all failed. Lambert proved that π was irrational in 1761, and by then it was suspected that the problem was insoluble. This was proved by Lindemann in 1882, when he showed that π is a transcendental number. We prove π's irrationality later; a proof of its transcendental nature is given by Klein[2].

2. Some formulae for π

Vieta gave one of the first instances of an infinite product in 1593; it is

$$\frac{2}{\pi} = \frac{\sqrt{2}}{2} \cdot \frac{\sqrt{(2 + \sqrt{2})}}{2} \cdot \frac{\sqrt{\{2 + \sqrt{(2 + \sqrt{2})}\}}}{2} \cdots$$

154

Successive factors converge to 1 with reasonable speed, the 25th being $1 - 10^{-15}$. Proof follows from the formulae

$$\sin\theta = 2\cos\frac{\theta}{2}\sin\frac{\theta}{2}, \quad \cos\frac{\theta}{2} = \frac{\sqrt{(2+2\cos\theta)}}{2}. \tag{1}$$

Repeated use of the first gives

$$1 = \sin\frac{\pi}{2} = 2^{n-1}\cos\frac{\pi}{4}\cos\frac{\pi}{8}\ldots\cos\frac{\pi}{2^n}\sin\frac{\pi}{2^n}.$$

Then the second formula is used to evaluate the cosines giving

$$\frac{1}{2^{n-1}\sin\pi/2^n} = \frac{\sqrt{2}}{2}\cdot\frac{\sqrt{(2+\sqrt{2})}}{2}\cdot\ldots\cdot\frac{\sqrt{\{2+(\sqrt{2}+\ldots)\}}}{2},$$

the last term containing $(n-1)$ nested root signs. Letting n tend to infinity and replacing sine by angle gives Vieta's formula.

Our next result, another infinite product, was discovered by Wallis (1616–1703). It reads

$$\frac{2}{\pi} = \frac{1\cdot3}{2\cdot2}\cdot\frac{3\cdot5}{4\cdot4}\cdot\frac{5\cdot7}{6\cdot6}\cdots\cdot\frac{(2n-1)(2n+1)}{2n\cdot2n}\cdots.$$

Convergence is very slow, the 50th pair of products giving $1 - 10^{-4}$. To prove the result we first obtain a certain reduction formula by integrating by parts. We have

$$I_n = \int_0^{\frac{1}{2}\pi}\sin^n\theta\,d\theta = [-\cos\theta\,\sin^{n-1}\theta]_0^{\frac{1}{2}\pi} + \int_0^{\frac{1}{2}\pi}(n-1)\sin^{n-2}\theta\cos^2\theta\,d\theta$$

$$= (n-1)(I_{n-2}-I_n),$$

on replacing $\cos^2\theta$ by $1-\sin^2\theta$.

This gives

$$I_n = \frac{n-1}{n}I_{n-2}. \tag{2}$$

Noting that $I_0 = \frac{1}{2}\pi$, $I_1 = 1$, we obtain

$$I_{2n} = \frac{2n-1}{2n}\cdot\frac{2n-3}{2n-2}\cdots\frac{1}{2}\cdot\frac{\pi}{2},$$

$$I_{2n+1} = \frac{2n}{2n+1}\cdot\frac{2n-2}{2n-1}\cdots\frac{2}{3}\cdot1.$$

Since $\sin\theta < 1$ we have

$$I_{2n} > I_{2n+1} > I_{2n+2},$$

for the integrand decreases as n increases. From (2) we see that, with increase of n, the ratio I_{2n}/I_{2n+2} tends to 1. The result just proved shows that this is also true of I_{2n}/I_{2n+1}, and this ratio yields the Wallis product.

Brouncker's continued fraction for π is of about the same period; it is

$$\frac{4}{\pi} = \cfrac{1}{1 + \cfrac{1^2}{2 + \cfrac{3^2}{2 + \cfrac{5^2}{2 + \dots}}}}.$$

It is easily derived from the continued fraction for $\tan^{-1} x$, given in Chapter 2. However, this result was found by Euler a century later, and we do not know what method Brouncker used.

Gregory's series (1671),

$$\tan^{-1} x = x - \tfrac{1}{3}x^3 + \tfrac{1}{5}x^5 - \dots,$$

can be derived by expanding $1/(1 + x^2)$ by the Binomial Theorem, and integrating the result. Putting $x = 1$ we obtain the Leibnitz series

$$\tfrac{1}{4}\pi = 1 - \tfrac{1}{3} + \tfrac{1}{5} - \tfrac{1}{7} + \dots.$$

Convergence is so slow as to make the result useless for computational purposes. However, the addition formula for $\tan (x + y)$ enables us to prove that

$$\tfrac{1}{4}\pi = 4 \tan^{-1} \tfrac{1}{5} - \tan^{-1} \tfrac{1}{239}. \tag{3}$$

Gregory's series converges rapidly for the second term, and is practicable for the first. This formula of Machin's has been used in several attempts to find π to many decimal digits.

Euler in about 1745 considered the problem of summing the inverse squares of the integers, thus discovering a set of series formulae for π. His reasoning, discussed by Polya[4], was on the following lines. Seeking an algebraic equation with roots of the form $1/n^2$, he was led to assume

$$\sin x = Ax \left(1 - \frac{x^2}{\pi^2}\right)\left(1 - \frac{x^2}{2^2 \pi^2}\right)\dots. \tag{4}$$

As $\sin (x)$ vanishes for $x = 0, \pm \pi, \pm 2\pi, \dots$ an extension of the Remainder Theorem makes this result plausible. For small x, the product is close to x, so that $A = 1$.

Taking logs of both sides, after dividing by x and replacing $\sin (x)$ by its Taylor series expansion about zero, gives

$$\log \left\{1 - \frac{x^2}{3!} + \frac{x^4}{5!} - \dots\right\} = \sum_{j=1}^{\infty} \log \left(1 - \frac{x^2}{j^2 \pi^2}\right).$$

The logarithmic expansion is then applied to both sides, it being assumed that this is justifiable for the double series that results from the right-hand side. We get

$$-\frac{x^2}{6}-\frac{x^4}{180}-\cdots = -\frac{x^2}{\pi^2}\sum_{j=1}^{\infty}\frac{1}{j^2}-\frac{x^4}{2\pi^4}\sum_{j=1}^{\infty}\frac{1}{j^4}-\cdots,$$

and comparison of powers of x yields the series

$$\frac{\pi^2}{6} = \frac{1}{1^2}+\frac{1}{2^2}+\frac{1}{3^2}+\cdots,$$

$$\frac{\pi^4}{90} = \frac{1}{1^4}+\frac{1}{2^4}+\frac{1}{3^4}+\cdots.$$

The first solved Euler's problem, the second converges rapidly enough to provide a useful way of determining π. It can be improved by noting that

$$1-\frac{1}{2^4}+\frac{1}{3^4}-\frac{1}{4^4}+\cdots = \frac{\pi^4}{90}\left(1-\frac{2}{16}\right) = \frac{7\pi^4}{720},$$

since the even terms of the original series add up to give $\frac{1}{16}$th of the full sum. The nth term of the new series is of the same magnitude as in the old; however, the alternation of signs is valuable. It is easy to prove that, because of the steadily decreasing terms, when the series is truncated the error incurred is of smaller magnitude than the first term omitted.

3. Three geometrical approximations

While π, a transcendental number, cannot be constructed by ruler and compass, attempts to square the circle have led to many geometrical approximations. The following was discovered by Kochansky, a Polish monk, in 1685. In Figure 2 the circle is of unit radius, and we construct angle BOC equal to $30°$. The line CD is 3 units long, and AD is very nearly equal to π. We have $CB = 1/\sqrt{3}$ so that

$$AD = \sqrt{\{2^2+(3-1/\sqrt{3})^2\}} = \sqrt{(\tfrac{40}{3}-\sqrt{12})} = 3\cdot141533,$$

contrasting with the true value $3\cdot141593$.

Before dealing with the second approximation we digress to obtain a continued fraction expansion for π with unit numerators. The decimal part, multiplied by 10^8 is first divided into 10^8. The remainder now becomes a divisor, and so on. This is of course the process for finding a highest common factor.

$$14,159,265)100,000,000(7$$
$$99,114,855$$
$$885,145)14,159,265(15$$
$$13,277,175$$
$$882,090)885,145(1$$
$$882,090$$
$$3,055$$

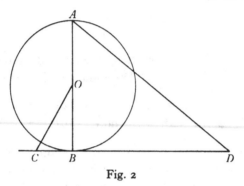

Fig. 2

The result is the continued fraction

$$3 + \cfrac{1}{7 + \cfrac{1}{15 + \cfrac{1}{1 + \dots}}}.$$

Had more terms been required we would have started with more digits in π. The next term is $\frac{1}{292}$, and its small size indicates that neglecting it, together with all subsequent fractions, causes only a small error in the approximation. Truncating at successively lower levels we find

$$3 = 3 \cdot 0,$$
$$\tfrac{22}{7} = 3 \cdot 1429,$$
$$\tfrac{333}{106} = 3 \cdot 141509,$$
$$\tfrac{355}{113} = 3 \cdot 14159292.$$

These values are to be compared with $3 \cdot 14159265$, and the last gives excellent agreement.

Figure 3 illustrates Gelder's construction based on the ratio $\tfrac{355}{113}$. The circle is of unit radius, and $OB = \tfrac{7}{8}$, $AC = \tfrac{1}{2}$, CD is perpendicular to

158

AO, while CE is parallel to BD. Then AE is near $(\pi - 3)$. The proof is straightforward, we first note that $AB = \sqrt{(113)}/8$. By similar triangles

$$\frac{AD}{AO} = \frac{AC}{AB} \quad \text{and} \quad \frac{AE}{AD} = \frac{AC}{AB},$$

the 'product' of these expressions gives

$$\frac{AE}{AO} = \frac{AC^2}{AB^2} = \frac{16}{113} = \frac{355}{113} - 3.$$

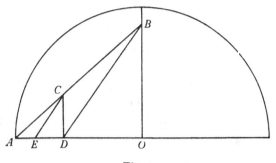

Fig. 3

The construction of Figure 4 is due to Snell, the discoverer of the law of refraction: $\sin (i) = \mu \sin (r)$. It gives an approximation to the length of circular arcs of small angle. To construct a line of approximately the

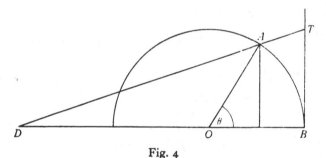

Fig. 4

same length as arc AB, produce the unit length BO to D so that $BD = 3$. The line DA cuts the tangent at B in T, and BT is nearly equal to θ. By similar triangles

$$\frac{BT}{3} = \frac{\sin \theta}{2 + \cos \theta} = \frac{1}{3} \left(\theta - \frac{\theta^5}{180} + \dots \right).$$

The use of Taylor's series, followed by simple division, leads to the

159

latter expression. The neglect of all but the first term is justified for small θ. Applied to an angle $\frac{1}{16}\pi$, readily determined by repeated bisection, the construction gives a good approximation to π.

4. The irrationality of π

If π is rational, then π^2 is equal to p^2/q^2 and is rational too. We assume the latter result, and arrive at a contradiction.

As a preliminary we consider

$$f(x) = \frac{x^n(1-x)^n}{n!} = u(x).v(x) \quad \text{with} \quad u(x) = \frac{x^n}{n!}.$$

We are interested in the values of the derivatives of $f(x)$ at $x = 0$ and 1. Because of symmetry about $x = \frac{1}{2}$ those of even order are equal, those of odd order differ only in sign. We shall prove that all derivatives at these points are signed integers or zero. The derivatives of u are

$$u^r(0) = 0 \quad (r \neq n),$$

$$u^n(0) = 1.$$

Using the Leibnitz formula for the derivatives of a product it is easy to show that

$$f^r(0) = 0 \quad (r < n \quad \text{or} \quad r > 2n),$$

$$f^r(0) = \binom{r}{n} v^{r-n}(0) = (-1)^{r-n} \binom{r}{n} n(n-1)\ldots(2n-r+1) \quad (n < r \leqslant 2n)$$

$$= 1 \quad (r = n).$$

Thus the result has been proved.

We next consider the expression

$$I = \int_0^1 f(x) \sin \pi x \, dx,$$

this is twice integrated by parts to give

$$I = \frac{f(0)+f(1)}{\pi} - \frac{1}{\pi^2} \int_0^1 f''(x) \sin \pi x \, dx.$$

As $f^{2n+2}(x) \equiv 0$, repeated application of this result leads to

$$\pi I = \left[f(0)+f(1) - \frac{f''(0)+f''(1)}{\pi^2} + \ldots + (-1)^n \frac{f^{2n}(0)+f^{2n}(1)}{\pi^{2n}} \right].$$

160

Our assumption of rational $\pi^2 = a/b$ implies

$$\pi a^n I = a^n [f(0)+f(1)] - a^{n-1} b [f''(0)+f''(1)] + \dots$$
$$+ (-1)^n b^n [f^{2n}(0)+f^{2n}(1)], \quad (5)$$

and the right-hand side is an integer.

As both $\sin(\pi x)$ and $x(1-x)$ lie between 0 and 1 we have

$$0 < \pi a^n I < \pi a^n \int_0^1 \frac{dx}{n!} = \frac{\pi a^n}{n!}.$$

If n is large enough, this quantity is less than 1, so the left-hand side of (5) cannot be an integer. The previous result is contradicted, and we must abandon the assumption that π^2 is rational.

This proof is due to Niven; he also uses similar methods to show that various trigonometric quantities are irrational (Niven[3]).

5. Buffon's needle problem

We conclude with an account of a statistical method of measuring π. This was devised by Buffon, a naturalist associated with investigations into the geometry of the bee's cell. A needle of length l is dropped at random on to a plane ruled with parallel lines at unit intervals. If $l < 1$, the probability that the needle cuts at least one line is $2l/\pi$; otherwise it is

$$\frac{2l}{\pi}(1-\cos \phi_0) + \left(1 - \frac{2\phi_0}{\pi}\right), \quad \text{where} \quad \sin \phi_0 = \frac{1}{l}. \quad (6)$$

Repeated experiments with a count of favourable cases leads to a Monte Carlo determination of π. Kahan[5] gives an interesting account of a carefully designed experiment. This includes a modified procedure designed to reduce the effect of random variations on the answer. Thus we can replace the parallel grid by a square grid with (6) suitably modified. This divides the effect of random errors by nearly 4, and further improvements are possible.

Let the centre of the needle be at distance x from the nearest line, and its direction make an acute angle ϕ with it, as in Figure 5(a). With $l < 1$, Figure 5(b) illustrates the rectangle over which (x, ϕ) ranges; we assume all points in this area occur with equal probability. Points leading to an intersection have $2x$ less than $l \sin \phi$, corresponding to the shaded area. The probability of an intersection is the area ratio

$$\int_0^{\frac{1}{2}\pi} \int_0^{\frac{1}{2}(l \sin \phi)} dx\, d\phi \Big/ \frac{\pi}{4} = \frac{2l}{\pi}.$$

If l exceeds unity sketch (c) shows that the probability is

$$\frac{4}{\pi}\int_0^{\phi_0}\int_0^{\frac{1}{2}(l\sin\phi)} dx\,d\phi + \frac{4}{\pi}(\tfrac{1}{2}\pi - \phi_0)\cdot\tfrac{1}{2},$$

which reduces to (6) above.

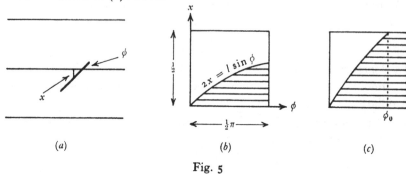

(a) (b) (c)

Fig. 5

This is a classical result in the theory of geometrical probability. Kendall and Moran[1] give an interesting account of this relatively neglected field. We shall use the above result for a needle of length l to deduce the probability that a convex curve of maximum breadth less than or equal to unity will intersect a line of the parallel set.

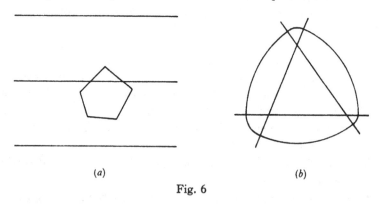

(a) (b)

Fig. 6

In Figure 6(a) we illustrate a convex polygon of n sides and of maximum breadth less than or equal to 1. The lengths of these sides are l_1 to l_n. We assume that the probability of cutting side j is $p(j)$, while the probability of cutting both j and k is $p(j, k)$. If side j is cut, then one other side is also cut, so that

$$p(j) = \sum_{(k)}{}' p(j, k), \tag{7}$$

162

the dash denoting the absence of $p(j, j)$ in the sum. If the polygon intersects a line then just two sides are cut, so the probability P of this happening will be

$$P = \Sigma\Sigma' p(j, k),$$

where the sum is taken over all possible different pairs j, k. If we sum (7) over all sides we get this latter sum counted twice, for $p(j, k) = p(k, j)$. By the result proved earlier, $p(j) = 2l_j/\pi$, so that

$$P = \tfrac{1}{2}\Sigma p(j) = \tfrac{1}{2}\Sigma \frac{2l_j}{\pi} = \frac{C}{\pi},$$

where C is the polygon's circumference. By replacing a convex curve by an approximating polygon, and proceeding to the limit, the result is seen to hold for curves too.

If the curve is of breadth I in all directions intersection is certain and P is unity. Moreover, by the above formula, $C = \pi$ for all such curves, a result due to Barbier that we shall meet again in the next chapter. Figure 6(b) is made up of 6 circular arcs with centres at the 3 vertices of the triangle. It is of constant breadth, and provides a direct confirmation of Barbier's theorem, since we see readily that its circumference is π times the constant breadth.

References

1. M. G. Kendall and P. A. P. Moran. *Geometrical Probability* (Griffin, 1963).
2. F. Klein. *Famous Problems of Elementary Geometry* (Dover, 1956).
3. I. Niven. *Irrational Numbers*, Carus Monograph no. 11 (Wiley, 1956).
4. G. Polya. *Mathematics and Plausible Reasoning*, vol. 1 (Princeton, 1954).
5. B. C. Kahan. 'A practical demonstration of a needle experiment designed to give a number of concurrent estimates of π.' *J. R. Statist. Soc.* (A), vol. 124, part 2, p. 227 (1961).

15

Rotors and curves of constant breadth

1. *Curves of constant breadth*

The distance between parallel tangents to a circle is constant, but it is not the only curve with this property. We consider a closed convex curve and a pair of parallel support lines (see Figure 3 of Chapter 13). The

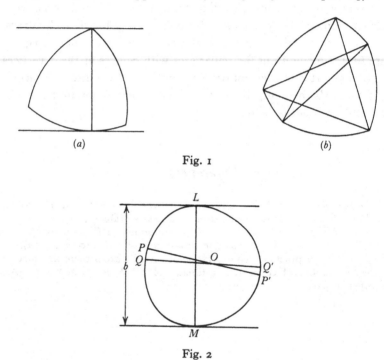

(a)   (b)

Fig. 1

Fig. 2

distance between them is the curve's breadth in the direction of these lines. We shall investigate curves for which the breadth is the same in any direction, and start with some simple examples.

In Figure 1 (*a*) the curve is defined by 3 circular arcs with centres at the vertices of an equilateral triangle whose side equals their radii. This is the Reuleaux triangle; in Figure 1 (*b*) it is generalised by starting with a crossed pentagon, not necessarily regular, but with sides of equal length. It can be expanded by a constant amount to remove discontinuities in slope; the result for the Reuleaux triangle is seen in Figure 3(*b*). Figure 6(*b*) of Chapter 14 shows how a 6-arc curve of constant breadth can be formed from any triangle. By using a pentagon a 10-arc curve can be constructed, and so on.

So far the curves discussed have been composed of circular arcs, but we shall now find a far more general procedure. There cannot be two points inside or on a curve of constant breadth *b* further apart than *b*. If there were, we could draw lines through them perpendicular to their join, and the curve's breadth in this direction would exceed *b*. The line *LM* joining points of contact of parallel support lines must therefore be perpendicular to them both, for *LM* cannot exceed *b*.

If we take *P* and *Q* close together and draw the normals there to meet the curve again at *P'* and *Q'*, the lines *PP'*, *QQ'* will cut at *O*, the centre of curvature at both *P* and *P'*. If the arc *LPM* is arbitrary, but has the property that all circles of radius *b* touching it enclose the curve, we can complete a curve of constant breadth as follows. Along the normal at *P* measure distance *b* to find *P'*, then the locus of *P'* is the other part of our curve. By our assumption about tangent circles, the centre of curvature at *P* lies inside *PP'*, and the resulting curve is convex. This heuristic proof can be supplemented by Rademacher and Toeplitz's[3] rigorous discussion.

We now consider the circumference *C* of our curve. Let angle POQ be $\delta\theta$, then we have

$$PQ = OP.\delta\theta, \quad P'Q' \backsimeq OP'.\delta\theta, \quad C = \int_0^{\pi} (OP + OP')d\theta = b\pi.$$

This surprising result is due to Barbier; Lyusternik[2] gives a rigorous proof. A proof based on geometrical probability was given at the end of Chapter 14. A direct verification is easy for the curves composed of circular arcs discussed above. We shall prove analogous results by a quite different method later.

It was proved in Chapter 13 that any convex curve possesses a circumscribing equiangular hexagon with central symmetry. If the curve is of constant breadth this hexagon is regular, a fact Eggleston[1] uses in proving the Blaschke–Lesbegue theorem. This states that, of all curves of given constant breadth *b*, the Reuleaux triangle has the smallest area. The curve of largest area is the circle; this follows from the fact that its

165

circumference is determined by b. Among all curves of given circumference the circle has greatest area, and it is also a curve of constant breadth.

2. *Mechanical applications*

These curves have a number of mechanical applications. In Figure 3(a) we enclose the curvilinear triangle in a square, and it can be rotated so as to maintain contact with the sides of the square. Curves of constant breadth are rotors for a square; we discuss equilateral triangle or Δ-rotors later. This property led to the invention of the Watts drill for

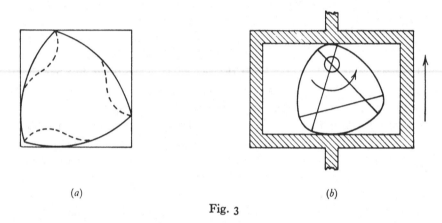

(a)   (b)

Fig. 3

square holes. A cylinder whose cross-section is a Reuleaux triangle has cutting edges formed on it by removing the regions indicated by dots in the figure, and the end is pointed. If rotated inside a square template and fed forward it drills a square hole, except for small rounded areas at each corner. The centre of rotation is changing throughout the motion, so a floating chuck is needed to impart the rotary movement. For mechanical reasons it is not feasible to use a similar technique for equilateral triangles; but, by using appropriately shaped rotors, drills for holes of 5, 6 and 8 sides have been made.

Figure 3(b) illustrates a cam of constant breadth, in this case a modified Reuleaux triangle, rotating about one of its three original vertices. The follower, shown shaded, is stationary over two arcs each of 60°; in the diagram it is on one of these portions, and a small further rotation in the direction shown will cause it to rise. Such periods of halted motion are needed in the feed mechanism for projectors of cine-film. By a suitable choice of cam shape a wide variety of motions can be generated. One of the mechanical advantages of such cams is that the drive on the follower is a positive one in both directions, in contrast to the spring loaded plunger needed with ordinary cams.

The rotating piston of the Wankel non-reciprocating engine is another example of a rotor, this time within a region of relatively complicated shape.

3. *Solids of constant breadth*

The simplest solid of constant breadth is obtained by rotating a Reuleaux triangle about an axis of symmetry, as shown in Figure 4(a).

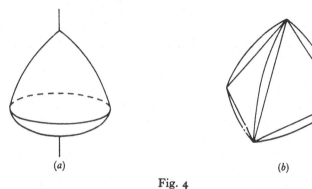

(a)   (b)

Fig. 4

We shall obtain two other such solids by a different generalisation of the Reuleaux triangle. Figure 4(b) illustrates the solid formed by constructing four spherical caps of radii equal to the sides of a regular tetrahedron, and with centres at its vertices. In the diagram it is assumed that the tetrahedron is opaque, while the spherical caps are transparent. This is not a solid of constant breadth unless suitably modified. It occurs in nature in the skeleton of the nassellarian, a tiny sea creature about 0·1 mm. in diameter. D'Arcy Thompson[5] has a beautiful drawing of it. It has, besides the curvilinear tetrahedron, six planes going through its centre and edges. Three of these, together with one of the four caps, can be seen in Figure 5(a). The angles between plane faces and between planes and spherical caps are all 120°.

In order to form the two solids of constant breadth shown in Figure 6, three edges of Figure 4(b) have to be rounded off. In the first solid these

167

edges meet at an apex, in the second they belong to one face. The new spindle-like surfaces are formed by rotating a circular arc about its chord through 60°.

In Figure 5(b) the spherical cap opposite A cuts the plane face ACD in the lower arc CD. The cap opposite B cuts BCD in the upper arc CD. Rotating one into the other generates the required surface opposite the

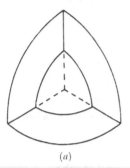

(a)

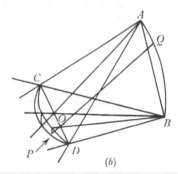

(b)

Fig. 5

unmodified arc AB. There are two possible types of contact of the solid with parallel support planes. In the first, one passes through an apex, the other touches the corresponding cap at a distance equal to the side

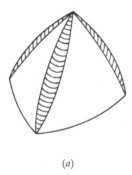

(a)

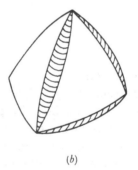

(b)

Fig. 6

of the tetrahedron. In the second, one plane touches the spindle-like surface at P. The normal there is POQ, cutting the opposite arc at Q, the point of contact of the other plane. If we rotate POQ about O so that Q moves into B, it is easy to see that the line in its new position equals BD in length.

Much less is known about solids of constant breadth than about the curves discussed above. There is no direct analogue of Barbier's theorem. However, Minkowski pointed out that their shadows by orthogonal projection are of constant circumference.

4. The addition of curves

We now describe an addition operation that provides a powerful tool for questions concerned with rotors. In Figure 7(a), we add A_1 of region C_1 to A_2 of C_2 to get the sum point (A_1+A_2). The process is one of vector addition with a fixed origin O. The sum total of all points like (A_1+A_2) is a new region (C_1+C_2), the sum of C_1 and C_2. This sum is unaltered, except for a parallel translation, if either C_1 or C_2 is moved parallel to itself. In Figure 7 (b), the arrowed vector indicates how A_2 moves into A_2'. The sum point (A_1+A_2) moves into (A_1+A_2') by the

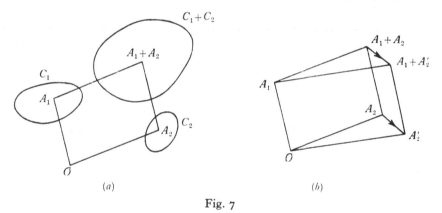

(a) (b)

Fig. 7

same vector. Hence (C_1+C_2) also gives (C_1+C_2') by the same translation. A change of origin is equivalent to the same parallel shift applied to C_1 and C_2, and, by the previous argument, produces no alteration in the form of (C_1+C_2). Thus we may select O at any convenient point.

In Figure 8(a) we add non-parallel lines l_1 and l_2, taking O at one end of l_1. Adding l_1 to point P of l_2 is equivalent to moving l_1 until O lies on P, and it assumes the position PQ. As P takes all possible positions on l_2, PQ sweeps out the parallelogram (l_1+l_2). In Figure 8 (b) the lines are parallel, and the sum is a line parallel to both, and with length equal to the sum of their lengths.

We now prove that, if C_1 and C_2 are convex, so is their sum. Take P, Q any two points of the sum region, then P is the sum of points P_1 and P_2

169

in C_1 and C_2 respectively. The same is true of Q, Q_1 and Q_2. As the base curves are convex all points of P_1Q_1 lie in C_1 and of P_2Q_2 in C_2. Thus all points of the sum of P_1Q_1 and P_2Q_2 lie in (C_1+C_2). This sum is a parallelogram with diagonal PQ, so all points of PQ lie in (C_1+C_2), i.e. it is a convex region.

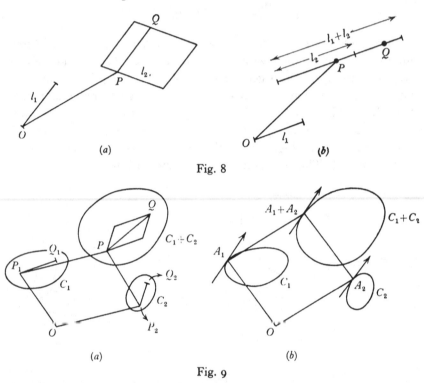

Fig. 8

Fig. 9

Next we take A_1 and A_2 on C_1 and C_2 so that the corresponding support lines are parallel, and directed in the same sense as in Figure 9(b). All points of C_1 and C_2 lie to the right of the corresponding support lines. These support lines add up to give a parallel line through (A_1+A_2). As all points of (C_1+C_2) lie on its right and (A_1+A_2) lies on it, it is the support line there, thus parallel support lines are additive.

In Figure 10 we take O on C_1. Adding C_1 to point P of C_2 is equivalent to a parallel translation that moves O on to P.

Fig. 10

170

Thus $(C_1 + C_2)$ is generated as the envelope of C_1' when P moves around C_2. In two positions of P, support lines of C_2 are parallel to the support line of C_1 at O. Thus we see that the breadth of $(C_1 + C_2)$ in this direction is the sum of the breadths of C_1 and C_2. As O can take any position on C_1, it follows that breadths in any direction are additive. Thus curves of constant breadth add to give a new curve with the same property. When adding a circle it is convenient to take O at its centre. We see that the curve of Figure 3 (b) is obtained by moving the centre of a circle round a Reuleaux triangle, an instance of the additive property just proved.

5. The additive property of circumferences

We now prove the remarkable theorem that circumferences are also additive. In doing so we derive another result of considerable use in the sequel.

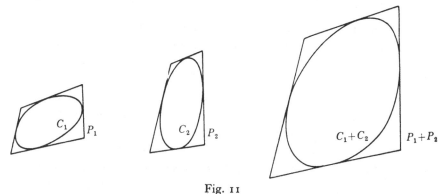

Fig. 11

In Figure 11 we have curves C_1 and C_2 inscribed in polygons P_1 and P_2 with corresponding sides parallel. The sum of these polygons is another with sides parallel to both. The length of a side of this new polygon is the sum of the corresponding sides in P_1 and P_2; moreover, the sum $(C_1 + C_2)$ is inscribed in it. These results all follow at once from the properties deduced above, and they are valuable in their own right. We next assume that the number of sides increases so that perimeters of polygons approximate to circumferences of inscribed curves. Since perimeters of polygons are additive, we see that the circumference of $(C_1 + C_2)$ is the sum of the circumferences of C_1 and C_2.

171

6. Rotors in the equilateral triangle

We shall describe such triangles as Δ-triangles for short in what
follows. In Figure 12, ABC is a Δ-triangle and AO is parallel to BC,
the arc LM is drawn with O as centre
to touch BC. Its mirror image in LM
completes a 'Δ-biangle' shown shaded
and with 60° angles at its vertices.

Draw LN perpendicular to OA,
meeting CA in N. Then the circle,
centre O, goes through N, and angle
LOM is twice angle LNM. This latter
angle is 30° so that LOM is 60°, and
LM is of fixed length. Thus the biangle will turn with arc LM touching

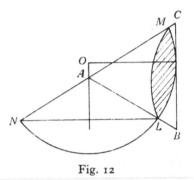

Fig. 12

BC. Moreover, it is easily seen that the two arcs at M include 60°
between their tangents. So when the biangle is rotated until M lies at C,
its arcs touch AC, BC there. Further rotation brings an arc in contact
with CA, and M on to BC. In short, we can rotate the Δ-biangle inside
ABC.

It can be proved that the perpendiculars to BA, CA at L and M meet
on the perpendicular from O to BC. Their common point is the biangle's instantaneous centre of rotation.

A simpler rotor is the Δ-triangle's incircle. It can be proved that, for
a given triangle, this is the rotor with greatest area, while the Δ-biangle
has the least area. In Figure 16(a) four circular arcs of radius d are
drawn with centres at the vertices of a square of side d. The curvilinear
quadrilateral so formed is another Δ-rotor; Yaglom and Boltyanskii[4]
give a proof.

In Figure 13 we add a Δ-biangle to a circle to produce a new convex
figure. In position B the biangle touches the sum-curve, in B' one of its
vertices lies on a circular arc of the curve. If we draw parallel Δ-triangles
around B and C, their sum is another Δ-triangle circumscribing $(B+C)$,
thus Δ-rotors are additive, and $(B+C)$ is a new Δ-rotor. This property is
analogous to the similar one for curves of constant breadth, i.e. rotors in
a square.

There is also an analogue of Barbier's theorem; all Δ-rotors in a given
Δ-triangle are of the same circumference.

172

In Figure 14 we start with the Δ-triangle T_1 containing a rotor. This configuration is rotated twice through 120° about a point O on one of its axes of symmetry, giving three Δ-triangles in all. The three triangles add up to a new Δ-triangle; the three rotors give a rotor inscribed in it. The sum figures have been enlarged in the diagram for the sake of clarity. We shall prove that the new large rotor is a circle. It then follows

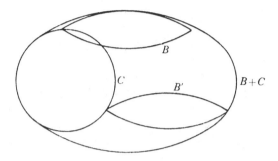

Fig. 13

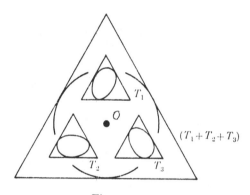

Fig. 14

that the circumference of each of the three small rotors is one-third of the circumference of this circle, and so depends only on the size of the Δ-triangle T_1.

The sum-rotor is unaltered by a rotation of 120° about O, for this leaves the constituent figures unchanged. Any Δ-triangle circumscribed about it is also invariant under this rotation. For it is the sum of three Δ-triangles that are merely interchanged by the rotation. Therefore its sides must be equidistant from O. As this is true of all circumscribing triangles, and they are of the same size, the sum-rotor is a circle with centre at O. The Barbier theorem for curves of constant breadth can be proved in the same way. Such curves are rotors in a square and four rotations, each of a right angle, lead to a similar proof.

7. Some generalizations

The Δ-rotor theory extends in a straightforward way to other regular polygons. However, of more interest is the possibility of rotors, other than circles, in non-regular polygons. A remarkable theorem, due to Fujiwara, states that, for $n \neq 4$, the necessary and sufficient conditions for the existence of such rotors are that all angles of the polygon are rational multiples of π and that it possesses an in-circle. For $n = 4$ the first condition is not necessary, thus all curves of constant breadth b constitute rotors for any rhombus with parallel sides a distance b apart.

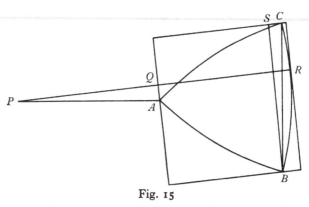

Fig. 15

Figure 15 leads to another quite different generalization. Here ABC is equilateral and $PA = AB$, the circular arc BC has P as centre, and so on for arcs CA and AB. Any rectangle is drawn circumscribing the curvilinear triangle ABC. If PQR is drawn perpendicular to two of its sides, and to BS, then angles APQ, SBC lie between 2 pairs of perpendicular lines, and so are equal. As triangles APQ, SBC are also right-angled, and $BC = PA$, they must be congruent. Thus $PQ = BS$, and the fixed length PR is the sum of sides of the circumscribing rectangle. It follows that all these rectangles have a fixed perimeter.

In Figure 16(b) the 90° biangle is another curve with circumscribing rectangles of fixed perimeter; Yaglom and Boltyanskii[4] give a proof. The addition of two curves C_1 and C_2, each with circumscribing rectangles

174

of fixed perimeter, is a curve $(C_1 + C_2)$ with the same property. To see this we consider parallel rectangles R_1 and R_2 circumscribing C_1 and C_2 respectively. The sum of R_1 and R_2 is a rectangle circumscribing $(C_1 + C_2)$, and its perimeter is the sum of the perimeters of R_1 and R_2. This sum is constant as R_1 and R_2 rotate with their sides remaining parallel.

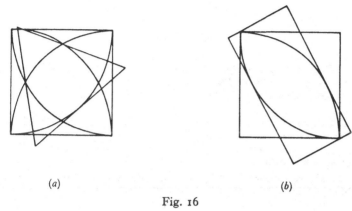

(a)   (b)

Fig. 16

Finally, these curves have circumferences depending only on the fixed perimeter of circumscribing rectangles, i.e. a generalized Barbier's theorem again applies. The proof follows the lines of that for Δ-rotors, depending on four rotations of $\frac{1}{2}\pi$, followed by addition of four equal figures.

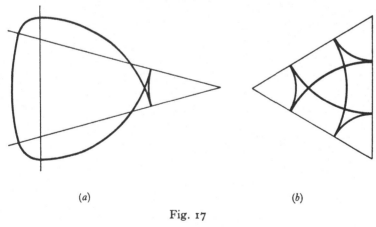

(a)   (b)

Fig. 17

So far all the curves considered have been convex. If the curvature condition on the arbitrarily specified arc LPM used in the construction of Section 1 is violated, the result will be a non-convex curve. It is

175

simpler to construct an example using circular arcs, and two are shown in Figure 17. The distance between parallel tangents is constant, so in this sense they are still curves of constant breadth, although having double points and cusps. Such a curve satisfies Barbier's theorem provided certain arcs have their lengths counted as negative in the 'circumference' of the curve.

References

1. H. G. Eggleston. *Convexity*, Cambridge Tract no. 47 (Cambridge, 1963).
2. L. A. Lyusternik. *Convex Figures and Polyhedra* (Dover, 1963).
3. H. Rademacher and O. Toeplitz. *The Enjoyment of Mathematics* (Princeton, 1957).
4. I. M. Yaglom and V. G. Boltyanskii. *Convex Figures* (Holt, Rinehart and Winston, 1961).
5. D'Arcy W. Thompson. *On Growth and Form*, abridged edition (Bonner) (Cambridge, 1961).
6. M. Goldberg. *Rotors in Polygons and Polyhedra* (Maths of Comp., **14**, No. 71 (1960).

Index

addition of curves, 169
  of lines, 169
  of circumferences, 171
aeroplane wing, 61
algebraic curves, 100, 111
  numbers, 135
Alhambra, 129
anchor ring, 77, 80
Andreini tessellations, 92
Apollonius, problem of, 45
approximation, Stirling's, 37, 72
  to binomial distribution, 67
Arbelos, 44
arc sine law, 72
Archimedean polyhedra, 9, 92
Archimedes and π, 153
  axiom of, 131
Arnold, 79, 87
astigmatic lens, 59
  surface, 58
asymptote, curvilinear, 104
axiom of Archimedes, 131

Ball, 9, 11, 94, 95, 99
Barbier's theorem, 163, 165, 172, 175
Bellman, 21
Bertelsen, 33
Bertrand, 35
Besicovitch, 96
Biblical references to π, 153
Binet's formula, 13
binomial distribution, 65
biological growth, 12, 17
Blaschke–Lesbegue theorem, 165
Blaschke's theorem, 144
Boltyanskii, 89, 99, 152, 172, 174
Bolyai, 88
Borel, 139
Brianchon's theorem, 58
Brouncker, 156
Brownian motion, 65
Buffon, 161

Cadwell, 31, 89, 99
cam of constant breadth, 166
Cantor, 138
Casey's theorem, 49
Cayley, 76
centring theorem, 143
ceratoid cusp, 109

chains of circles, 47
  of spheres, 48
characteristic of a surface, 79
charm, dodecahedral, 2
Chebyshev, 36
Cherwell, 40
Chung, 72
circle chains, 47
coaxal circles, 50
Cohn–Vossen, 57, 63, 77, 79, 87, 129
coin tossing, 64
combinatorial geometry, 142
compound solids, 5
constant breadth, curves of, 163, 164
  solids of, 167
continued fractions, 15, 130, 156
convergents, 15
Courant, 37, 40, 137, 141
covering problems, 147
Coxeter, 4, 8, 9, 10, 11, 13, 17, 21, 44
crossed polygon, 5, 29
  polytope, 8
cube, 1, 9, 89, 92
  in n-space, 7
cuboctahedron, 9, 93
Cundy, 9, 11, 44, 51
curves, algebraic, 100, 111
  of constant breadth, 163, 164
curvilinear asymptote, 104
cusp, ceratoid, 109
  rhamphoid, 108
cuspidal edge, 61
cyclic constructions, 22

de la Vallée Poussin, 36
de Moivre, 67
Debrunner, 152
Dehn, 89
decimal fractions, 133
delta bi-angle, 172
  rotor, 172
Descartes, 100
developable surface, 61
dice, icosahedral, 2
Dirichlet, 14, 34, 133
discrete groups, 116
dissection of the cube, 89
  of rectangles, 88
dodecahedral charm, 2

177

179